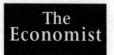

E-TRENDS

The Economist

E-TRENDS

THE ECONOMIST IN ASSOCIATION WITH
PROFILE BOOKS LTD

Published by Profile Books Ltd
58A Hatton Garden, London EC1N 8LX
www.profilebooks.co.uk

The greatest care has been taken in compiling this book.
However, no responsibility can be accepted by the publishers or compilers
for the accuracy of the information presented.

Typeset in EcoType by MacGuru
info@macguru.org.uk

Printed in Great Britain by
St Edmundsbury Press, Bury St Edmunds

A CIP catalogue record for this book is available
from the British Library

ISBN 1 86197 386 1

Contents

List of contributors

Frances Cairncross is *The Economist*'s management editor. Her book, *The Company of the Future: How the Communications Revolution is Changing Management*, will be published in February 2002 by Harvard Business School Press in the United States and by Profile Books in Britain.

Emma Duncan is the editor of *The Economist*'s Britain section.

Simon Long writes for *The Economist*'s electronic commentary, *Global Agenda*. Before that he was the paper's finance editor.

John Peet is *The Economist*'s business affairs editor.

Mark Roberts, who contributed to Part 7, is *The Economist*'s correspondent for America's west coast.

Ludwig Siegele is *The Economist*'s US technology correspondent, based near Silicon Valley.

Tom Standage is *The Economist*'s technology correspondent in London.

Matthew Symonds is an associate editor of *The Economist*. He is currently writing a book about Larry Ellison and Oracle Corporation, to be published by Simon & Schuster in spring 2003.

Pam Woodall is *The Economist*'s economics editor

All the surveys in this collection were edited by **Barbara Beck**, *The Economist*'s surveys editor, who also selected the material in this book and wrote the introduction.

Full texts of all the articles from *The Economist* are available to both print and electronic subscribers through the paper's website, www.economist.com .

Preface

THIS BOOK IS a collection of surveys and articles that appeared in *The Economist* between the spring of 2000 and the autumn of 2001; the date of publication is given at the end of each contribution. In some instances the author has added a postscript, but by and large the articles remain as pertinent as when they were first written. Together they demonstrate how profoundly the communications revolution is changing all our lives.

Introduction

BACK IN 1995, *The Economist* took a small gamble: it published a 15-page survey on a relatively obscure phenomenon called the Internet. True, the net had been around for 25 years, but for most of that time it had been the arcane domain of computer scientists and academics. Recently, though, technological advances had made it more accessible, and the number of users worldwide had been doubling every year, to the proud figure of about 20m at the time of publication. The survey agreed with the critics that the Internet was chaotic, slow, frustrating and intimidating, but insisted that it was here to stay, because it allowed people better to exercise one of their most basic desires: to communicate.

A handful of years on, and a dotcom boom and bust later, that conclusion has been triumphantly vindicated. A combination of Moore's Law (which states that computing power and capacity double every 18 months) and Metcalfe's Law (which says that the value of a network grows roughly in line with the number of users squared) has ensured that in many developed countries, and particularly in America, the Internet has now become a near-utility. The learning curve has been extremely steep. As yet, few people have figured out how to make money from Internet ventures, and many have lost their shirts. But there can be no doubt that the net increasingly permeates every aspect of our lives.

This book gathers together a series of surveys and other articles published in *The Economist* over the past couple of years which illustrate the relentless onward march of the Internet in different spheres. The first part, on the so-called "new economy", examines the impact of this communication revolution on the economy as a whole, and compares it with the effect of earlier technological innovations, such as steam power, electricity, the telegraph and the telephone. If those are anything to go by, the benefits of such innovations take many years to materialise in full, and although the Internet and computers are being adopted more rapidly than many earlier technologies, the effects are only just beginning to come through. One of the biggest claims made for the Internet is that it is capable of raising the productivity of entire economies, by allowing both labour and capital to be used more efficiently.

Economists disagree on whether it has actually done so to date, but evidence from America suggests that some of the recent growth in productivity has been structural rather than cyclical, and may be related to the use of information technology.

Governments have come late to the Internet, partly because they are not subject to the competitive pressures that have driven many businesses to become early adopters. But they have now taken up the challenge (part 2), in the hope of becoming more efficient, accessible, transparent and accountable in their citizens' eyes – and saving lots of money too. This is not a risk-free venture: for example, widespread use of the Internet by government may exacerbate the "digital divide", the growing gap between the electronic haves and have-nots. Moreover, an increasingly efficient, plugged-in government may end up knowing far more about its citizens than is good for either privacy or civil liberty. On the plus side, though, e-government is capable of giving individuals a much bigger say in the way their country is run.

The e-subject that has attracted most comment over the past few years, and that has had the biggest effect on people's wallets, is e-commerce (part 3). It seems only yesterday that eager investors in dotcom ventures were learning about the different forms of e-commerce, such as B2B (business-to-business), B2C (business-to-consumer) and the rest. Anyone who had half an idea for a dotcom launch could make a mint. It seemed too good to be true, and it was: the dotcom bubble duly burst. A couple of years on, a few lessons have been learnt. One is that devising business models which extract profits from the Internet, whether for pure-play (electronic-only) ventures or for mixed businesses, is extraordinarily difficult. So far, the vast majority of dotcom ventures have failed. Another lesson is that conventional old-economy businesses, so despised for a while, often have the best chance of harnessing the new technology in profitable ways. All in all, the main beneficiary from e-commerce may turn out to be not business at all but the consumer, for whom the extra competition may generate more choice, lower prices and better service.

The most important change the Internet has brought about in many businesses is in the way they are managed (part 4). Communications both inside and outside the company have become incomparably easier, which affects relationships within the enterprise as well as links with suppliers and customers, and ultimately the shape of the company itself. The conventional wisdom, first put forward in 1937 by Ronald Coase, an economist who later won a Nobel prize, was that people banded together into companies because it was cheaper to transact business jointly rather than separately. But the Internet has reduced those transaction costs, and some – perhaps most – of a company's activities can be contracted out or spun off. Such changes are bound to be disruptive, so the people in charge not only have to be good managers in the old-fashioned sense but need a raft of extra e-skills as well.

Management techniques apart, some industries have already thoroughly embraced the Internet for the conduct of their main business. Take financial services (part 5). Activities such as share-trading and, in some parts of the world, financial retailing are increasingly being carried out on the Internet. Against that, some financial markets, particularly on the wholesale side, are highly automated already, but rely on proprietary systems, and have no great incentive to move to the web. For the moment, many existing large financial institutions seem well placed to survive the Internet and continue to prosper. They have trusted brands, large customer bases and often plenty of money to invest in whatever technology is required. But there are nagging worries that in the long run they will be "disintermediated" away: that Internet-based businesses will bypass them in payments, clearing and settlement.

The mobile-phone industry is facing a different problem (part 6). The impending shift to third-generation (3G) mobile phones will enable huge numbers of people to gain access to the Internet on the move, opening up a host of new business opportunities. But many mobile-network operators have paid vastly over the odds for licences to operate 3G networks, and may have trouble finding the money to develop those networks. They have also encountered big technical problems that have already delayed the start of 3G services. On the more cheerful side, consumers seem eager to get access to mobile-Internet services, and are used to paying call charges, so will probably stump up.

Getting paid is something that the electronic-entertainment industry (part 7) has yet to resolve. Advertising, even before the downturn, was unlikely to make services pay on its own, and persuading consumers to pay subscriptions is hard. For example, they expect to be able to download music from the web free. Among the few things they will shell out on are e-games and e-pornography. Leaving aside a few niche businesses, the digital revolution has been a huge disappointment for the entertainment industry. But for those in search of education rather than entertainment, the Internet has proved a success, both in schools and for educating and training grown-ups.

In many respects, the Internet has opened up the world. Users can send and receive e-mails or log on to a website from anywhere on the globe, enjoying near-instant communication at negligible cost. Increasingly, too, the software they use on their computers will come out of the ether (part 8). Many experts believe that in future a vast cloud of software will be available on the Internet from which users can choose precisely what they need at any particular moment, rather than buying a piece of software in a box, putting it on their computer and waiting for it to become obsolete, or finding they need something they haven't got. At the same time, oddly, the Internet itself is becoming less cloud-like and global by the day. When it started, it was thought to be everywhere and nowhere. But governments dislike the thought of a hovering cloud that can transport, say, an unwanted political idea or a libellous article instantly round the world. Technology that gives away an Internet user's geographical location is becoming increasingly effective, allowing local law-enforcers to home in on offenders. The Internet is no longer free as a bird.

Nor has it kept all the promises that were made for it in its first few wild and wonderful years. Nicholas Negroponte, an American technology guru, once declared that, thanks to the Internet, the children of the future were not going to know what nationalism was. Perfect communications, he assumed, would speedily resolve all misunderstandings between people everywhere. Alas, he was wrong.

But in another sense it is still true that, as the early pioneers claimed, "the Internet changes everything". Look at the world ten years ago, and again now, and spot the myriad differences the Internet has already made. Then cast your mind ten years forward, and imagine all the changes it might yet bring about. You'll hardly recognise the place.

I

THE NEW ECONOMY

Untangling e-conomics

Will the economic benefits of information technology match those of earlier technological revolutions? Quite probably, but the laws of economics will still apply

"**E**VERYTHING THAT CAN be invented has been invented." With these sweeping words, the Commissioner of the United States Office of Patents recommended in 1899 that his office be abolished, so spectacular had been the wave of innovation in the late 19th century. History is littered with such foolish predictions about technology. The lesson is that any analysis of the economic consequences of the current burst of innovation in information technology (IT – computers, software, telecoms and the Internet) should proceed with care. At one end, the Internet's boosters have boldly proclaimed it as the greatest invention since the wheel, transforming the world so radically that the old economics textbooks need ripping up. At the other extreme, sceptics say that computers and the Internet are not remotely as important as steam power, the telegraph or electricity. In their view, IT stands for "insignificant toys", and when the technology bubble bursts, its economic benefit will turn out to be no greater than that of the 17th-century tulip bubble.

The first programmable electronic computer, with a memory of 20 words, was built in 1946, but the IT revolution did not really start until the spread of mainframe computers in the late 1960s and the invention of the microprocessor in 1971. The pace of technological advance since then has been popularly summed up by Moore's Law. Gordon Moore, the co-founder of Intel, forecast in 1965 that the processing power of a silicon chip would double every 18 months. And so it has, resulting in an enormous increase in computer processing capacity and a sharp decline in costs (see chart 1). Scientists reckon that Moore's Law still has at least another decade to run. By 2010 a typical computer is likely to have 10m times the processing power of a computer in 1975, at a lower real cost.

Over the past 40 years global computing power has increased a billionfold. Number-crunching tasks that once took a week can now be done in seconds. Today a Ford Taurus car contains more computing power than the multimillion-dollar mainframe computers used in the Apollo space programme. Cheaper processing power allows computers to be used for more and more purposes. In 1985, it cost Ford $60,000 each time it crashed a car into a wall to find out what would happen in an accident. Now a collision can be simulated by computer for around $100. BP Amoco uses 3D seismic-exploration technology to prospect for oil, cutting the cost of finding oil from nearly $10 a barrel in 1991 to only $1 today.

The capacity and speed of communications networks has also increased massively. In 1970 it would have cost $187 to transmit *Encyclopaedia Britannica* as an electronic data file coast to coast in America, because transmission speeds were slow and long-distance calls expensive. Today the entire content of the Library of Congress could be sent across America for just $40. As bandwidth expands, costs will fall further. Within ten years, international phone calls could, in effect, be free, with telecoms firms charging a monthly fee for unlimited calls.

As communications costs plunge, more and more computers are being linked together. The benefit of being online increases exponentially with the number of connections. According to Metcalfe's Law, attributed to Robert Metcalfe, a pioneer of computer networking, the value of a network grows roughly in line with the square of the number of users. The Internet got going properly only with the invention of the World Wide Web in 1990 and the browser in 1993, but the number of users worldwide has already climbed to more than 350m, and may reach 1 billion within four years.

Between the extremes

IT is revolutionising the way we communicate, work, shop and play. But is it really changing the economy? The ultra-optimists argue that IT helps economies to grow much faster, and that it has also eliminated both inflation and the business cycle. As a result, the old rules of economics and traditional ways of valuing shares no longer apply. Cybersceptics retort that sending e-mail, downloading photos of friends or booking holidays online may be fun, yet the Internet does not begin to compare with innovations such as the printing press, the steam engine or electricity. Some even say that America's current prosperity is little more than a bubble.

Whom to believe? The trouble is that IT commentators go over the top at both extremes. Either they deny that anything has changed, or they insist that everything has changed. This survey will argue that both are wrong, and that the truth – as so often – lies somewhere in the middle. The economic benefits of the IT revolution could well be big, perhaps as big as those from electricity. But the gains will be nowhere near enough to justify current share prices on Wall Street. America is experiencing a speculative bubble – as it has done during most technological revolutions in the past two centuries.

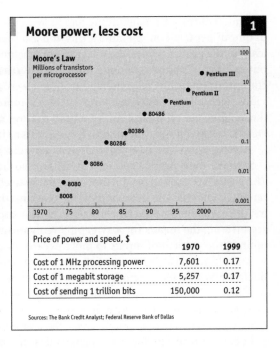

Moore power, less cost 1

Moore's Law
Millions of transistors per microprocessor

Price of power and speed, $	1970	1999
Cost of 1 MHz processing power	7,601	0.17
Cost of 1 megabit storage	5,257	0.17
Cost of sending 1 trillion bits	150,000	0.12

Sources: The Bank Credit Analyst; Federal Reserve Bank of Dallas

The Internet is far from unique in human history. It has much in common with the telegraph, invented in the 1830s, as Tom Standage, a journalist on *The Economist*, explains in his book *The Victorian Internet*. The telegraph, too, brought a big fall in communications costs and increased the flow of information through the economy. But it hardly turned conventional economic wisdom on its head.

Extra brain-power

The value of IT and the Internet lies in their capacity to store, analyse and communicate information instantly, anywhere, at negligible cost. As Brad DeLong, an economist at the University of California at Berkeley, puts it: "IT and the Internet amplify brain power in the same way that the technologies of the industrial revolution amplified muscle power." But is IT really in the same league as previous technological revolutions? There are several tests.

First, how radically does it change day-to-day life? Arguably, the railways, the telegraph and electricity brought about much more dramatic changes than the Internet. For instance, electric light extended the working day, and railways allowed goods and people to be moved much more quickly and easily across the country. Yet the inventions that have the biggest scientific or social impact do not necessarily yield the biggest economic gains. The printing press, seen by some as the most important invention of the past millennium, had little measurable effect on growth in output per head. In scientific terms, the Internet may not be as significant as the printing press, the telegraph or electricity, but it may yet turn out to have a bigger economic impact. One reason is that the cost of communications has plummeted far more steeply than that of any previous technology, allowing it to be used more widely and deeply throughout the economy. An invention that remains expensive, as the electric telegraph did, is bound to have a lesser effect.

A second test of a new technology is how far it allows businesses to reorganise their production processes, and so become more efficient. The steam age moved production from the household to the factory; the railways allowed the development of mass markets; and with electricity, the assembly line became possible. Now computers and the Internet are offering the means for a sweeping reorganisation of business, from online procurement of inputs to more decentralisation and outsourcing.

The ultimate test, however, is the impact of a new technology on productivity across the economy as a whole, either by allowing existing products to be made more efficiently or by creating entirely new products. Faster productivity growth is the key to higher living standards. After years when people puzzled over the apparent failure of computers to boost productivity, there are signs at last that productivity growth in America is accelerating. The question is whether that faster growth is sustainable. Undeniably, though, America's economy has had a fabulous decade in which it achieved both faster growth and lower inflation, and some part of that is due to IT.

And whatever the impact of IT so far, there is more to come. Paul Saffo, who heads the Institute for the Future, in California, believes that the IT revolution has only just begun, in terms of both innovation and the adoption of new technologies. Corporate America's R&D has increased by an annual average of 11% over the past five years, which suggests that innovation will go on. As yet, only 6% of the world's population is online; even in the rich world, the figure is only 35%. Only a third of American manufacturing firms are using the Internet for procurement or sales. All technologies follow an S-shaped path (see chart 2). They are slow to get going, but once they reach critical mass the technology spreads fast. The world may already be half-way up the curve for computers, but for the Internet it is only at the bottom of the steep part, from where it is likely to take off rapidly. Moreover, IT is only one of three technological revolutions currently under way. Together with fuel-cell technology, and genetics and biotechnology, it could create a much more powerful "long wave" than some of its predecessors.

Even so, predictions about future growth must be kept in perspective. Those who claim that technology has created a new growth paradigm that will allow America's GDP to keep expanding at well over 4% a year do not realise just how bold their forecasts are. That sort of annual rate implies growth in GDP per head of more than 3%. For that to

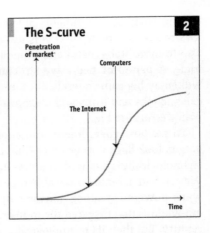

materialise, computers and the Internet would need to be a far more important engine of growth than steam, railways or electricity. Through most of the 19th century America's GDP per head grew by less than 1.5% a year, and in the 20th century by an average of just under 2%. In truth, many current expectations for American growth are probably unrealistic.

On the other hand, global growth may well turn out to be faster than in the past. America has been the first to embrace the IT revolution and the new economy, which is why so much of the evidence in this survey is concentrated in that country. But it is no longer alone. A later section of the survey will argue that if the rewards from IT are significant in America, the gains in Europe, Japan and many emerging economies could be even bigger. If so, this could yet prove to be the biggest technological revolution ever for the world as a whole.

So is it true that the "new economy" is making a nonsense of the laws of economics? It is argued that rules for, say, monetary and antitrust policy that worked in the age of steel and cars no longer apply now that computers and networks hold sway. But as Carl Shapiro and Hal Varian neatly put it in their book *Information Rules*: "Technology changes, economic laws do not." The business cycle has not really been eliminated; if economies grow too fast, inflation will still rise; share prices still depend on profits; and governments still need to remain on their guard against the abuse of monopoly power.

Don't burn the textbooks

But perhaps the most important economic rule of all is that new technology is not a panacea that cures every economic ill. To reap the full benefits from IT, governments still need to pursue sound policies. America's recent economic success is not due to new technology alone, but also to more stable fiscal and monetary policies, deregulation and free trade. A period of pervasive structural change lies ahead. Economies will enjoy big gains overall, but these will not be evenly spread. Many existing jobs and firms will disappear. In this environment, the risks of policy errors are high.

To see how governments can choke the economic benefits of innovation, look back 600 years to China, which at that time was the most technologically advanced country in the world. Centuries before the West, it had invented moveable-type printing, the blast furnace and the water-powered spinning machine. By 1400 it had in place many of the innovations that triggered the industrial revolution in Britain in the 18th century. But then its technological progress went into reverse, because its rulers kept such tight control on the new technology that it could not spread. It is a warning that the fruits of the IT revolution should not be taken for granted.

Elementary, my dear Watson

How information technology can boost economic growth

IN THE 1940S Thomas Watson, then chairman of IBM, predicted that the world market for computers would add up to five; he simply could not foresee any commercial possibilities. Today there are around 300m active computers in the world, so the economic impact of IT will turn out to be somewhat bigger than Mr Watson might have guessed. But how big?

People nowadays take it for granted that they will grow richer year by year. Yet for most of human history, growth in world output per head averaged little more than 0.1% a year. It was not until the late 18th century that growth accelerated, to an average of 1.2% a year over the past 200 years (see chart 3), thanks to a spurt in technological innovation. Since then, the world has seen four main waves of innovation. The first, from the 1780s to the 1840s, was the industrial revolution in Britain, fuelled by steam power; the second, from the 1840s to the 1890s, was the railway age; the third, from the 1890s to the 1950s, was driven by electric power and the car. Now we are in the information age.

People are often frightened of technological change. Yet the world would be much more frightening without innovation. Economies have limited resources of capital and labour. So without better ways to use these resources, growth would soon run out of steam.

Traditional models of growth developed in the 1950s concentrated largely on inputs of capital and labour, and had nothing to say about technological change. It was seen as exogenous, something that rained down from heaven. But a new theory, developed in the 1980s by Paul Romer and others, put technological change at centre stage. This "new growth theory" regards knowledge creation as endogenous, responding to market incentives such as improved profit opportunities or better education. Rather than raining down at a steady rate, the pace of technological change depends partly on governments and firms. Mr Romer argues that the economic incentives for innovation have strengthened in recent years. Raising finance for innovation has become easier, and a bigger global market has increased the likely return. Global R&D as a share of GDP has increased. It is claimed, strikingly, that about 90% of all the scientists who have ever lived are alive today. The pace of innovation

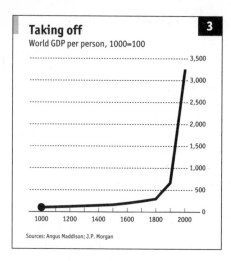

Taking off `3`
World GDP per person, 1000=100

- 3,500
- 3,000
- 2,500
- 2,000
- 1,500
- 1,000
- 500
- 0

1000 1200 1400 1600 1800 2000

Sources: Angus Maddison; J.P. Morgan

does not just seem to be faster: it really has increased.

Taking the plunge

A good gauge of the pace of technological change is the rate of decline in the cost of a new technology. Over the past three decades, the real price of computer processing power has fallen by 99.999%, an average decline of 35% a year. The cost of telephone calls has declined more slowly, but over a longer period. In 1930, a three-minute call from New York to London cost more than $300 in today's prices; the same call now costs less than 20 cents – an annual decline of around 10%.

These price plunges are much bigger than those in previous technological revolutions. The first steam engines were little cheaper than water power. By 1850 the real cost of steam power had fallen by only 50% from its level in 1790. The building of the railways reduced freight rates across America by 40% in real terms between 1870 and 1913, an annual decline of only 3%. The introduction of the telegraph hugely reduced the time it took to send information over long distances, but the service remained expensive. In the 1860s, a transatlantic telegram cost $70 a word in today's prices. Over the next decade the cost fell, but a 20-word message still cost the equivalent of around $200 to send. Today a 20-page document can be e-mailed for a mere cent. Electricity prices fell more steeply, but still by an average of only 6% a year in real terms between 1890 and 1920.

Thanks to rapidly falling prices, computers and the Internet are being adopted more quickly than previous general-purpose technologies, such as steam and electricity. It took more than a century after its invention before steam became the dominant source of power in Britain. Electricity achieved a 50% share of the power used by America's manufacturing industry 90 years after the discovery of electromagnetic induction, and 40 years after the first power station was built. By contrast, half of all Americans already use a personal computer, 50 years after the invention of computers and only 30 years after the microprocessor was invented. The Internet is approaching 50% penetration in America 30 years after it was invented and only seven years since it was launched commercially in 1993.

In addition to plunging prices, computers and the Internet have four other noteworthy features:

- IT is pervasive: it can boost efficiency in almost everything a firm does, from design to marketing to accounting, and in every sector of the economy. The productivity gains of steam, electricity and railways were mainly concentrated in the manufacture and distribution of goods. This could be the first technological revolution to boost productivity in services, from health care and education to finance and government. That would be no small matter: services account for nearly three-fifths of America's GDP.
- By increasing access to information, IT helps to make markets work more efficiently. Economists at UBS Warburg suggest that the "new economy" should really be called the "nude economy" because the Internet makes it more exposed and transparent. The Internet allows consumers to seek the lowest price, and firms to get quotes from more suppliers; it also reduces transaction costs and barriers to entry. In other words, it moves the economy closer to the textbook model of perfect competition, which assumes abundant information, many buyers and sellers, zero transaction costs and no barriers to entry. IT makes these assumptions a bit less far-fetched. (However, it also seems to increase monopoly power in some industries, which will be discussed in a later section of this survey.)

  Better-informed markets should ensure that resources are allocated to their most productive use. Farmers can get instant information on weather, prices and crop conditions in other regions. Manufacturers can track changes in demand more closely via direct links to electronic scanners in shops.
- IT is truly global. More and more knowledge can be stored as a string of zeros and ones and sent anywhere in the world at negligible cost. Information technology and globalisation are intimately linked. By reducing the cost of communications, IT has helped to globalise production and capital markets. In turn, globalisation spurs competition and hence innovation, and speeds up the diffusion of new technology through trade and investment.

◪ IT speeds up innovation itself, by making it easier and cheaper to process large amounts of data and reducing the time it takes to design new products. Thanks to ever more powerful computers, the mapping of the human genome, completed in 2000, took much less time than first expected.

Net gains

Many economists believe that although computers are undoubtedly useful on their own, it will take the Internet to unlock their full economic potential. E-commerce still accounts for only 1% of total sales in America, but it is growing rapidly. Dotcom firms, such as Amazon and eBay, have become household names, but far more important from an economic point of view will be business-to-business (B2B) e-commerce, linking buyers and sellers electronically along the supply chain. The Gartner Group, a consultancy, forecasts that global B2B e-commerce will reach $4 trillion by 2003, compared with less than $400 billion of online sales to consumers.

The best way to analyse the impact of the Internet on the economy is as a fall in the cost of an input, in this case information. Expressed diagrammatically, this pushes the aggregate supply curve (an economy's productive potential) out to the right (see chart 4), in exactly the same way as the invention of the wheel or electricity did in the past. Assuming no change in aggregate demand (D1), the equilibrium level of production rises from Q1 to Q2, and the price level falls from P1 to P2.

B2B e-commerce can cut firms' costs in several ways. First, it reduces procurement costs, both by making it easier to find the cheapest supplier and through efficiency gains. It is much cheaper to place an order online, and there are likely to be fewer errors in orders and invoicing. That may seem trivial, but Cisco reports that a quarter of its orders used to have to be reworked because of errors in its phone and fax ordering system. When it switched to online ordering, the error rate fell to 2%, saving the company $500m. British Telecom claims that buying goods and services online reduces the cost of processing a transaction by 90% and cuts the direct costs of goods and services it buys by 11%.

A second possible saving is from much lower distribution costs for goods and services that can be delivered electronically, such as financial services, software and music. The marginal cost to a bank of a transaction over the Internet is a mere cent, compared with 27 cents via a cash machine, 52 cents by telephone and $1.14 by bank teller. Online commerce also allows more efficient supply-chain management, cutting out layers of middlemen. And lastly, better information reduces the need for firms to keep large stocks. Dell Computer's build-to-order model completely eliminates inventories, and is being widely copied.

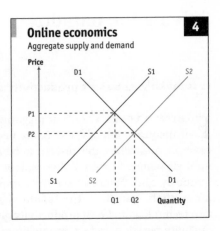

Online economics 4
Aggregate supply and demand

The B2B exchanges being set up by car, steel, construction and aerospace firms will provide a more efficient marketplace for buyers and sellers to exchange products. Such exchanges are likely to spring up in most industries. GM, Ford, Daimler-Chrysler and Renault-Nissan plan to move all their business to a joint electronic exchange with a turnover of $250 billion and 60,000 suppliers. According to one estimate, dealing with suppliers online could reduce the cost of making a car by 14%.

The biggest savings are likely to come in procurement. A report by Goldman Sachs, an investment bank, estimates that online purchasing could save firms anything from 2% in the coal industry to perhaps 40% in electronic components. As a result of such cost savings, Goldman Sachs reckons, B2B e-commerce could boost the level of output in the rich economies by an average of 5% over time. More than half of that would come through within ten years, an increase of 0.25% a year in the rate of growth over the next decade. Add in the potential indirect cost savings from the Internet as firms reorganise the way they do business, and the total gains could be considerably bigger.

The popular distinction between the old and the new economy completely misses the point. The most important aspect of the new economy is not the shift to high-tech industries, but the way that IT will improve the efficiency of all parts of the economy, especially old-economy firms. This distinction will be examined further in a later section of this survey. But first those productivity gains have to materialise – and economists find it impossible to agree on how far IT has already started to lift America's productivity growth.

Solving the paradox

IT is making America's productivity grow faster, but for how long?

EIGHTEEN YEARS AGO *Time* magazine declared the computer "person of the year". But as for so many people and firms feted on the front cover of magazines, this proved to be a curse: computers failed to live up to their billing. In 1987, Robert Solow, a Nobel laureate in economics, famously said: "You can see the computer age everywhere but in the productivity statistics." The failure of massive investment in information technology to boost productivity growth became known as the productivity paradox. In fact, productivity growth slowed sharply in most countries in the 1970s and 1980s. The surge in America's productivity growth since the mid-1990s has therefore been seized upon with relish. Has the productivity paradox now been solved?

This is no trivial question, for productivity growth is the single most important economic indicator. It determines how fast living standards can grow. The reason why the average American today is seven times better off than his counterpart at the turn of the century is that he is seven times as productive. And faster growth not only lifts living standards, it also boosts tax revenues and makes it easier to pay for tomorrow's pensions.

Spending on IT equipment and software now accounts for about half of all investment by American firms. So why has it taken so long for that investment to show up in faster productivity growth? History suggests that there were also long lags before both steam power and electricity boosted productivity. Work by Paul David, an economist at Oxford University, shows that productivity growth did not accelerate until 40 years after the introduction of electric power in the early 1880s. This was partly because it took until 1920 for at least half of American industrial machinery to be powered by electricity. But firms also needed time to figure out how to reorganise their factories around electric power to reap the efficiency gains.

Mr David suggests that a technology will start having a significant effect on productivity only when it has reached a 50% penetration rate. American computer use has reached the 50% mark only recently, and other rich economies still lag behind (see chart 5). That puts IT at roughly the same stage now that electricity had reached in 1920. Almost exactly on cue, growth in labour pro-ductivity in America's business sector has increased to an annual average of 2.9% since 1996, from an average of 1.4% in 1975–95 (see chart 6). In the year to the second quarter of 2000, productivity surged by 5.2%.

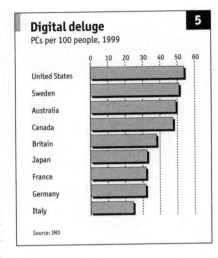

Digital deluge
PCs per 100 people, 1999

Source: IMD

But economists disagree on whether this increase in produc-tivity growth is sustainable. To complicate things, there are two different measures on offer: labour productivity (output per man-hour), and total factor productivity (TFP, which takes account of the efficiency with which both capital and labour inputs are used). To most people, it is labour productivity that matters, because this is what ultimately determines living standards. Economists, though, get more excited about TFP growth, which they see as a costless way of boosting growth without increasing scarce inputs. Faster TFP growth automatically increases labour-productivity growth.

IT can boost growth in labour productivity in three ways: by increas-ing the amount of capital deployed per worker (ie, capital deepening), as firms invest in IT; by speeding up TFP growth in IT-producing industries, thanks to technical advances; and by increasing TFP growth in sectors that use IT. Nobody can deny that productivity in the sector that pro-duces IT goods has surged, with growth in the 1990s averaging 24% a year. The disagreement is about the effect of IT on the rest of the economy.

Of the flurry of studies on the spurt in American productivity that have appeared over the past year, one of the most optimistic is by Stephen Oliner and Daniel Sichel at the Federal Reserve in Washington. It probably also comes closest to the current thinking of the Fed's chair-man, Alan Greenspan. The two economists conclude that IT has been the key factor behind America's improved productivity growth, and they expect a substantial portion of it to persist. They estimate that nearly half of the acceleration in productivity growth between the first and second halves of the 1990s was due to capital deepening as firms invested in IT. The other half was due to faster TFP growth, of which

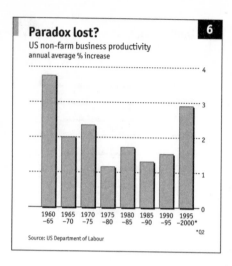

Paradox lost? `6`

US non-farm business productivity
annual average % increase

Source: US Department of Labour

two-fifths came from efficiency gains in computer
production. The authors conclude that roughly two-thirds
of the increase in labour productivity was due directly to
the production of or investment in computers.

Another study, by Dale
Jorgenson, at Harvard University, and Kevin Stiroh, at
the New York Fed, reaches
similar conclusions: heavy
investment in computers and
faster productivity growth in
the computer industry have
substantially boosted labour-productivity growth. However, the authors
worry that although TFP growth outside the computer sector has
increased, there is little evidence that this is linked to IT. Indeed, the sectors which have invested most in IT have generally seen smaller productivity gains. This could be due to measurement problems, but for the
moment it casts some doubt on the argument that IT is boosting TFP
growth throughout the economy.

Nevertheless, Messrs Jorgenson and Stiroh conclude that labour-productivity growth of around 2.3% a year could be sustained over the next
decade. That would allow America's GDP to grow at an average rate of
almost 3.5% a year without pushing up inflation, compared with an average growth rate of around 3% in the two decades to 1995.

Tools or toys?

However, Robert Gordon, an economist at Northwestern University and
an outspoken new-economy sceptic, is less impressed by America's productivity "miracle" than other economists. He reckons that the entire
increase in total factor productivity outside the computer sector is due to
the economic cycle. At times of rapid growth, firms work employees
harder, so productivity rises; but then it falls again in the next downturn.
Moreover, he finds that after excluding the manufacture of all durable
goods as well as of computers, there has been absolutely no increase in
labour productivity in the remaining 88% of the economy, after adjusting for the cycle. Yet this is where most of the investment in computers

has taken place. He concludes that the productivity paradox is alive and well.

Mr Gordon is not surprised that IT has failed to lift TFP growth throughout the economy. Computers and the Internet, he says, do not rate as an "industrial revolution", as did electricity and the car. Much Internet activity, he argues, is merely a substitute for things that are already being done. For example, downloading music simply replaces buying a CD; it does not create new products, in the way that electricity prompted the invention of the vacuum cleaner and the fridge. Indeed, the Internet can even reduce productivity in the workplace. The traffic on many consumer-oriented websites, he notes, peaks in the middle of the working day, not in the evening.

The main reason why Mr Gordon's conclusions differ from those of other researchers is that he adjusts productivity growth for the effects of the economic cycle. This is a reasonable thing to try to do, because falling unemployment shows that output has been growing faster than trend. But many economists are unhappy about the way he has done it. They are convinced that at least part of the increase in productivity growth is structural, if only because it is highly unusual for productivity to accelerate so late in an economic boom. The usual pattern is for it to slow down at that point.

Mr Greenspan recently dismissed the notion that the increase in productivity growth is largely cyclical. He also argued that the underlying rate of productivity growth was still accelerating. A study by economists at the OECD seems to give his view some support. After adjusting for the economic cycle, it concludes that labour-productivity growth and TFP growth both increased significantly in America in the 1990s.

A second point of difference is that Mr Gordon's test of the economic importance of IT – whether it has boosted TFP growth in sectors that use IT – is tougher than that of the other studies, which merely ask whether IT has lifted labour-productivity growth in the economy as a whole.

Why does the composition of the increase in labour-productivity growth matter? If faster labour-productivity growth is largely due to capital deepening, growth will remain high only if the price of IT equipment continues to fall. If technological progress in the IT sector were to slow, then overall productivity growth would be hit by a double whammy: the rate of TFP growth in the IT industries would fall, and the pace of investment in IT in the rest of the economy would slow.

However, scientists are confident that the rapid rate of innovation, and hence the fall in prices, will continue for at least another decade. If so, capital deepening will persist for some time. In any case, the split between capital deepening and TFP growth is very sensitive to the method of calculation. Using a wider measure of TFP, the OECD estimates that cyclically adjusted TFP growth has been growing faster in recent years than Mr Gordon suggests.

Mr Gordon's wider dismissal of IT is also somewhat unfair. He concentrates mainly on consumer use of the Internet, yet the biggest economic effect is likely to flow from B2B e-commerce. This has only just got going, so any productivity gains would not be expected to show up yet. And although IT may not yet have created many new products, it has opened up many new opportunities. Genetics and biotechnology, mobile phones, online auctions and financial derivatives would all be impossible without low-cost computer processing power.

The evidence from aggregate economic data may be mixed, but studies that look at individual firms suggest that computers have yielded substantial gains. In an analysis of 600 big American firms between 1987 and 1994, Erik Brynjolfsson at MIT and Lorin Hitt at the University of Pennsylvania found that investment in computers appeared to boost annual TFP growth by 0.25–0.5%. The productivity gains got bigger over longer periods, confirming that it takes time for firms to reorganise their business before they reap the full benefits of IT. Their research also shows that firms that coupled IT investment with changes in their organisational structure, such as decentralisation, enjoyed the biggest productivity gains from IT.

Messrs Brynjolfsson and Hitt argue that much of the benefit of IT comes in the form of improved product quality, time savings and convenience, which rarely show up in official macroeconomic data (see "virtual guesswork" on page 21). Microeconomic studies are able to identify these gains because firms whose products offer such intangible benefits will enjoy higher revenues.

Outside America, much less research has been done on the economic effects of IT. Comparisons are difficult because different countries use different methods to measure IT and to allow for quality improvements. All things considered, it seems likely that official figures understate European productivity growth relative to America's. IT investment rose strongly in all the G7 economies in the 1990s, but its contribution to growth is much less significant in Japan and most European economies than in America, largely because IT equipment accounts for a much smaller share of the total capital stock: only 3% in Japan and Germany against 7% in America.

By far the best international study has been done by Andrea Bassanini, Stefano Scarpetta and Ignazio Visco, at the OECD. They find that after adjusting for the economic cycle, annual TFP growth increased by at least half a percentage point in the 1990s in Australia, Canada and the Scandinavian economies as well as America, but it fell in Japan and the big European economies.

Two technologies 7
US business productivity
annual average % increase

	Labour productivity	Total factor productivity
Electricity era		
1909–19	2.1	1.3
1919–29	2.3	2.0
IT era		
1985–95	1.4	0.5
1995–2000	2.9	1.5*
Sources: US Bureau of Economic Analysis; OECD		*Estimate

Ranking revolutions

America has invested more, and earlier, in IT than the other big economies, so the economic benefits would be expected to emerge there first. It is still too early to judge how IT lines up against previous industrial revolutions, but it is possible to compare some of the growth forecasts now being made for the next decade with actual growth rates during the eras of steam and electricity.

Suppose, optimistically, that America's average rate of labour-productivity growth in the late 1990s were to be sustained for the next couple of decades as IT and the Internet continue to transform the way business is done. This would allow GDP per head to grow by around 3% a year, much faster than during the first industrial revolution's peak in the mid-19th century, when GDP per head grew by an average of around 1.5%, or during the electricity age, when growth accelerated to just over 2% in the 1920s.

In fact, previous technological revolutions resulted in more modest rates of overall productivity growth than most people realise. In its prime years in the 19th century, the world's first industrial revolution produced average labour-productivity growth in Britain of barely 1% a year. Electricity provided a bigger spark, with America's labour-productivity growth in manufacturing jumping to more than 5% a year in the 1920s. But the productivity growth rate across the whole economy was a less impressive 2.3% (see table 7).

New-paradigmers who suggest that rates of productivity growth of 3–4% a year are sustainable for the next decade or so are really saying that IT will have a far bigger economic impact than electricity, telephones

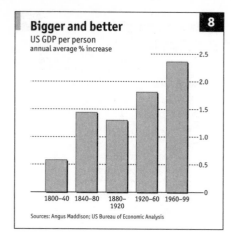

Bigger and better 8

US GDP per person
annual average % increase

Sources: Angus Maddison; US Bureau of Economic Analysis

and cars. That is very ambitious. More likely, America's long-term trend rate of labour productivity and hence per capita growth might be lifted to an annual 2.5%. That might not sound much, but it would make IT at least as significant as electricity.

Over the past 200 years, growth in GDP per head has gradually accelerated from 0.6% a year in 1800–40 to 2.3% in 1960–99 (see chart 8). A growth rate in GDP per head of around 2.5% over the next few decades would fit this trend. Paul Romer, the father of new growth theory, believes that the rate of growth has increased over time because of increasing returns to knowledge. Knowledge builds upon itself: the more that mankind discovers, the better it gets at the process of discovery. That rings true, but as Jack Triplett at the Brookings Institution points out, simply to keep productivity growth constant, the pace of introduction of new technology and new products needs to keep increasing.

The recent spurt in labour-productivity growth in America is almost certainly exaggerated by the current economic boom. But at the same time the official figures will probably understate the likely full effect of IT on the structural rate of productivity growth, for two reasons. First, official statistics significantly understate growth. Second, many economists believe that the Internet will trigger faster productivity growth by prompting firms to reorganise from top to bottom. Bigger gains may therefore lie in the future. But will they be big enough to justify the current level of share prices?

Virtual guesswork

IT IS IRONIC that in this information age there is a distinct dearth of statistics with which to measure the true economic impact of computers. This is not just a matter of a lack of data on new businesses, such as e-commerce. The main problem is that statistics designed for the industrial age are ill-equipped to measure output in the 21st century.

A breakdown of productivity growth by industry shows that in many of the sectors that use computers most intensively, notably banking and education, productivity actually declined in the 1990s (see chart 9). So is IT a waste of money? Not necessarily: more likely, the figures are flawed. The service sector has always been hardest to measure, and IT and the Internet have exacerbated the problem because much of their benefit comes in the shape not of cost savings, but of increased product quality, convenience and customer service. Such gains rarely show up in GDP figures.

When the quality of a product is changing rapidly, number-crunchers find it tricky to work out which part of an increase in nominal spending is due to higher prices and which part to gains in real output. If statisticians had their way, products would all be like Henry Ford's model-T car, introduced in 1908 and basically unchanged for two decades; but in practice change is becoming ever more rapid, and the faster the rate of product innovation, the bigger the statistical error is likely to be. If quality improvements are ignored, productivity growth will be understated. For example, the introduction of ATMs and online banking has substantially improved the service offered to customers, yet the official statistics take little account of these benefits. One study of American banking puts the increase in output between 1977 and 1994 at 7% a year after allowing for technology-related improvements in the quality of service, whereas the official statistics recorded an annual increase of only 1.3%.

Health-care statistics look particularly fishy. According to official figures, productivity in America's health-care industry has fallen by more than 20% over the past decade; indeed, total factor productivity is now almost 40% down on 1960. Yet there is little doubt that medical care is much better today than it was 40 years ago. Diagnosis is more accurate, patients need to spend less time in hospital, and new medical equipment allows less invasive treatment. All this implies big cost savings, as well as greater convenience for the patient. Again, however, official

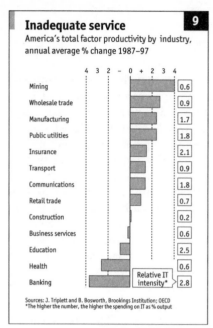

Inadequate service 9

America's total factor productivity by industry, annual average % change 1987–97

Industry	Value	Relative IT intensity*
Mining		0.6
Wholesale trade		0.9
Manufacturing		1.7
Public utilities		1.8
Insurance		2.1
Transport		0.9
Communications		1.8
Retail trade		0.7
Construction		0.2
Business services		0.6
Education		2.5
Health		0.6
Banking		2.8

Sources: J. Triplett and B. Bosworth, Brookings Institution; OECD
*The higher the number, the higher the spending on IT as % output

output figures, based on indicators such as the number of doctors and hospital beds, fail to take account of higher quality, so growth is understated.

Official statistics also fail to capture the benefit of the increasing choice consumers enjoy. In Henry Ford's day customers could have any colour they wanted, so long as it was black. Since then choices have proliferated. In the early 1970s Americans could buy five types of running shoes; in 1999 they could pick from 285. There used to be four types of milk; now there are 19. IT also makes it easier and cheaper to offer personalised goods and services. CDs, computers and even golf clubs can be customised for an individual buyer. Michael Cox and Richard Alm, at the Federal Reserve Bank of Dallas, argue that mass production was all about producing more stuff at lower cost; mass customisation is about producing the right stuff. Consumer satisfaction has increased, but GDP fails to capture such intangible benefits.

Lost in cyberspace

There are good reasons, therefore, to suspect that official figures understate productivity growth – and by more than they may have done in the past. To be fair, America's Department of Commerce has made big strides over the past decade in improving its numbers. It has pioneered new methods for measuring changes in the quality and price of computers, and since 1999 it has included software as part of IT investment. Most other countries are some way behind America in measuring the importance of IT. Morgan Stanley, a bank, reckons that if Europe used the same price deflator for IT as America to calculate its GDP, its growth rate might turn out to be half a percentage point higher.

In this statistical fog, only one thing seems certain: the productivity of official statisticians has seriously declined.

Bubble.com

All technological revolutions carry risks as well as rewards

IF INFORMATION TECHNOLOGY is lifting America's rate of growth, surely that justifies the current lofty heights of share prices? Sadly, no. Wall Street still looks dangerously high relative to likely future profits. Every previous technological revolution has created a speculative bubble, and there is no reason why IT should be different.

New-economy fanatics argue that in this new world of rapid technological change, old methods of share valuation have become irrelevant. Profits are for wimps. But both economic theory and history suggest otherwise. In his book "Irrational Exuberance", Robert Shiller, an economist at Yale University, tracks the price/earnings (P/E) ratio of America's S&P 500 over 120 years, a period that covers huge technological change: America's railway boom, electricity, telephones, radio and cars. With each wave of technology, share prices soared and later fell. Ominously, though, prices now are higher in relation to profits than they have ever been before (see chart 10).

The inventions of the late 19th century drove p-e ratios to a peak in 1901, the year of the first transatlantic radio transmission. By 1920 shares prices had dropped by 70% in real terms. The roaring twenties were also seen as a "new era": share prices soared as electricity boosted efficiency and car ownership spread. After peaking in 1929, real share prices tumbled by 80% over the next three years.

There are many similarities between the Internet today and Britain's railway mania in the 1840s. Would-be rail millionaires raised vast sums of money on the stockmarket to finance proposed lines. Most railway companies never paid a penny to shareholders, and many went bust, largely because over-investment created excess capacity. The Great Western Railway was for decades the most admired railway company in Britain, yet anyone who had bought shares at its launch in 1835 (at a fraction of their peak in 1845) and held them until 1913 would have seen an annual return of only 5%. Even so, the railways brought huge economic benefits to the economy long after share prices crashed. The lesson is that although IT may be causing a bubble, it may still produce long-term economic gains. But many investors could lose their shirts.

Current valuations of dotcom shares seem to assume that they are

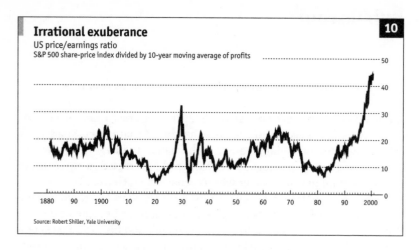

Irrational exuberance `10`
US price/earnings ratio
S&P 500 share-price index divided by 10-year moving average of profits

Source: Robert Shiller, Yale University

going to grab a huge chunk of market share from existing firms. Yet history suggests that the gains from technological revolutions often go to unexpected quarters. The biggest winners from America's railway boom were small firms and farmers who benefited from the opening up of the continent. It is a sobering thought that 99% of the 5,000 railway companies that once existed in America are no longer around. The same is true of 2,000 car firms. And according to a study by Goldman Sachs, profits and share prices of the early electricity firms were disappointing, despite the industry's profound effect on the economy.

Goldilocks and the three myths

This time, it is said, will be different. Three popular explanations are offered to justify the high level of share prices: that profits will grow faster; that the economy and hence equities have become less risky; and that lower, more stable inflation will reduce real interest rates. True or false?

Faster productivity growth should indeed boost profits. But even if America's future average economic growth is as steep as optimists believe, say just over 4% a year, the current level of share prices implies that profits will rise even faster. Looking back over time, the share of profits in America's national income has been fairly stable. During the electrification of American industry, profits actually fell slightly as a share of GDP: although firms' costs came down, strong competition ensured that savings were passed on to consumers.

Profits can outpace GDP for a period if companies build more monopoly power, but if IT reduces barriers to entry and increases competition, profit margins are more likely to shrink than widen. B2B e-commerce will cut costs, but one company's lower costs are another's lower prices and profits. Moreover, by increasing price transparency, the Internet shifts power from producers to consumers, and so is likely to squeeze average profit margins. Consumers, not producers, will reap the lion's share of the rewards of IT. If the Internet increases competition, it cannot also increase the share of profits in GDP. It has to be one or the other.

What about the argument that the equity-risk premium (the premium that investors demand over risk-free assets such as government bonds) has fallen close to zero because of greater economic stability? Not only have fiscal and monetary policies become more prudent, it is said, but IT also helps to smooth the economic cycle. B2B e-commerce allows firms to hold fewer stocks and, by providing better information about changes in demand, it also reduces the risk of overstocking. But the business cycle has been declared dead many times in the past, usually just before a recession. A hard landing remains a real risk for the American economy.

Another flaw with this theory is that stockmarket volatility remains high. Valuing individual shares becomes more difficult in periods of rapid change because current revenues and profits may be poor predictors of future performance. As in all technological revolutions, some new firms will make fortunes but most will fail, which implies a greater dispersion of returns from individual shares. This surely makes equities more risky, not less.

The proposition that the real rate of interest (used to discount future profits) will be lower in future because of lower and more stable inflation is another myth. If the IT revolution increases profitable investment opportunities, then the equilibrium real interest rate must rise in order to encourage households to save more to finance the higher level of investment.

Taking account of all this, share prices seem to assume there will be an implausible rate of growth in profits. Martin Barnes of The Bank Credit Analyst, a research group based in Canada, calculates that assuming the equity-risk premium is 2% (well below its historical average of 5%), then the current level of America's S&P 500 index implies profits growth of over 6% a year in real terms over the next three decades, well above likely GDP growth.

A crash in share prices would make a serious if temporary dent in

America's economy, even though the underlying economic benefits of IT would continue. Stockmarkets in many other economies are over-valued too, but a bursting of the bubble would claim many more victims in America than in Japan or Europe, partly because far more people own shares and partly because in recent years American households and companies have borrowed huge sums in the expectation that share prices will continue to climb. Sooner or later they are likely to discover their mistake.

Monetary matters

Technological revolutions and financial bubbles seem to go hand in hand, but has the Fed done all it could to prevent a bubble inflating? Judged by the traditional test of inflation, the Fed has done a superb job, combining relatively low inflation with rapid growth. Setting monetary policy in the new economy is no easy task, because the old relationship between growth and inflation seems to have broken down. No central banker believes the popular claim that inflation is dead. In the long run inflation is determined by monetary conditions; the Internet can affect only relative prices, not the overall rate of inflation. But if America's productivity growth has indeed increased, it is safe for the Fed to allow the economy to grow a bit faster before touching the brakes. The snag is that nobody knows what the new rate of productivity growth is, and if the economy grows too fast, it will still overheat.

In the face of increased uncertainty about underlying productivity growth, many economists now argue that a central bank should not try to restrain an expansion until there is visible evidence that inflation is rising. A premature tightening could stifle investment and innovation. The Fed has, more or less, taken this advice. But, as a result, it has encouraged share prices to move even higher.

Should central banks worry more about the surge in share prices that has accompanied the IT revolution? In theory, there is a strong case for central banks to try to prick bubbles before they get too big. Letting out some air early on can help to prevent a painful crash later on. The problem is that it is hard for a central bank to know whether it is dealing with a financial bubble or a new era of much faster growth. Jan Hatzius, an economist with Goldman Sachs in New York, argues that when share prices and investment are booming as they were in America in the late 1990s, the correct policy is to raise interest rates regardless. If it turns out to be a bubble, then the central bank needs to raise interest rates to let out some air. If it is a new era of faster growth and new investment opportunities, then the equilibrium real interest rate (the rate at which monetary policy neither boosts nor restrains the economy) would rise, so the central bank would be right to move interest rates towards that level.

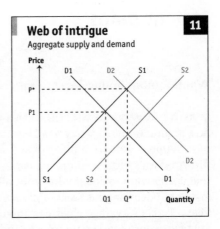

Web of intrigue   **11**
Aggregate supply and demand

Many people believe that the Internet makes a central banker's job easier by helping to hold down prices. But initially, the IT revolution might actually increase inflationary pressures, as shown in chart 11. In the long term, IT will shift the economy's aggregate supply curve from S1 to S2, but this will happen gradually. Meanwhile, investors will anticipate faster future growth in output and profits, pushing up share prices. This will boost households' wealth and encourage them to spend more, even before the increase in supply has materialised. As a result, the demand curve will shift to the right, from D1 to D2, pushing up the price level to P*. The risk is that if the increase in demand outstrips the increase in supply, inflation will rise unless the central bank raises interest rates. This could describe America today, as Mr Greenspan hinted in a speech early in 2000.

Central bankers clearly have their work cut out in these uncertain times. Yet for shouldering this onerous responsibility, Mr Greenspan is paid a comparatively modest $140,000 a year. Central banking is one of the very few knowledge jobs that have not benefited from fat pay rises in recent years.

Labour pains

Wanted: more brains, less brawn

GO TO WWW.PAYWATCH.ORG, a website set up by trade unionists
to monitor bosses' pay at all big companies, and tap in your annual
pay. Suppose you are a blue-collar worker employed by General Electric, America's biggest company, earning $25,000 a year. You will learn
that if you want to equal what your ultimate boss, Jack Welch, got paid
last year (including his stock options), you will have to work for 3,663
years. Today's average chief executive earns 475 times as much as the
average factory worker, up from a ratio of 42 in 1980. The new economy
is rewarding some more handsomely than others.

Until a few years ago, workers' main worry was that new technology
would create mass unemployment as people were replaced by computers and robots. America's boom during the past few years has muted
such fears: unemployment has fallen to 4%, the lowest rate for 30 years.
Ever since Britain's Luddites in the early 19th century smashed the
power looms and spinning jennies that threatened their livelihood,
people have feared that technological change would increase unemployment. Yet during two centuries of huge technological progress,
employment has risen almost continuously. Millions of jobs have been
destroyed, but even larger numbers of new jobs have taken their place.
For blacksmiths and coachmen, read car mechanics and airline pilots.

But that is little consolation if you are one of those whose job is
destroyed by new technology. A steel worker cannot easily get a job as
a computer programmer. Most of the jobs being lost as a result of IT are
concentrated among the low-skilled, whereas many of the new jobs
require good education and skills. As the demand for brains has risen
relative to the demand for brawn, so wage differentials have widened
in favour of the better-educated. Since 1979, average weekly earnings of
college graduates in America have risen by more than 30% relative to
those of high-school graduates (see chart 12), increasing the wage gap to
its widest for at least 60 years. The wage gap between college graduates
and high-school drop-outs has grown by twice as much. Since average
real wages rose relatively slowly for much of this period, the real pay of
the least educated has actually fallen over the past 20 years.

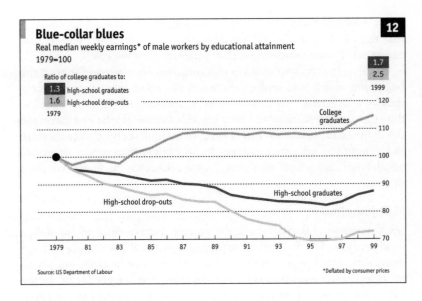

Blue-collar blues
12

Real median weekly earnings* of male workers by educational attainment
1979=100

Ratio of college graduates to:

1.3 high-school graduates
1.6 high-school drop-outs
1979

1.7
2.5
1999

Source: US Department of Labour

*Deflated by consumer prices

Why life isn't fair

Economists have put forward four main explanations for the increase in wage inequality: technological change; increased imports from low-wage developing economies; higher immigration of low-skilled workers; and the waning power of trade unions. All four have probably played a role, but most economists reckon that new technology is by far the most important factor. Trade by itself is simply not large enough to be the major culprit, and the timing is wrong. America's trade with developing economies grew much faster in the 1990s than in the 1980s, yet most of the rise in wage inequality dates from the 1980s.

There are two reasons why computers might increase the relative demand for better-educated and more highly skilled workers. First, low-skilled, routine jobs, done by clerical and production workers, can be automated and replaced by computers more easily than professional or managerial jobs. Second, computers complement skilled workers, increasing the return on the creative use of information, whether in designing a car, trading bonds or managing a company.

To shed some light on the factors behind rising wage inequality, Larry Katz, an economist at Harvard University, has examined the changes in wage differentials and skill levels over the 20th century. Educational standards have increased continuously over time, yet despite a larger supply of educated workers, the wage premium enjoyed by the better educated has increased in every decade since 1950. This suggests that the increase in the relative demand for skills started well before computers became widespread, but accelerated in the 1980s.

Even so, the pay gap today is still considerably narrower than at the beginning of the 20th century. Between 1900 and 1939, wage differentials by educational level were severely compressed. Factory electrification, like IT, also eliminated many unskilled manual jobs and increased the demand for skills, but this was more than offset by a huge increase in the supply of educated workers. In 1910 fewer than 10% of American youths had high-school diplomas; by the mid-1930s the figure had risen to 40%.

Skill premium

Mr Katz suggests that although the greater relative demand for skilled workers caused by IT must have played a part in the widening of wage inequality over the past two decades, slower growth in the supply of more educated workers may have been an even bigger factor. In the 1970s, the supply of educated workers surged in America as the baby-boom generation entered the workforce and college enrolment rose. But since then the education level of the workforce has improved much more slowly.

A comparison of the United States with Canada supports this argument. During the 1980s and 1990s the ratio between the earnings of university graduates and high-school graduates rose sharply in America, but fell in Canada. In both countries the demand for skills rose by similar margins, but the supply of educated workers rose much more rapidly in Canada than in the United States.

Looking at a wider range of countries, Mr Katz finds that where wage differentials between skilled and unskilled workers have widened the most, growth in the supply of better-educated workers has generally slowed down. In contrast, in France, Germany and the Netherlands, where wage differentials have not increased over the past two decades, the supply of educated workers has grown rapidly. Static wage differentials in continental Europe are usually explained by factors such as powerful trade unions and high minimum wages. But it is possible that faster expansion in the supply of well-educated workers is more important. This suggests that the real culprit behind rising inequality in America is not IT, but the government's failure to improve education and training.

Since the mid-1990s, wage inequality in America has flattened or even narrowed slightly, and growth in the relative demand for college-educated workers seems to have slowed, despite the continuing spread of computers. Will IT continue to favour better-educated workers in future?

Some economists argue that part of the increased wage premium enjoyed by skilled workers reflects the fact that they tend to be more flexible, so all technological change increases the relative demand for skills during a transitional period. But as technologies mature, the advantage of the better-educated wanes. If this is true, the wage premium for skill or education depends on the pace of innovation. As a technology matures, the skill premium will narrow. Perhaps this is happening now. However, a more likely explanation of why inequality has stopped rising in recent years is that America's economic boom has reduced the unemployment rate to a historic low and pushed up wages at the bottom end of the labour market.

An alternative, and more persuasive, theory about technology and jobs argues that each technological innovation favours different skills. Electricity and computers have both increased the relative demand for skilled workers, whereas the mechanisation of factories during the steam age in the 19th century increased the relative demand for unskilled workers. Highly skilled craftsmen, such as weavers, were replaced by machines and unskilled labour. So perhaps what matters is not the pace of innovation, but the type. IT, the current driver of change, favours better-educated workers, so during this particular wave of innovation the demand for such workers will go on growing.

Even so, if the government puts more effort into increasing the supply of well-educated workers, then America's wage inequality could narrow in future years. However, that still leaves the question of the growing divide between the information haves and have-nots. Richer and better-educated people are more likely to have a computer and access to the Internet (see chart 13). In 1998, 60% of Americans with incomes above $75,000 used the Internet, compared with under 20% of those with incomes below $25,000. More than 60% of college-educated workers, but only about 15% of high-school drop-outs, used the Internet. Poorer, less educated people are therefore at a double disadvantage. They have less access to information which might help them to get a better job, and they are shut out of e-commerce and the opportunity to seek lower prices that could most benefit the less well-off.

But the Internet also has a more direct effect on the labour market: a growing number of jobs are being advertised online. About 400 of the

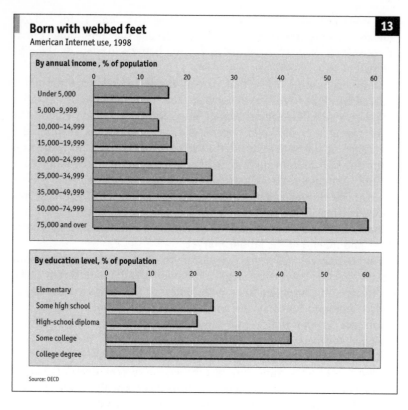

Born with webbed feet — **13**
American Internet use, 1998

By annual income , % of population

By education level, % of population

Source: OECD

world's 500 biggest firms (and over 90% of American ones) use websites for recruitment, and half accept applications online. Employment websites, such as Monster.com, can improve information and reduce search costs in the labour market. By better matching of vacancies and job seekers, this may reduce the level of unemployment consistent with stable inflation. But what will the Internet do to pay structures?

One view, based on standard trade theory, holds that by making it easier for workers to keep informed about job opportunities and pay rates, the Internet will help to create a uniform market for skills and reduce pay variations within occupations, just as Amazon.com has caused a move towards uniform book pricing across America.

A competing theory suggests that as the Internet expands the market for individuals with special talents, small differences in ability will lead to huge differences in rewards. This idea is based on the work of Sherwin Rosen, an economist at the University of Chicago. In a classic paper written in 1981, "The Economics of Superstars", he explained why in sectors such as sports and films a few top stars are paid vast sums whereas the runners-up lag far behind. This "winner-takes-all" principle already applies to an increasing number of occupations, including bond dealers, doctors, lawyers and chief executives. In these jobs, being slightly ahead of the pack is hugely important. An ambitious investment bank does not want the second-best bond dealer, so it pays well over the odds for the best. IT has expanded the market for such skills to a global scale, so the premium for superstar talent has increased.

David Autor, at MIT, suggests that the Internet could cause both effects in different parts of the labour market. In routine occupations, such as cashiers and clerks, pay is likely to become more uniform as technology reduces regional wage differentials. On the other hand, the superstar effect could spread to more occupations, such as teaching and software engineering, as the Internet increases the power of talented individuals to serve a bigger market.

But if winners take all in the labour market, does the knowledge economy reward companies in the same way?

Knowledge is power

Do we need a new competition policy for the new economy?

HOW MANY MICROSOFT computer programmers does it take to change a light bulb? None. Bill Gates will simply call a press conference and announce that the new standard is darkness. Every revolution has its enemies, and this time Microsoft has been cast as the main villain of the information economy. Whatever the outcome of the battle between Microsoft and the American government, some economists worry that because of the very nature of information and knowledge, which form the building bricks of the new economy, more and more monopolies like Microsoft are likely to emerge. But how does that fit with the accepted wisdom that IT and the Internet will make markets more efficient, and will therefore boost competition?

Economies are increasingly based on knowledge. Finding better ways of doing things has always been the main source of long-term growth. What is new is that a growing chunk of production in the modern economy is in the form of intangibles, based on the exploitation of ideas rather than material things: the so-called "weightless economy". In 1900 only one-third of American workers were employed in the service sector; now more than three-quarters are (see chart 14). More and more goods, too, from Mercedes cars to Nike trainers, have an increasing amount of knowledge embedded in them, in the form of design or customer service.

Economists have a problem with knowledge because it seems to defy the basic economic law of scarcity. If a physical object – a spade, say – is sold, the seller ceases to own it. But when an idea is sold, the seller still possesses it and can sell it over and over again. However much knowledge is used, it does not get used up. Yet the market system as described by Adam Smith 200 years ago was based on the notion of scarcity, including a cost structure in which it is more expensive to produce two of anything than one.

Traditional economic theory assumes that most industries run into "diminishing returns" at some point because unit costs start to rise, so no one firm can corner the market. But an increasing number of information products (anything that can be transformed into a string of zeros and ones), such as software, books, movies, financial services and websites, have "increasing returns". Information is expensive to produce, but cheap to reproduce. High fixed costs and negligible variable costs give these industries vast potential economies of scale. A new software program might cost millions of dollars to develop, but each extra copy costs next to nothing to make, especially if it is distributed over the Internet.

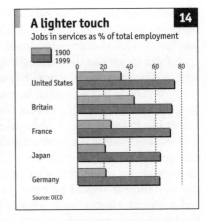

A lighter touch

Jobs in services as % of total employment

1900
1999

Source: OECD

There is nothing new about increasing returns. Alfred Marshall, a British economist, discussed them at length in 1890. Gas, electricity and railways were subject to increasing returns long before the information age. But increasing returns may be more prevalent in information goods because of their cost structure. Besides, economies of scale have increased. In the days of Standard Oil early in the 20th century, if a firm was twice as big as its rivals, its average unit costs might be 10% lower. Today, if a software firm is twice as big as its competitor, its average unit costs might be up to 50% lower. This makes it harder for new entrants to break into a market.

In such circumstances, the natural market structure therefore becomes a monopoly. An added complication with information goods is that economies of scale may apply not just on the supply side but on the demand side as well, thanks to network effects (which economists call "network externalities"). The value of many information goods, such as fax machines or software packages, increases as more people use them. Microsoft's Windows is valued by customers precisely because it is so widely used. Network effects can thus create strong barriers to entry. If everybody you know uses Microsoft Word, then you will find life easier if you use it too.

Once again, network effects are not new. A century ago the Bell System, later to become AT&T, took advantage of network externalities to become dominant in the telephone business. However, Carl Shapiro of the University of California at Berkeley reckons that economies of networks have become more important relative to traditional economies of scale. The combination of demand-side and supply-side economies of scale in many information industries can be extremely powerful. Higher sales not only reduce production costs, but they also make the product even more valuable to other users. In such markets, one firm tends to become dominant.

A third factor can then strengthen a leader's grip on the market: the lock-in effect. Once a customer has learned how to use a computer program, say, he is loth to switch because of the hassle of learning a new program. Users gain big benefits from common standards, so a newcomer has to show a huge advantage to persuade consumers to switch.

This suggests that the antitrust authorities will be kept busy. But some commentators suggest that the old competition rules are no longer appropriate for the information economy. In particular, they argue that the government should go easy on high-tech companies. With rapid technological change and vigorous competition, they say, current market share means little; monopolies will prove only temporary. Furthermore, breaking up a monopoly could actually hurt consumers. A traditional monopoly maximises profits by restricting supply and raising the price. But in information goods, a firm facing demand- and supply-side economies of scale will do the exact opposite: it will increase output and reduce the price. There is a risk, therefore, that if antitrust policy limits the market share of high-tech firms, prices could rise. So perhaps there is a case for greater tolerance of monopolies to allow them to reap full economies of scale, in the knowledge that rapid innovation within the industry will always keep them on their toes. If they become inefficient, they will quickly be displaced by sharper rivals.

A game of monopoly

Some economists have argued, therefore, that by clamping down on Microsoft, the Department of Justice has made a mistake. In a speech in 2000 Larry Summers, America's Treasury secretary, seemed to sympathise with the view that natural monopolies may be good, not bad for the consumer: "The only incentive to produce anything is the possession of temporary monopoly power ... without that power, the price will be bid down to marginal cost and high fixed costs cannot be recouped. So the constant pursuit of that monopoly power becomes the central driving thrust of the new economy." In other words, the economics of information requires an imperfect market, so that innovators can recoup their investment. This follows from the ideas of Joseph Schumpeter, an early-20th-century economist. His theory of "creative destruction" suggested that monopoly can actually stimulate innovation and growth, because it increases the incentive to innovate when a firm can capture more of the gains without being copied by rivals.

However, all the talk about the need for new competition rules misses the point. The government's case against Microsoft was not that it has a monopoly, or that big is bad, but that it protected and extended its dominant position through anti-competitive behaviour. As Joel Klein, head of the antitrust division at the justice department, says, "The legitimate and illegitimate ways of acquiring and maintaining market power have not really changed in 100 years." The core principles of competition policy are still relevant. Microsoft used its monopoly in operating systems to squeeze out rival software firms, thus making it more difficult for new technologies to enter the market. The key issue for competition policy in the new economy is not market share, but the abuse of market power by a dominant firm to discourage innovation by others.

Arguing that Microsoft broke the law is one thing. Much trickier is the issue of whether splitting up the firm will make consumers any better off. If network effects exist, then consumers benefit from the biggest network. That is one reason why the government rejected the idea of splitting the Windows monopoly into three new firms. Instead, it wants to break Microsoft into two: a Windows company and an applications company which would own Internet Explorer, Word and such like. Each company would be free to compete in all lines of business, developing products that compete with each other. The applications company would have an incentive to develop office software not just for Windows, but for other operating systems, helping them to grow. In turn, Windows would have more incentive to co-operate with makers of rival applications. Other software firms would have more incentive to innovate, with less fear of being squashed. This should spur innovation and competition.

Paul Romer, a leading advocate of the importance of innovation for growth, firmly believes that competition is more likely than monopoly to encourage innovation. Mr Romer rejects the idea that because technological change today is faster, antitrust enforcement is less important. That, he says, rests on the false notion that technological change is exogenous, simply raining down from heaven. If that were true, faster technological change would indeed tend to undermine monopoly power more quickly, at a lower cost to society. Instead, he says, the pace of technological innovation is influenced by economic incentives. And since new technology would undermine the monopoly power of an incumbent, it has a huge incentive to try to stifle innovation to protect its monopoly position, and thereby discourage new firms from innovating.

If Mr Romer is right, that makes competition policy more, not less important during times of rapid technological change, because there are many more opportunities for blocking innovation.

Competition authorities will therefore need to watch out, but only up to a point. Monopolies are unlikely to pop up all over the place in the new economy. High-tech industries in which network effects loom large account for only 8% of America's GDP. Microsoft is the exception rather than the rule. There may be a tendency towards monopoly in pure information goods, but in most of the economy IT will help to increase competition.

Broadly speaking, the Internet reduces barriers to entry, because it is cheaper to set up a business online than to open a traditional shop or office. The Internet also makes it easier for consumers to compare prices. Both these factors increase competition. It does not matter if only a small fraction of goods is being sold online so far; traditional firms will still find it harder to push up prices.

Don't do it yourself

Most previous technological breakthroughs have increased the optimal size of firms either by reducing production costs and increasing economies of scale, as with electricity and steam, or by reducing transport costs, as with railways, thus favouring concentration. By contrast, outside the digitisable sectors such as software the Internet reduces economies of scale in most of the economy by increasing the opportunities for outsourcing and by lowering fixed costs.

More than 60 years ago Ronald Coase, a Nobel-prize-winning economist, explained why firms are vertically integrated (as opposed to individuals buying and selling goods and services at every stage of production). The main reasons, he said, were imperfect information and the need to minimise transaction costs. A firm can either produce component parts or services itself or buy them from a supplier. They will probably be cheaper if bought in the marketplace, but against that the firm will have to spend time and money on finding what is available, and on ordering the products.

In the past, these transaction costs were high, so firms often preferred to do lots of things in-house, which made them bigger. Vertical integration solved the problem of imperfect information. But as the Internet increases access to information and reduces transaction costs between firms and suppliers, it makes it more attractive for firms to concentrate on what they are best at and buy in other goods and services from outside. This reduces their optimal size. A small firm can now use accounting software rather than employing an accountant, a word processor instead of a typist, and e-mail or voicemail instead of a telephone receptionist.

The Internet offers small and medium-sized firms many of the advantages of large, diversified firms. It gives them access to the same information as big firms, and makes it easier for them to get into international markets. Many big firms have been using electronic-data-interchange systems for years to communicate with their bigger suppliers. The Internet does the job much more easily and cheaply, making such things accessible to firms of all sizes.

It is true, therefore, that IT both diminishes and increases competition, but it is not really much of a paradox. In industries where network externalities are important, IT will favour giants to exploit economies of scale, both on the supply and the demand side. In the rest of the economy tiddlers will thrive.

The end of taxes?

NOTHING IN THIS world is certain, said Benjamin Franklin, except death and taxes. Even the Internet cannot prevent death, but some people predict that it will make it harder for governments to collect taxes, forcing them to take an axe to their welfare states. More likely, however, the composition of the tax burden will change.

The Internet will make the taxman's job harder in three ways:

- It makes it easier to avoid paying sales tax. If a German buys a CD from a local shop, he automatically pays VAT at a rate of 16%. If he buys it from an American online retailer who delivers it by post, he may well escape the tax he is supposed to pay, because the taxman cannot open every package entering the country. However, the most serious problem arises over products downloaded from the net, such as music, software and videos. It is almost impossible for the taxman to track digital products.

  In America Internet buyers rarely pay tax. Sales taxes are usually collected by the seller at the point of sale, but under American law mail-order firms do not have to collect taxes on sales in other states. In the past the tax loss was modest, but if B2C e-commerce really takes off, the loophole could cause serious damage. Studies show that people living in states with high sales taxes are much more likely to buy online.

- The Internet increases the mobility of firms and certain kinds of skilled workers. Businesses or individuals operating over the net can move to low-tax countries or to tax havens. For example, many British gambling firms have recently set up online operations offshore. The Internet makes it harder to pinpoint the identity and location of individuals or businesses engaged in taxable activities. A domain name may give no clue to the location of a site. And if plans to develop anonymous e-money bear fruit, potential taxpayers will become even harder to identify.

- The Internet could also make life harder for the taxman by cutting out retailers, bankers and other middlemen who currently play an important role in collecting taxes or providing useful information.

At present, e-commerce accounts for only a tiny fraction of total spending, so it is unlikely to cause a serious erosion of the tax base in the near future. The size of the potential drain on tax revenue also tends to be exaggerated: most products cannot be digitised and distributed over the net; most people do not want to move abroad to work; and where a company decides to set up in business depends on many other factors besides tax. Nevertheless, the problem is likely to grow. European governments, which raise more revenue from consumption taxes than America, are likely to be squeezed most.

The European Commission is worried about losing tax revenues through the net. It has proposed that foreign companies with annual online sales of more than €100,000 in the EU should register for VAT in at least one EU country and then collect the tax on all services downloaded from the Internet. But this would be almost impossible to enforce.

It will be no bad thing if the Internet does lead to increased tax competition between economies, forcing governments to reduce tax rates. But do not expect overall tax burdens to decline by much. The Internet will certainly increase the pressure on governments to reduce taxes on company profits and on consumption, but their most likely response is to shift the tax burden to people and things that can be pinned down more easily – such as ordinary workers, property and energy use.

In 1831 a British member of Parliament asked Michael Faraday, a pioneer of electrical theory, what use his discovery might be. Mr Faraday replied that he did not know, but he was sure governments would one day tax it. The Internet may be rather harder to tax, but someone, somewhere will find a way.

Catch up if you can

Europe and Japan cannot afford to miss the boat

SO FAR THE "new economy" has largely been an American spectacle, with little sign of an increase in productivity growth in Japan or the big European economies. Many Americans expect it to stay that way because the tired "old" economies of Europe and Japan lack the necessary innovation and entrepreneurial flair. But historically, the biggest economic gains from a new technology have come not from its invention and production, but from its exploitation. Over the coming years, as IT and B2B e-commerce spread all over the globe, America's economic lead should narrow.

Growth in America spurted ahead of the other big rich economies during the 1990s. Since 1995 the country has enjoyed average annual GDP growth of 4.2%, compared with 1.8% in Germany and only 1.2% in Japan. At least some of this can be explained by their different positions in the economic cycle. The gap in growth rates of GDP per head is also smaller because population growth in America has been faster. But by any measure America has outperformed the others. America's lead in IT and the Internet, it is argued, will give it a big advantage for many years to come, and present European and Japanese high-tech firms with formidable barriers to entry. After all, Europe lacks world-class technology producers like Microsoft, Cisco or Dell. Of the world's 50 biggest IT companies by revenue, 36 are American, nine Japanese and only four European. IT production accounts for 7% of America's GDP, 6.5% of Japan's and 4% of Europe's .

But this gloomy view ignores the vital point that it is the use of IT that will do most to lift productivity, not the making of IT products. Many (though not all; remember Robert Gordon) of the studies that have dissected the recent increase in America's productivity growth suggest that IT production accounted for only about one-quarter of the increase in labour-productivity growth in the second half of the 1990s. A more important factor was investment in IT. Europe and Japan currently spend less on IT as a share of GDP than does America (see chart 15), but their investment is starting to catch up.

Europe and Japan do not need to create cutting-edge technology to close the productivity gap with America. They can make their

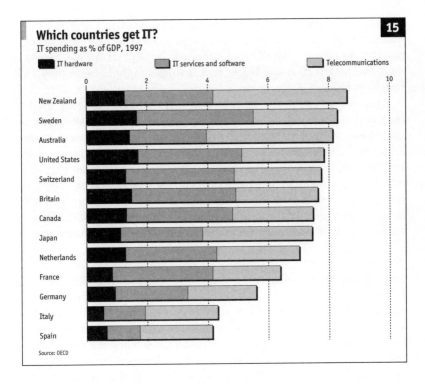

Which countries get IT? `15`

IT spending as % of GDP, 1997

■ IT hardware ■ IT services and software □ Telecommunications

Source: OECD

economies more productive simply by adopting or imitating American technology and B2B e-commerce. For all the talk about first-mover advantage, there are actually several advantages to being a follower. Catching up is much cheaper than trail-blazing. A Japanese or European firm buying IT equipment today will pay much less than it would have had to a few years ago. For example, in 1993 American firms invested $143 billion in IT, but the same level of computer processing power, estimates Paul Donovan, an economist at UBS Warburg in London, could now be had for perhaps $15 billion, thanks to falling computer prices.

Second-movers are also able to wait and see what works. They can cherry-pick the best bits and avoid the mistakes of American firms. As a result, Mr Donovan predicts that over the next decade growth in GDP per head could well be faster in Europe and Japan than in America. However, this does not mean that America will lose its economic supremacy; merely that its lead will narrow.

There are plenty of archaic business practices left in Europe and Japan that keep prices high and productivity low. IT and the Internet will make inroads into these by increasing transparency and competition. Corporate America, after a decade or so of vigorous restructuring, has already cut out a lot of economic waste. Japan and Europe, by contrast, can look forward to bigger cost savings simply because they are further behind.

E-commerce could also help to transform Japan's famously inefficient and expensive distribution system. A recent study by McKinsey, a consultancy, found that although productivity in many Japanese manufacturing sectors was higher than in America, in retailing it was only half as high. If consumers can buy much more cheaply from abroad – which the Internet will help them do – domestic producers and retailers will be forced to reduce their prices. The Internet also offers Japanese firms a way to cut out the swathes of inefficient middlemen. The longer the supply chain, the bigger the potential gains from B2B e-commerce.

An obstacle race

Yet although the potential cost savings from IT in Japan and Europe may be bigger than in America, there remain some big obstacles to realising that potential. Inflexible labour and product markets could hinder the shift of labour and capital that is needed to unlock productivity gains. For example, strict employment-protection laws block the swift reallocation of workers from old to new industries. America's better economic performance over the past decade may owe something to its flexible, competitive markets as well as to its use of technology.

Starting a company is also harder in Europe and Japan, because venture-capital markets are less developed and new businesses get entangled in red tape. It can take ten times as long and cost four times as much to start a new business in continental Europe as in America. Recent research by the OECD has found some evidence that the benefits of IT may indeed be amplified by flexible labour and product markets. Economies with the most flexible arrangements have adopted IT more swiftly, and have also seen a better performance in TFP growth in the 1990s (see chart 16).

All this means that to close the productivity gap, it is not enough for Europe and Japan simply to invest in IT: they must also shake up their labour and product markets. On past experience, that could mean a long wait, but things are changing. European governments are embracing tax reform and greater labour-market flexibility more swiftly than anybody expected a few years ago. The venture-capital industry is growing briskly, and many European businesses are starting to adopt a management style

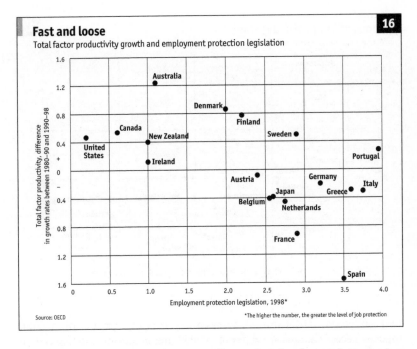

Fast and loose `16`

Total factor productivity growth and employment protection legislation

Total factor productivity, difference in growth rates between 1980–90 and 1990–98

Employment protection legislation, 1998*

Source: OECD

*The higher the number, the greater the level of job protection

closer to America's. The Internet is itself a catalyst for change, by exposing restrictive rules and helping to remove barriers to trade. European and Japanese consumers need only click on their screens to see what a raw deal they are getting. Likewise, high tax rates become harder to maintain when people and firms can up sticks and move to a lighter tax regime (see "The end of taxes", page 40).

In the cyber-age, national restrictions on shop-opening hours or strict rules on pricing and promotion are becoming a nonsense. In 2000, the German economics minister said he was planning to abolish some retail legislation that gets in the way of e-commerce, such as the law that forbids shops to discount prices by more than 3% below the manufacturer's recommended price. In Japan, the Internet will force structural change by bringing stronger competition to the supply chain. It strikes at the heart of the *keiretsu* system, the network of shareholding relationships that link manufacturers with their preferred suppliers and retailers.

Structural rigidities will continue to slow the rate at which the productivity gap is being closed, but at least Europe and Japan are starting to catch up in their use of IT. In 1997, 48% of American employees but only 28% of German ones used the Internet, a 20% gap. By 1999 both countries

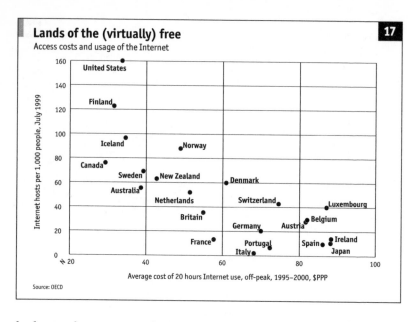

Lands of the (virtually) free

17

Access costs and usage of the Internet

Internet hosts per 1,000 people, July 1999

160 — United States

140

120 — Finland

100 — Iceland · Norway

80 — Canada

60 — Sweden · New Zealand · Denmark

Australia · Netherlands · Switzerland · Luxembourg

40 — Britain · Belgium

Germany · Austria

20 — France · Portugal · Spain · Ireland

Italy · Japan

0

₦ 20 · 40 · 60 · 80 · 100

Average cost of 20 hours Internet use, off-peak, 1995–2000, $PPP

Source: OECD

had moved on, to 65% and 58% respectively, leaving a gap of only 7%. But Internet access charges are generally still higher in Europe and Japan than in America, not least because governments have been slow to liberalise and introduce competition into the "local loop", the final link between the telephone network and homes and offices. Countries where the cost of access is low are generally heavier Internet users (see chart 17).

There is one sector, however, where Japan and many European countries lead America: in the use of mobile phones, which some think could one day become the main gateway to the Internet. Proportionally, many more Japanese than Americans have mobile phones, and one in three Japanese users already has Internet capability. America's fragmented market of operators and operating standards has held back the use of mobiles across the country.

America also lags behind in another crucial area: education. At the very top, it has the best universities in the world. But for the bulk of the population, Japan and Europe provide a better education. This could seriously hold back America's economy in the years ahead. Japan and Europe may have been slower to adopt IT, but once they get round to it, the economic pay-off could be bigger than in America. As Paul Saffo, at California's Institute for the Future, puts it, "The early bird may catch the worm, but it is always the second mouse that gets the cheese."

Falling through the net?

For the developing world, IT is more of an opportunity than a threat

WITH INFORMATION TECHNOLOGY now claimed to be the main engine of growth over the next couple of decades, many people worry that developing economies, which have far fewer computers and Internet connections than the rich world, will get left behind. The income gap between rich and poor countries will widen further. But such fears about a "digital divide" seem to be based on a misunderstanding of the nature of growth as well as of the nature of IT. If IT can boost growth in the rich economies, why should it not do the same trick in emerging economies?

Pessimists point out that the rich countries account for only 15% of the world's population but 90% of global IT spending and 80% of Internet users. Much of the developing world is too poor to buy computers or telephones. In Bangladesh a computer costs the equivalent of eight years' average pay. The 2 billion people living in low-income economies (with average incomes below $800 per head) have only 35 telephone lines and five personal computers for every 1,000 people, compared with 650 phone lines and 540 computers in America. One in two Americans is online, compared with only one in 250 Africans.

Not only are developing countries less wired (see chart 18), but the Internet may cause the gap between rich and poor nations to widen further, worries Avinash Persaud, an economist at State Street Bank. He has three particular concerns. The first is that the "network externalities" helping first-movers to establish a dominant position will favour American giants, so that local firms in emerging economies will be frozen out of e-commerce. The second is that the shift in power from sellers to buyers which the Internet inevitably entails – the next supplier is never more than a mouse-click away – will harm poor countries. Since emerging economies, especially commodity producers, tend to come low down in the supply chain, he fears that their profit margins will be squeezed by rich-country firms. And lastly, Mr Persaud argues that high-tech shares in rich economies have offered investors a much more attractive combination of risk and return than emerging economies, so poorer countries will enjoy less inward investment than they might

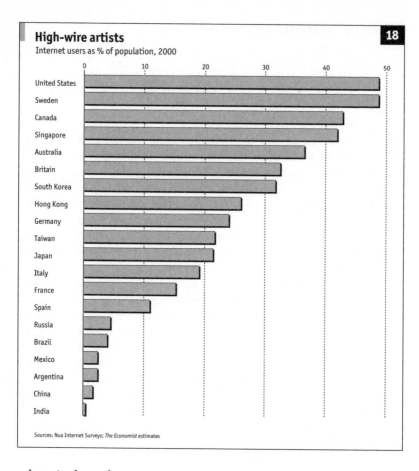

High-wire artists

Internet users as % of population, 2000

United States
Sweden
Canada
Singapore
Australia
Britain
South Korea
Hong Kong
Germany
Taiwan
Japan
Italy
France
Spain
Russia
Brazil
Mexico
Argentina
China
India

Sources: Nua Internet Surveys; *The Economist* estimates

otherwise have done.

Each of these arguments has a grain of truth in it, but there are good reasons for hoping that many emerging economies could nevertheless gain even more from IT than the rich world. In rich economies, the only way to sustain rapid growth is to increase productivity by devising new technologies or better management methods. Poor countries, by contrast, start off with much less capital per worker. For example, in the mid-1990s the average amount of capital deployed per worker in Thailand was only one-eighth of that in America. So developing countries have huge scope to grow rapidly by buying rich countries' technology and copying their production methods. This allows them to grow faster than developed economies, even if they start with fewer computers. As latecomers, poorer countries do not need to reinvent the wheel or the computer, but merely to open their economies to ideas from the rich world.

The key question, therefore, is how fast technology diffuses across borders to poorer economies, and the answer is cheerful. Computers, modern telecommunications and the Internet all reduce communications costs and break down geographical borders, so they are bound to speed up the global diffusion of knowledge. Previous technologies such as railways and electricity took decades to spread to developing countries, but IT is advancing in leaps and bounds. OECD figures show that IT spending in developing economies has been growing more than twice as fast as in developed ones over the past decade (though admittedly from a low base).

IT can even allow developing economies to leapfrog old technologies, for example, by skipping intermediate stages such as copper wires and analogue telephones. New wireless technologies require less fixed investment and maintenance than traditional wire-based ones, so they are more effective in countries with sparse populations and tricky terrain. Mobile phones can extend communications to areas that copper wires might have taken decades to reach, allowing remote villages to tap into the global store of knowledge.

The Internet offers virtually free access to a huge amount of information and expert advice on subjects from engineering and plant cultivation to birth control and health care. A single Internet connection can be shared by many, giving schools access to the world's top libraries when they previously did not even have books. Distance learning gives students the chance to be taught by better teachers. The African Virtual University, which is partly financed by the World Bank, uses satellites to broadcast televised courses to students in 15 African countries, who communicate with teachers by e-mail, fax and telephone.

Another good thing about IT, from the emerging countries' point of view, is that it reduces the optimal size of a firm in most industries. Firms in emerging economies are typically smaller than in rich countries, and the Internet, argues Andy Xie, an economist at Morgan Stanley in Hong Kong, allows such small firms to sell direct into global markets at lower cost. The Internet makes it possible for a tailor in Shanghai to handmake a suit for a lawyer in Boston, then FedEx it to him. A women's weaving co-operative in a remote village in Guyana is selling hammocks over the Internet for $1,000 each. Firms in Africa can bid online for procurement contracts tendered by America's General Electric.

Furthermore, by bringing down the cost of communicating with someone on the other side of the world, IT makes it easier for multinational firms to move production to emerging economies to take advantage of low labour costs, but ensure close contact with head office. That should help poorer countries to attract more foreign direct investment. IT also allows some previously untradable services to be traded just like physical goods. Any activity that can be conducted via a screen and telephone can be carried out anywhere in the world. Computer programming, airline revenue accounting, insurance claims and call centres have all been outsourced to developing economies.

Roadblocks

But before we get carried away with rosy visions of IT bringing huge prosperity to emerging economies, some big caveats need to be added. There is nothing automatic about the process of economic catch-up. IT will certainly increase the opportunities for emerging economies to narrow the income gap with rich countries, but wiring the country is only the beginning. IT is not a panacea that allows governments to avoid doing all the hard stuff, such as opening up markets to foreign trade and investment, liberalising telecommunications, protecting property rights, improving education, and ensuring an effective legal system and efficient financial markets. Indeed, IT makes it even more important for governments to do all these basic things, because it increases the rewards for doing so. For example, open markets help to speed up technology transfer, and education increases a country's ability to absorb knowledge. There is little point in spending millions of dollars connecting villages to the Internet if most people cannot even read and write.

Many developing countries, especially in Africa, are at a huge disadvantage so long as telecommunications services remain in the hands of an inefficient, state-controlled monopoly. Because of inadequate investment, waiting lists for telephones are long and charges have not fallen as fast as in the rich economies. Developing countries pay, on average, three times more than rich-country users to access the net. According to an UNCTAD study, 20 hours of access a month costs $90 in Mexico, equivalent to 15% of average income, compared with only $25 in America, a mere 1% of average income. In Africa, average access charges top $200 a month.

This suggests that although IT may help many emerging economies to catch up with the developed world, it will also lead to a further widening in economic performance within the developing world itself. Matti

Pohjola, a Finnish economist, has analysed the relationship between IT investment and growth in 39 countries over the period 1980–95. He found that whereas IT investment appears to boost growth in developed economies, the same is not true in developing countries. It

Eastern divide		19
per 1,000 population, 2000	China	India
Telephone main lines	110	30
Mobile-phone subscribers	54	3
Personal computers	10	5
Internet users	16	4

Sources: Pyramid Research; IMD, Nua Internet Surveys

would seem that to reap the economic benefits from IT investment, developing countries need to put in place other policies as well.

The way not to do it is to throw a lot of public money at developing high-tech industries sheltered by trade barriers. When Brazil tried (unsuccessfully) to develop a national computer industry behind strong protectionist walls, it imposed huge costs on the whole economy: computer prices were much higher than they otherwise would have been because of import restrictions. The money would have been much better spent on increasing numbers at secondary school, which only one-third of Brazilian children attend at present. Likewise, Malaysia's much-hyped Multimedia Supercorridor, which the government has built at vast expense to create an Asian Silicon Valley and lure foreign high-tech firms, does not really make sense. The comparative advantage of emerging economies lies in applying new technology developed in rich economies, not trying to invent it.

Bangalore in India is a popular – and misleading – example of how IT can affect emerging economies. It has a thriving software industry, thanks to English-speaking workers with good technical skills and low labour costs. Programmers are paid a quarter of what they would earn in America. Software exports have been growing by 50% a year, reaching almost $6 billion in 1999. But this is not a quick-fix, high-tech route to development. A few sophisticated Indians are creating a lot of wealth for themselves, increasing the gap between India's rich and poor, but there have been few positive spillovers to the rest of the economy. Thanks to over-regulation and a lack of competition in the telecoms sector, which is only now being tackled, most Indians have no access to telephones or the Internet, and capacity bottlenecks mean that phone lines are often jammed. A better example of a low-income country that is wholeheartedly embracing IT is China, which has four times as many telephone lines and Internet users per 1,000 people as India, and 18 times as many mobile phones (see table 19).

Asia.com

The East Asian economies are likely to benefit more from IT than Africa or Latin America. Africa lacks many of the economic and legal institutions needed for a thriving information economy, and Latin America is well behind Asia on educational standards. East Asia has not only adopted many of the right policies to enable it to gain from the use of IT, but as a big manufacturer of IT equipment it could also do very well out of the global IT investment boom. Electronics account for one-third of the region's exports.

Bernie Eschweiler, an economist at J.P. Morgan in Singapore, argues that the biggest gains from the Internet in Asia will be in services. Asian manufacturing is famously efficient, but many service industries, being tightly regulated and closed to competition, are surprisingly flabby. The Internet can help to change this by giving consumers more power.

American firms such as Wal-Mart and J.C. Penney insist that suppliers abroad deal with them over the net, which has forced their Asian trade partners to take to e-commerce sooner rather than later. In Thailand, all importers and exporters have to be online because the government has passed a law requiring all trade documentation to be provided on the web.

South Korea has embraced the Internet more fervently than most. By 2001, 20m of its 48m people were expected to be wired. South Korea has the foundations of a knowledge-based economy, with higher investment in education as a proportion of GDP than in many developed economies. But a knowledge-based economy also needs a more competitive environment that provides incentives for the efficient use of knowledge. The government is using the web as a way of forcing change within South Korea's economy. It ordered all state-owned firms to make 50% of their purchases online by 2001. All government procurement is to be on the net by 2002. The hope is that increased transparency and open competition will help to transform the country's traditional cosy and opaque business relations. B2B e-commerce will break the links between the *chaebol* (conglomerates) and their suppliers, and so help to loosen the stranglehold of the *chaebol* on the economy.

Sun Bae Kim, an economist at Goldman Sachs in Hong Kong, reckons that over the next decade the efficiency gains from IT and e-commerce will be bigger in emerging Asia than in the rich economies. Goldman Sachs has tried to estimate the size of these gains. In addition to the direct savings in procurement costs from B2B e-commerce, the bank has attempted, more ambitiously, to estimate the spillover benefits of IT in reducing barriers to competition and eliminating inefficiencies in the supply chain.

Mr Kim suggests that the effects of the Internet will be similar to those of an economy opening up to trade. Economic studies suggest that a one-percentage-point increase in trade as a share of GDP boosts the level of productivity by 0.5–2%. Making the brave assumption that opening up an economy to e-commerce will give a similar spur to productivity, and adding in the direct cost savings from procurement, the bank estimates that over time the Internet will boost the level of GDP by amounts ranging from 5% in Indonesia to an impressive 12% in Singapore. This gain will be spread over a couple of decades, so annual growth rates might increase by between 0.2% and 0.8% over the next ten years.

The Internet can help developing countries catch up with developed ones, so lack of access will clearly prejudice their growth prospects. But the same might have been said about any previous technology, from telephones to electricity. Developing countries may have fewer computers and Internet connections than the rich economies, but it does not automatically follow that they will grow more slowly. Fears that the "digital divide" will widen the income gap between rich and poor countries are exaggerated. Indeed, perhaps the biggest risk is that governments, businesses and aid agencies in emerging economies will get distracted by the Internet and concentrate all their efforts on getting wired, but fail to tackle deeper economic obstacles to development.

The Internet will assist development, but it is not a magic drug for growth. Opening markets, breaking up telecoms monopolies and improving education are far more important concerns. Those economies that get left behind should blame themselves, not technology.

The beginning of a great adventure

Globalisation and IT were made for each other

THIS SURVEY HAS outlined many common fallacies about the new economy. The new-economy fanatics, who believe that the economic benefits of the Internet will be far bigger than those of any previous technological revolution, have been shown to be just as misguided as the sceptics who consider the IT revolution as nothing more than a stockmarket bubble. The economic benefits of IT are likely to be big, perhaps even as big as those from electricity, but not big enough to justify the current dizzy heights of share prices. Two other ideas – that the IT revolution is uniquely American, and that the information economy requires all the rules of economics to be rewritten – have also been knocked firmly on the head. But one more fallacy still needs to be deleted: the proposition that the information economy can thrive completely free of government interference.

It is true that IT brings the economy closer to the textbook model of perfect competition, but governments still retain an important role in ensuring that the opportunities offered by IT are fully exploited. Well-functioning markets for labour, products and capital are important, but on their own they are not enough. Investment in education, too, will be crucial, to ensure that the workforce is equipped for the information economy. America, in particular, needs to get moving on that. Governments also have a role to play in encouraging innovation. Studies suggest that the social return from R&D is at least twice as big as the private return because of spillover benefits to other firms. Companies may invest too little in research because they are unable to capture all the benefits, which suggests there is a case for government support for R&D, especially basic science research. The Internet may now be synonymous with free markets, but in the beginning it was itself the product of government funding. Yet government tax credits and research grants for R&D may not be enough.

Paul Romer argues that such subsidies increase the demand for scientists and engineers, but if the supply does not expand, the entire increase in spending may be squandered on higher pay instead of boosting innovation. In the long run higher pay should encourage more young people to study science and engineering, but the education system suffers from serious time-lags. In particular, it is notoriously bad at switching resources between university departments, because every head of department is determined to maintain the status quo.

Between the mid-1980s and mid-1990s, the number of degrees awarded in engineering, mathematics and computer sciences in America actually fell, despite the unfolding of the computer revolution. Mr Romer suggests that to increase the number of graduates in science and engineering, the government should spend some of the money it now uses to subsidise R&D on grants and fellowships that would boost the supply of scientists directly.

Keep it global

"I'm all for progress; it's change I don't like." Mark Twain's words probably sum up the feelings of many of those who took to the streets of Seattle in 1999 to protest against the meeting of the World Trade Organisation. Globalisation is an easier villain to blame than technology (the Internet, after all, helped the protesters to organise their demonstrations), but in reality IT and globalisation are closely related. By reducing the cost of information and communication, IT has helped to globalise production and capital markets. In turn, globalisation amplifies the economic gains from IT. Perhaps the most important role of governments in the information economy, therefore, is to keep markets open. A retreat from globalisation would seriously hurt the new economy, which needs the free flow of trade and capital to maximise the benefits of IT.

Today's globalisation is not just the inevitable result of technological change; it is also driven by trade liberalisation, which could be reversed. In some ways the world economy at the start of the 20th century was just as globalised as it is now, but after the first world war governments imposed trade barriers and capital controls, so economies turned inwards.

Could globalisation be undone again? Not so easily, because it goes much deeper today than last time round. Many more economies are now part of the global market, and economies and multinationals are much more interconnected. A mobile phone might be designed in London and made in China from parts produced in Canada, America and Sweden, on the orders of a headquarters in Finland. A hundred years ago foreign investment by multinationals was much less important than it is now.

IT has itself encouraged faster growth in international trade, especially of products that are delivered over the net. Even if governments were to turn protectionist, trade in such products would be almost impossible to block. Thanks to IT, information is also much more global than it was a century ago. As consumers find out more about products on offer abroad, and discover the benefits of being part of the global economy, political pressure to open borders to trade is likely to intensify. It would therefore be far harder for globalisation to be reversed now. But it would not be out of the question.

The information-technology revolution has barely begun, but it is spreading fast. A century ago, technological innovations took decades to make their way around the world. Today, developing countries have almost immediate access to new knowledge, and the faster pace of diffusion of technology is itself boosting global growth. The economic benefits of electricity in the 1920s and 1930s were partly undermined by growing protectionism. That prompts the tantalising thought that IT, combined with the benefits of globalisation, might deliver even bigger economic gains to the world economy as a whole than electricity did all those years ago. A retreat into protectionism, on the other hand, would have bigger costs today than ever before in human history.

The material on pages 3–56 first appeared in a survey written by Pam Woodall in *The Economist* in September 2000.

POSTSCRIPT

SINCE THIS SURVEY was published, both the ultra-optimists and the cybersceptics seem to have been proved wrong. America's IT boom has turned to bust and many of the claims of the new economy have been exposed as false, especially the notions that the business cycle was dead and traditional methods of valuing shares were irrelevant. Spending on IT has plunged and the high-tech share-price bubble has burst. Share prices have fallen sharply from their peak in 2000, but in late 2001 still remained historically high in relation to profits.

The most important pillar of the new economy, faster productivity growth, has been badly dented, but not demolished. Government figures for average American productivity growth in 1996–2000 were revised down to 2.5%, from an original estimate of almost 3%. As the economy slid into recession in 2001, the country's productivity growth slowed, but remained faster than in any previous recession during the past quarter-century. The latest best guess is that America's sustainable rate of productivity growth is around 2–2.25% – well below estimates of 3% widely believed possible back in 2000, but considerably faster than the 1.4% average during the two decades to 1995.

What's left?

The fall in American productivity in 2001 might be seen as the death knell of the "new economy". But its obituary would be premature

ONE BY ONE, the claims of the IT-powered new economy to be special are being exposed as myths. The new-economy sceptics, who long argued that computers and the Internet did not rate in the same economic league as electricity or the car, are now grinning smugly. They have a long list of unfulfilled promises to point to.

- **The cycle.** Top of the list is the idea that the traditional business cycle is dead. Now that America's economy is slowing sharply and unemployment is rising, everybody is trying to deny that they ever made such a claim. Instead, they argue that IT helps to smooth the cycle. But even that claim looks suspect: if America's GDP growth in 2001 slows to the average forecast of 1.5%, then the decline in growth from 5% in 2000 would be very abrupt indeed. And because of the excesses that have built up during the boom years – such as too much investment and too little saving – there is a high risk that the downturn could be much deeper. If America's GDP growth were to fall below 1% in 2001, it would be the sharpest slowdown between any two years since the 1974–75 oil crisis. So much for smoothing.
- **Inventories.** One reason why IT was supposed to smooth out the economic cycle was that fancy computer systems and instant information would allow firms to ensure that production never got too far ahead of sales, and would thus avoid an excessive build-up of inventories. Yet despite spending millions on cutting-edge software, firms failed to spot the slowdown late in 2000.

  Across the economy as a whole, "just-in-time" inventory management has reduced the ratio of inventories to sales. But it cannot guard against excessive stocks when shocks occur. It can only ensure that for any given shock, the error is smaller than it would otherwise have been. So far, the build-up in stocks is modest given how sharply demand has fallen. But it is far from negligible.

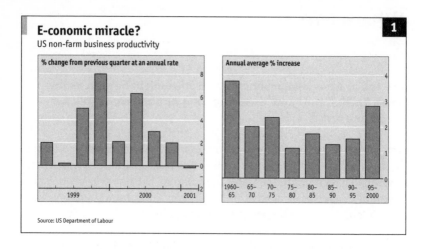

E-conomic miracle?
US non-farm business productivity

% change from previous quarter at an annual rate

Annual average % increase

Source: US Department of Labour

■ **Investment.** The third myth of the new economy was that IT spending was recession-proof. Demand for IT equipment, it was argued, would continue to grow briskly throughout any downturn in the old economy because the pace of innovation was so fast that existing IT equipment would rapidly become obsolete. Firms would be forced to keep investing merely to stay competitive. In reality, as profits have fallen, firms have decided that they can easily delay buying new PCs or upgrading their e-mail systems.

■ **Profits.** Myth number four was that corporate profits would continue to rise rapidly for years to come. Average profits look set to fall by at least 10% in 2001. Yet historically, profits have tended to increase broadly in line with nominal GDP. If anything, the ratio of profits to GDP should now be falling slightly, because IT is likely to trim profit margins by increasing price transparency and shifting power from sellers to buyers.

■ **Share prices.** The fifth (and silliest) claim was that in this new world of rapid technological change, old methods of share valuation had become irrelevant. Profits were for wimps, it claimed. Falls of 90–100% in the share prices of loss-making dotcom firms show that profits do matter after all. During previous technological revolutions (spurred by the invention of the railways, electricity or the automobile, for example) share prices similarly soared and then tumbled. But this time prices rose to a higher level in relation to profits than ever before in history.

The productivity puzzle

What about the most important claim of the new economy, namely that investment in IT has lifted productivity growth? The fall in productivity in 2001 appears to confirm that this was also a myth. But America's productivity gains cannot be dismissed so easily. Recent estimates of sustainable productivity growth were almost certainly too optimistic, but there is still reason to believe that some of the productivity gains will survive even as the IT boom turns to bust.

Labour productivity growth in America's non-farm business sector rose to an annual average of almost 3% over the five years to 2000, up from 1.4% between 1975 and 1995. This faster productivity growth has been the lifeblood of claims about the new economy. It helped to deliver faster GDP growth with low inflation, higher profits and a large budget surplus.

One of the hottest debates in economics for the past few years has been about how much of this increase in productivity growth was structural and how much of it was cyclical. During boom times firms tend to work their employees harder, producing a cyclical rise in productivity growth, which then declines during the subsequent recession.

Robert Gordon, an economist at America's Northwestern University, and one of the most outspoken new-economy sceptics, reckons that outside the manufacture of computers and other durable goods there has been no increase in labour productivity growth after adjusting for the effects of the economic cycle. As a result, he estimated in 2000 that America's structural productivity growth was somewhere around 2%.

At the other extreme, two studies published early in 2001 – one by the president's Council of Economic Advisers and the other by the Federal Reserve – both conclude that virtually all the increase in labour productivity growth since 1995 has been structural, putting the sustainable rate at close to 3%.

It is hard to believe that none of the increase in productivity growth was cyclical, given the strength of the recent economic boom. On the other hand, microeconomic studies appear to confirm that a sizeable chunk of the increase has been structural. A study by Kevin Stiroh, an economist at the Federal Reserve Bank of New York, examined productivity growth in individual industries. Mr Stiroh found that those industries which invested most in IT in the early 1990s saw the biggest productivity gains in the late 1990s. If the increase in productivity growth had been largely cyclical, the gains should have been more equally spread across industries.

If the fall in productivity growth since mid-2000 is only cyclical (and due to weaker demand) then, once the economy recovers, productivity growth should resume its faster trend. What is becoming increasingly clear, however, is that the productivity gains in recent years were not only exaggerated by the cyclical impact of the economic boom, they were also inflated by an unsustainable IT investment boom that has now turned to bust.

Investing in IT can increase labour productivity either by increasing the amount of capital employed per worker (ie, "capital deepening") or by speeding up total factor productivity (TFP – the efficiency with which both capital and labour are used). Unlike previous spurts in America's productivity growth, the recent one has been unusually dependent on capital deepening rather than on TFP. A 2000 study, by Stephen Oliner and Daniel Sichel at the Federal Reserve, concluded that almost half of the acceleration in productivity growth between the first and second halves of the 1990s was due to capital deepening. If that is so, a decline in IT investment will have worrying implications for future productivity growth.

Deepening concerns

The IT investment boom was unsustainable. The initial rise in productivity growth during the 1990s generated bumper profits, which encouraged firms to become over-optimistic about future returns. At the same time, the stockmarket bubble pushed capital costs down to virtually zero. The inevitable result was over-investment. Credit Suisse First Boston (CSFB) estimates that American firms have overspent on IT equipment to the tune of $190 billion since 1999.

Now, as profits plunge, firms are being forced to cut their investment plans. CSFB suggests that if firms try to eliminate the surplus over two years, real IT investment will need to fall by an average of 16% in 2001 and 2002. If firms try to extend the life of their existing IT equipment during the downturn, then new investment would have to fall by even more.

One year's dip in investment would have only a modest impact on structural productivity growth if capital spending were then to bounce back strongly. More important is the longer-term trend in investment. America's business capital stock has recently been growing at a real annual rate of over 5%, well in excess of the growth in GDP. Jan Hatzius, an economist at Goldman Sachs, estimates that the trend growth in the capital stock will now slow to 3–3.5%. If so, estimates Mr Hatzius, the

contribution of capital deepening to labour productivity
growth will be reduced to
0.75%, half its recent rate.

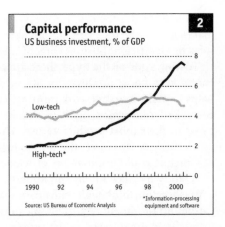

Capital performance
US business investment, % of GDP

Source: US Bureau of Economic Analysis

*Information-processing equipment and software

Mr Hatzius also expects the
growth in total factor productivity to slow as the cyclical
boost to productivity fades,
and as efficiency gains in
computer manufacturing di-
minish. One measure of the
latter is the pace of price deflation for IT equipment. Price
deflation accelerated in the
late 1990s, partly because of increased competition and partly because
of increased efficiency in the manufacture of chips. But it has since
slowed. Computer prices fell by 16% in the year to the end of March
2001, down from an average annual decline of 25% in the previous five
years.

Putting all this together, Goldman Sachs has reduced its estimate of
underlying productivity growth to a more modest 2.25%. That would be
barely half the rate of productivity growth in 2000, but it would still be
significantly higher than the 1.4% average annual rate recorded between
1975 and 1995. Only if investment stagnated for several years would
productivity growth revert to its pre-1995 pace.

And that seems unlikely. The IT boom was mainly driven by falling
IT prices that were following Moore's law, which states that the processing power of a silicon chip roughly doubles every 18 months. This
encouraged firms to substitute IT equipment for labour or other capital.
Scientists believe that Moore's law should hold for at least another ten
years, allowing IT prices to continue to fall and hence encouraging further IT capital deepening, albeit at a slower pace than in recent years.

Yet to be revealed

Not only is the pace of technological innovation likely to continue to
support productivity growth for at least another decade, but firms have
yet to exploit fully the potential of their existing IT. It is not just the
direct impact of computers and the Internet on productivity that matters, but also the ability of firms to organise their businesses more efficiently as a result. Just as the steam age moved production from the

household to the factory, railways allowed the development of mass markets, and electricity made possible the assembly line, so IT allows for a more efficient business organisation.

Tossing aside all the hype, there are sound reasons for expecting IT to continue to boost productivity. By increasing access to information, it helps to make markets work more efficiently and it reduces transaction costs. Better-informed markets should ensure that resources are allocated to their most productive use. The most important aspect of the new economy was never the shift to high-tech industries; it was the way in which IT could improve the efficiency of old-economy firms.

A study by Alice Rivlin and Robert Litan at the Brookings Institution, a Washington think-tank, considers ways in which the Internet might further lift productivity growth. It examines how leading-edge firms are using the Internet to reduce transaction costs, to boost the efficiency of supply-chain management, and to improve communications with customers and suppliers. It then projects best practice across each sector.

The study's tentative conclusion is that the economy as a whole can look forward to productivity gains from the Internet of 0.2–0.4% a year for the next five years. The potential cost savings look especially large in the health-care industry, where there is plenty of scope to improve the management of medical records and to communicate better with patients. Only 3% of Americans have so far corresponded with their doctor by e-mail; the rest still use more time-consuming methods, such as the telephone or a personal visit.

It is also worth remembering that not all of America's productivity growth in recent years has come from IT. Some of it has been the pay-off from earlier economic deregulation designed to make labour, product and capital markets work more efficiently. Whatever happens to IT spending, these benefits will endure.

High-tech investment and productivity growth are, however, unlikely to return to the giddy pace of 1990–2000. The best guess is that productivity growth will average 2–2.5% per year over the next decade, still well above the pace of the 1970s and 1980s. Those who claimed that 3–3.5% growth was sustainable will be disappointed, but their expectations were far too bold. They implied that IT would have a much bigger economic impact than cars and electricity did in the 1920s, when annual labour productivity growth in the non-farm business sector averaged 2.5%.

Slower-than-expected growth in productivity has two implications for policymakers, however. First, after taking account of labour-market changes, it means the rate at which the Fed can safely allow the economy to grow without pushing up inflation is 3–3.5%, much less than in recent years. Second, it implies a smaller budget surplus – and hence less room for tax cuts. Back-of-the-envelope calculations by Goldman Sachs suggest that if productivity growth falls to 2.25%, the cumulative budget surplus in the years 2002–11 would fall to just over $2 trillion, as against the $3.1 trillion estimated by the Congressional Budget Office.

In recent years, economists have tended to exaggerate about IT. Either they have denied that anything has changed, or they have insisted that everything has changed. The truth, as ever, lay somewhere in the middle.

Future productivity gains from computers and the Internet will probably not be enough to justify current share prices, even after 2001's slide, but they will still matter. An increase in productivity growth of 0.5–1.0 percentage points per annum may not sound very exciting, but it will lift future living standards and make it easier for the government to pay tomorrow's pensions. That, rather than the get-rich-quick culture of the dotcoms, is the true stuff of the new economy.

The material that appears on pages 57–63 was first published in a special report in *The Economist* in May 2001.

POSTSCRIPT

SINCE THIS ARTICLE was written, America's average productivity growth in the five years to 2000 has been revised down to 2.5%, from an original estimate of almost 3%. As the economy sank into recession in 2001, productivity growth slowed, but it still held up better than in previous recessions. This suggests that at least part of the gains of the new economy were structural, not cyclical.

Older, wiser, webbier

The greatest impact of the Internet looks like being found in old firms, not new ones

THE SPECTACULAR BURSTING of the Internet bubble has led some to question the very importance of the net. Not that long ago, it was said that this was the greatest change since the Industrial Revolution two centuries ago, and thus that it would have a greater effect on productivity and management than did electricity and the telephone in the first quarter of the 20th century. Executives queued up to attend e-business conferences in order to learn how to bring the magic of the web to their companies, and speakers vied to produce the best soundbites. "E-business or out of business" was one of the favourites. And now? "Delete or, insert and" would seem the right adjustment to that slogan.

Not so fast. Just as the hype in 1999 and early 2000 was hugely exaggerated, so is the gloom of today, with its increasingly common, peremptory dismissal of the Internet. The lesson of the boom and bust is that the Internet and associated new technologies do not magically bring eternal and rapid productivity growth. Nor was the sheer weight of venture capital and equity investment sufficient to change the behaviour of customers overnight, either in retail ("B2C" in the ghastly shorthand) or in wholesale (B2B) markets. People change rather more slowly than computers and telecoms might like them to; and they are less good at dreaming up new business models than venture capitalists might wish them to be.

But the lesson does not stop there. Where e-business has had a genuine and sometimes powerful effect is in the transformation of established companies. Even as the headlines have blared about the bust of first B2C and then B2B firms, these older giants have been quietly taking to new technology and the Internet with a new purpose.

Cut costs, demolish barriers

The Economist ran a series of e-strategy briefs or case-studies that examined how several large, older companies have responded. The spread was broad: from America (GE and Enron) to Japan (Seven-Eleven), from manufacturing (Valeo) to financial services (Merrill Lynch), from emerging markets (Cemex) to rich ones (Siemens). Lessons from these companies' use of the net have been equally varied: squeezing suppliers at GE, knowledge-sharing across a big conglomerate at Siemens, going international at Cemex, responding to an outside threat at Merrill Lynch, building your own electronic network at Seven-Eleven, reaching a wider group of customers at Valeo and exploiting the net's trading power at Enron.

Yet there are also some common threads that can be picked out from the experience of these companies. The most immediately important – and relevant, in these increasingly stringent economic times – is the huge scope for cost-cutting that the Internet offers. GE now does more business on its own private online marketplace than do all the public B2B exchanges put together. Siemens hopes to cut its annual costs in the medium term by 3–5%. The room for more is evident. One estimate suggests that, for routine office purchases, e-procurement costs only a tenth as much per order as does physical procurement. Low-cost airlines such as Ryanair have chopped their costs hugely by using the Internet to cut out travel agents and dispense with ticketing. And many companies have barely scratched the surface. A survey early in 2001 by the National Association of Manufacturers found that only one-third of American manufacturers were using the Internet to sell or to procure products or services.

A second point is that, contrary to one of its early myths, the Internet does not seem to offer huge "first-mover" advantages. Most of the companies studied took to the Internet relatively late and with some caution. Yet they do not seem to have suffered; indeed, they may have gained from being able to avoid both the mistakes and the huge spending of the pioneers. The sad fate of many "pure-play" Internet retailers confirms that established companies seem able to catch up relatively easily; and it also suggests that the Internet may be lowering not raising barriers to entry. The contrast between the news that Britain's biggest supermarket chain, Tesco, is selling its e-buying system to America's Safeway and the troubles of the most ambitious and best-capitalised online grocery chain, Webvan, is telling.

Third is the more intensified competition that the Internet is everywhere bringing. GE's Jack Welch famously took to the Internet after seeing his family do the Christmas shopping online in 1999; he then coined the address "destroyyourbusiness.com" as a graphic instruction to his divisions to embrace the Internet or risk being eaten up by competitors. That may have reflected some of the hype of the time, but there can be no doubt that in the coming months the pressure on costs, margins and prices will be intense in many markets. Valeo and Cemex illustrate well the effect of being able to extend a company's competitive reach globally thanks to the Internet, spreading their costs over a widening market.

The winning consumer

Does all this mean that business will, after all, be the main beneficiary of both the Internet and new technology more broadly? Maybe not. For although there seems to be plenty of scope for cost-cutting and even for productivity improvements, neither may end up feeding through into greater profits. Rather, greater competition, more transparency and lower barriers to entry suggest that the biggest beneficiaries may ultimately be consumers.

In this, as in so many other ways, the Internet may indeed resemble such earlier revolutions as electricity, the car and the telephone. It took companies some time to work out how best to use these new technologies; when they did so, the transformation of work habits and business structures was profound, especially in old companies; but the biggest winners were surely consumers, whose standard of living was hugely increased. That remains the promise that new technology and the Internet hold out today. It is a promise that it is well worth striving to deliver.

The material that appears on pages 64–66 was first published as a special report in *The Economist* in June 2001.

II

GOVERNMENT AND THE INTERNET

The next revolution

After e-commerce, get ready for e-government

IN DOWNTOWN PHOENIX, Arizona, people are queuing in a grubby municipal office to renew their car and truck registrations. They are visibly bored and frustrated, but what can they do? All over the world, people dealing with government departments and agencies are having to engage in dreary and time-consuming activities they would much rather avoid.

What is unusual about Arizona is that the locals have a choice. Since 1996, a pioneering project called ServiceArizona has allowed them to carry out a growing range of transactions on the web, from ordering personalised number plates to replacing lost ID cards. Instead of having to stand in a queue at the motor vehicle department, they can go online and renew their registrations 24 hours a day, seven days a week, in a transaction that takes an average of two minutes.

What is more, ServiceArizona has not cost taxpayers a cent to set up, and is free to users. The website was built and is maintained and hosted by IBM, which is being paid 2% of the value of each transaction – about $4 for each vehicle registration. But because processing an online request costs only $1.60, compared with $6.60 for a counter transaction, the state also saves money. With 15% of renewals now being processed by ServiceArizona, the motor vehicle department saves around $1.7m a year.

That allows Penny Martucci, the decidedly ungeeky grandmother behind the project, to devote extra resources to improving her department's offline service. But perhaps even more gratifying for her are the e-mails she gets from satisfied customers. In a recent survey, the Arizona motor vehicle department scored an 80% approval rating for its service, head and shoulders above other departments. John Kelly, the state's ambitious chief information officer, is pushing other departments in the same direction, and has put out to tender a contract for building a portal that will link them all together.

There is nothing spectacular about ServiceArizona, but it is a straw in the Internet wind that is beginning to blow through government departments and agencies all over the world. Within the next five years it will transform not only the way in which most public services are delivered,

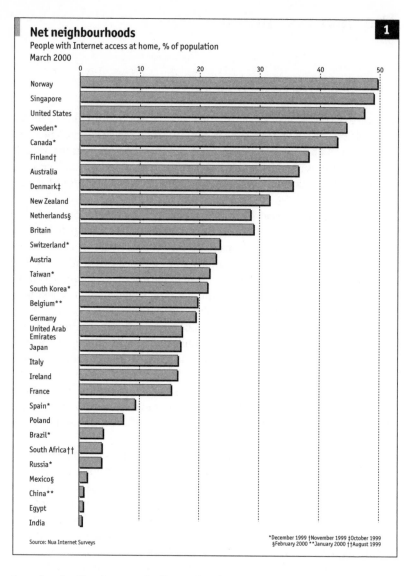

Net neighbourhoods 1
People with Internet access at home, % of population
March 2000

Norway
Singapore
United States
Sweden*
Canada*
Finland†
Australia
Denmark‡
New Zealand
Netherlands§
Britain
Switzerland*
Austria
Taiwan*
South Korea*
Belgium**
Germany
United Arab Emirates
Japan
Italy
Ireland
France
Spain*
Poland
Brazil*
South Africa††
Russia*
Mexico§
China**
Egypt
India

Source: Nua Internet Surveys

*December 1999 †November 1999 ‡October 1999
§February 2000 **January 2000 ††August 1999

but also the fundamental relationship between government and citizen. After e-commerce and e-business, the next Internet revolution will be e-government.

Little and late

With few exceptions, governments have come late to the Internet. Although the net was born out of a project sponsored by America's Department of Defence, most governments have seen their job as creating a benign environment in which the hoped-for economic and social benefits of the Internet could unfold, rather than actively harnessing the fancy new technology to their own ends.

There have been some good reasons for reticence, as well as bad ones. Governments, even more than commercial enterprises, were deeply worried about the potential impact of the millennium bug (known as Y2K) on their computer systems, and the social disruption that might follow. Much of the available IT funding and expertise was channelled into forestalling millennium disaster.

But even without the distraction of the Y2K threat, governments and their agencies would have lacked many of the private sector's incentives to adopt the Internet. As monopoly suppliers, they are not worried about being "Amazoned" – waking up one morning to find a new web-based competitor with the potential to destroy their business. Transactions with government are rarely a matter of choice, and agencies collecting tax or managing entitlement programmes do not see the Internet as a challenge to their existence. Nor are the people running government services likely to be rewarded – with share options for, say, devising an innovative Internet strategy.

There is also the question of access. Even in America, the proportion of people with an Internet connection at home is still under 50%, and in most of Europe it is less than half that (see Chart 1). Governments cannot choose their customers; the services they provide must be for everyone, and much of what they do involves dealing with the poor, the less-well-educated and the elderly – precisely the people least likely to be wired. Lastly, security and trust are even bigger concerns for government than for the private sector. Banks and insurance companies may know quite a lot about their customers, but nothing matches the range and detail of information that governments require from their citizens. Unless the integrity of that information can be guaranteed, the scope for governments to make constructive use of the Internet will remain limited.

But despite the late start, the pressure is now on to catch up fast. The spur may not be competition in the conventional sense, but talk to public servants and politicians almost anywhere, and the sense of urgency and excitement is tangible. Helped by the big IT vendors, governments are realising that by applying much the same technologies and principles that are fuelling the e-business revolution, they can achieve a similar transformation. Reinventing government, a fashionable but premature idea a decade ago, is at last being made possible by the Internet.

More prosaically, the starting point for most e-government projects is the desire to reduce costs and make tax revenues go further. The potential for savings comes from the sheer scale of public-sector spending and from the opportunities to make internal processes more efficient. American federal, state and local procurement spending on materials and services in 2000 was estimated to be around $550 billion. Some big private-sector companies are now achieving annual savings in the region of 20% by putting their supply chains on the web. If government services in the United States could replicate that, they could save $110 billion a year. In the European Union, where the member states' combined procurement spending is about €720 billion ($778 billion), savings could be of a similar order. As with commercial businesses, the benefits come from the way the web can slash purchasing and fulfilment cycles, lower administrative costs by up to 75% and halve stocks.

Reasons to be "e"

Governments are also under pressure to meet rising expectations of service. Not many people enjoy dealing with their government; they do it because they have to. But that does not mean the experience has to be as dismal as it usually turns out to be. As increasing numbers of consumers become used to the quality of service offered by the best web retailers and service providers, their willingness to accept slum standards in the public sector is coming under strain. If the same 24-hour, seven-days-a-week availability and convenience, fast delivery, customer focus and personalisation became the norm in the public sector, it would not just make life easier, it would fundamentally change the way that people view government itself.

One of the greatest problems for anyone who has dealings with government, whether as a citizen or a business, is its sheer complexity. The average government has between 50 and 70 different departments and agencies. Just finding out which is the right one for the task in hand can be hard enough. Worse, even for fairly straightforward matters such as licensing a business, selling a house or registering the birth of a child, a number of different agencies requiring a plethora of different forms may be involved. Moreover, they expect users to communicate with each of them in turn rather than being prepared to communicate with each other.

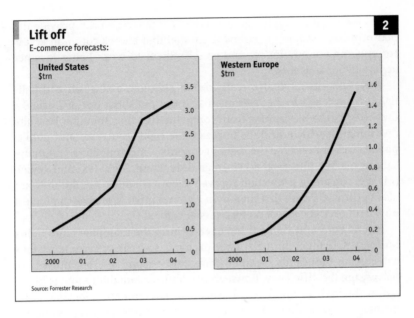

Lift off

E-commerce forecasts:

United States
$trn

Western Europe
$trn

2000 01 02 03 04

Source: Forrester Research

One of the basic reasons for public-sector inefficiency – "bureau-cracy" – is that, whereas departments are vertically organised, many of the services that they have to deliver require complex collaboration between employees across departments. The British government has for several years been preaching the need for "joined-up government", but has found that the underlying structures of government conspire against it.

The Internet offers a solution to both problems. Increasingly, governments are coming round to the view that they will need to construct Internet portals, similar to consumer portals such as Yahoo!, that can provide a one-stop shop for all of a citizen's needs. A central government portal of this kind has just been launched in Singapore; another is being developed in Austria. In Britain, BT has recently won the contract to build UK Online, a portal to offer government services that should, in a basic form, start up in the autumn.

These government portals are being designed to allow users to find what they are looking for by using questions such as "How do I ...?" or asking about so-called "life events", such as a change in marital or employment status. The citizen does not need to know about the organisational complexity behind the scenes because the portal will take him smoothly to where he wants to go. The same technology infrastructure that is built to link and integrate services for the citizen can provide the platform for a secure government intranet that allows common access for government employees and enables them to work better together across all departments.

Governments have also realised that although a sluggish and half-hearted approach to e-government will not put them out of business, they may not be as immune from competition as they thought. True, the first thing they have to do is to create the right regulatory and public-policy environment for the digital economy – a competitive communications market, universal access, digital signatures, light taxation, online privacy, consumer protection for web shoppers and so on. But they are also becoming aware that their own e-government strategies can have a powerful catalytic effect on business in general.

Just as Ford and General Motors can push their suppliers into doing business with them through online exchanges, so can governments, thus galvanising thousands of small firms into becoming e-businesses. By harnessing the efficiency, transparency and accountability that is inherent in the web to improve all aspects of government-to-business and business-to-government transactions, they can deliver a big economic boost. And by ensuring access for all to the Internet as the main channel of dealing with government, they can be a powerful force in bridging the "digital divide" between the haves and the have-nots and stimulating online education.

Last but not least, by improving the quality of their relationship with citizens, they can make a big difference to the attractiveness of their country, region or city as a place to live and work. This goes beyond the delivery of services through the Internet, and to the beginnings of digital democracy.

The first signs are already there: a Democratic primary in Arizona (again) in which online voting boosted voter turnout to six times its usual level; more accountability for elected officials, thanks to e-mail and web-casting; and online campaigning and fundraising, as in the 2000 American presidential election.

Yet making e-government a reality will be extraordinarily difficult. Britain's e-minister, Patricia Hewitt, points to one of the more obvious reasons: "If Jack Welch says that GE is going to become an e-business, it does, and pretty quickly. Government is different." Politicians can be as visionary as they like, but unless they can get the machinery of government to take notice, nothing much will happen. And persuading public servants to abandon their paper-shuffling ways and embrace change on a hitherto unimagined scale will require relentless pressure – as well as some of the carrots and sticks that keep people going in the private sector.

Jay Nussbaum, who heads Oracle's service industries business in Reston, Virginia, has a mantra for e-government: "Start small, scale fast, deliver value." In other words, it is important to bank some quick wins from smaller projects that achieve what they set out to do, like ServiceArizona, before moving on to bigger things like the all-embracing portals that cover every aspect of government activity.

The potential is enormous, but governments will need committed leadership, a full understanding of e-business principles and a clear strategy for overcoming the barriers to change: the departmental rivalries, the hostility of unions, the fears of individuals and the sheer size of the thing. For once, the technology – although crucial to making it all possible – is the least of the worries.

No gain without pain

Why the transition to e-government will hurt

TODD RAMSEY, IBM's worldwide head of government services, does not beat about the bush. "About 85% of all public-sector IT projects are deemed to be failures," he says. That does not mean they are total disasters, but that they usually take longer to implement, cost more and deliver less than was planned. For anyone who is getting overexcited about e-government, that is a sobering reflection.

At least the reasons why big government IT projects get into difficulties are well understood. The trouble often starts with the way contracts are awarded, typically by tender. This can be a nightmare. Steve Dempsey, Andersen Consulting's [now Accenture] e-government specialist in Britain, says tender documents often run to 1,000 pages, and picking the winner can take 18 months.

At that point, according to Mr Ramsey, the contract can become a battleground as both sides take out their frustrations on each other. Often the customer – ie, the agency or government department involved – deserves blame for poor project management and inflexibility, but vendors too must take their share for over-promising and under-bidding. They, in turn, would argue that this happens because public servants tend to award contracts on the basis of price rather than value, quality or past performance.

Another problem – not unique to the public sector, but especially prevalent there – has been a tendency for IT integration firms to over-customise (and thus over-complicate) applications. This is because officials are reluctant to ask a highly unionised workforce to change its working practices so that software can be used straight out of the box.

And changes could involve more than working practices; they could entail big job losses too. There may be scope for retraining some clerical and counter staff to improve offline customer service, but many existing workers will not adapt to new working patterns. Private industry can redeploy people more easily by changing its remit, but that is not an option for government departments and agencies whose goals and powers are usually mandated by statute.

With so much scar tissue around, the confidence shown in the prospects for e-government may seem surprising. But according to

Elizabeth Echols, a member of Al Gore's electronic commerce working group, "People on the Hill are desperate to link their names to anything with an 'e' on it." For Tony Blair, Britain's prime minister, the Internet is crucial to the modernisation of government which he believes must be carried out if non-conservative parties which support the idea of an active state are to have a long-term future. As for the vendors – the big infrastructure and software firms as well as the IT integrators – they keep saying that the Internet changes everything, so they had better believe it themselves.

On the technology front, at least, the Internet, or rather the success of Internet standards and protocols, has indeed changed things a great deal. It is the Internet's open standards, allowing everyone to connect with everyone else, which are the basis of its power. In past government IT projects, special programmes had to be written for every application to make it work with the assortment of hardware and software chugging away in every government department. But today's web applications, often written in the Java programming language, are designed to run straight from the Internet browser.

What is more, virtually all the technologies that will make e-government possible are already working for e-business. For example, applications for enterprise resource planning, customer-relationship management and supply-chain management, business intelligence and data-mining tools, Internet procurement and payment systems are all available now and need very little adaptation for public-sector use. In the same way, the security protocols, the multi-layered firewalls and the public key infrastructures needed for authentication and the protection of data are already available off the shelf. Even the vendors, always keen to offer their own patent solutions, agree that technology is not a barrier to the introduction of e-government.

In the past, government IT projects have been carried through from beginning to end within a single department or agency. That has meant that they have often merely streamlined existing work processes instead of redesigning them or getting rid of them altogether. But just as Internet technologies have forced many big businesses to change what they do as well as the way they do it, so they are about to transform the biggest and most change-resistant business of all – government.

Four far-from-easy steps
The way to e-government divides into four distinct stages. The first stage – which is as far as most governments today have got – involves

departments and agencies using the web to post information about themselves for the benefit of citizens and business partners. Thousands of such "one-way" communication sites are already up and running. In the second stage, these sites become tools for two-way communication, allowing citizens to provide new information about themselves – such as a change of address – instead of telephoning or writing. There are also plenty of these around, although many depend on e-mail.

During the third stage, things start to get more interesting. At this point, websites allow a formal, quantifiable exchange of value to take place. It might be renewing a licence, paying a fine or enrolling for an educational course. There are several hundred such sites, mostly operating at the state or local government rather than central government level. More sophisticated versions can guide applicants through making a claim for benefit or filing a tax return. Examples include the Pennsylvania Department of Labour's site and Britain's Inland Revenue site that went live in April 2000 (only to be taken down for repairs a week or so later). Such sites substitute an element of web-based self-service for work previously carried out by public servants, and need to be co-ordinated with offline channels. They begin to challenge traditional working practices and processes.

The final stage, of which more later, is a portal that integrates the complete range of government services, and provides a path to them that is based on need and function, not on department or agency. A single log-on and password allows users to get in touch with any part of government. Many governments have plans for such portals, but at present only two such sites are anything other than local: MAXI, operated by Australia's state of Victoria, and Singapore's eCitizen Centre.

The British government's experience in preparing to launch its UK Online portal illustrates some of the problems. On the face of it, Britain, with a highly centralised government, a parliamentary democracy and a powerful prime minister who sees himself as a "moderniser", should be well placed to move quickly. Its targets are highly ambitious: 25% of its services are meant to be available online by 2002, and 100% by 2005. But the difficulties are immense. According to Andersen Consulting's Steve Dempsey, nobody has really grasped the extent of the change that is called for or, for that matter, the sheer inefficiency within government that will be exposed by this project.

Mr Dempsey should know: Andersen Consulting has played a pioneering role in the management of big public-service IT projects. It won the contract to build a new system for Britain's National Insurance records (NIRS) that eventually went live in 1999. With 65m National Insurance accounts, more than $60 billion in contributions and over 100m online transactions each year, not to mention annual changes in the law that affect the whole system, it was a project of huge scale and complexity.

It was also the first big IT project carried out under the Private Finance Initiative, a scheme designed to shift funding and risk to the private sector. In return for designing and building the system, Andersen was to receive payments on a per-usage basis for seven years. Although NIRS eventually turned out to be a success, it illustrates the difficulties that bedevil work for the public sector, says Mr Dempsey. "There was simply no partnership with the Department of Social Security, and as a consequence the relationship soured."

UK Online has been split into two projects: one to design and build the site, the other to integrate the systems of all departments to create the single gateway to government that is the point of the portal. The prospect of working on the gateway – likely to be a £2 billion project – has vendors such as EDS and IBM licking their lips, but there is also trepidation.

Mr Dempsey frets that it is not clear who in government will ultimately be responsible for the project, or how and where it will be funded. The government's "e-envoy", Alex Allan, is based in the Cabinet Office, which has already said that each department should make its own preparations for integration. But somebody should be asking whether all of these departments will still make sense once services are delivered electronically. The Inland Revenue, Britain's tax-collection agency, has proposed that because it is much further down the e-road than any other department, it should make the decisions about vital technical standards that will allow data to be transferred between departments. As a precedent, it can cite Singapore's successful e-Citizen project, which has been led by the Ministry of Finance.

There are other worries too. The government's advisers are quite sensibly thinking about a public–private partnership that would, in effect, create a joint venture. But as IBM's Mr Ramsey points out: "At the end of the day, it's the government that has to control the channel." In other words, unless the government is committed to making the portal its preferred way of doing business, people may not use it.

Keep it bite-sized

Britain's push to turn itself into an e-government has won widespread acclaim. Al Edmonds, the retired American Air Force general who runs the government division of the world's biggest systems integrator, EDS, even describes Britain as "the template" for what other governments, including his own, are trying to do. There is nothing wrong with a project as ambitious as the gateway: Mr Ramsey says that when it comes to e-government portal architecture, "cheap and cheerful won't give you the foundation you need to grow." But the technology firms are saying something else as well: beware of trying to bite off more than you can chew.

Mr Ramsey feels that such projects need to be broken down into smaller segments, for example, getting every agency to agree to use just one system for procurement on the web. Oracle's Jay Nussbaum also warns against "The Big Bang of trying to do too much too fast." Mr Nussbaum claims that outlining a vision and strategy is the easy bit. Among other success factors that are needed, he suggests, are a profound understanding of the principles of e-business, and a realistic assessment of the organisation's readiness for what is about to hit it.

In common with other vendors, Mr Nussbaum also emphasises the importance of scoring some quick wins to boost morale. His highly practical mantra is: "Start small, scale fast, deliver value." David Kleinberg, a feisty deputy chief financial officer at the US Department of Transportation, who has been working with Oracle to put much of his operation on the web, took on some 20 "IT baronies" within his sprawling department. "By eating a slice at a time," he says, "they now see the value and they want to play – the costs just come oozing out of the system."

Britain's UK Online is much bigger than any integrated portal either running now or coming soon, and some anxiety can be sensed even within the ranks of government evangelists. John Clark of the Cabinet Office's Performance and Innovation Unit, who is preparing a report on electronic service delivery, reckons that if the portal is to be a success, there is a lot of work to be done that most people in government have not even thought of yet. The big organisational question remains largely unanswered: how do you motivate people to accept massive change in the absence of overwhelming market pressure? Mr Clark concluded a recent presentation with a salutary quote from George Bernard Shaw: "Reformers have the idea that change can be achieved by brute sanity."

E-government is definitely not for the politically timid or half-hearted. One-stop, non-stop e-government portals will revolutionise not just the way public services are delivered, but government itself as well. The trouble with revolutions is that they rarely go according to plan. They also have a nasty habit of eating their own young.

Quick fixes

IN THE LONG term, no self-respecting government moderniser would settle for anything less than a fully fledged portal complete with a zappy, citizen-centred interface. But most e-government projects start off much more modestly, as initiatives championed within a particular department or agency. Typically, they go after the things whose transfer to the web can make a noticeable difference without big changes in existing work practices or IT infrastructure. They mostly fall into one of three categories: intranet applications that allow data to be gathered, processed and shared in new, more efficient ways; extranets that link government to business suppliers, bringing discipline and cost savings to procurement; and public websites that give citizens and businesses a self-service channel for their dealings with government.

One example of the first kind is a web-based intranet application developed by EDS for the naval airbase at Corpus Christi in Texas. A large proportion of the helicopters based there, having come into service in the 1960s, were getting increasingly decrepit. Records on the parts most prone to failure and how best to fix them did exist, but in a form that made them almost useless: hand-written paper dockets stowed away in thousands of filing cabinets.

The solution was to take all the old service sheets and scan them into a database that could be searched by serial number or by key word, such as "leaking O-rings". Now an authorised mechanic anywhere in the world can both enter and retrieve information instantly, so the right spares can be held.

Another example is the work IBM has done with Emekli Sandigi, a Turkish government social-security organisation that collects the premiums and pays the health expenses of 2m public-sector pensioners and their dependants. Previously, pharmacists might have to wait up to two months to be paid for medicines, and had to process 15m prescriptions a year by hand, risking both error and fraud. Emekli Sandigi also maintains detailed health files on its members that must be regularly and laboriously updated manually.

By linking 17,000 pharmacies together by Internet and intranet, all the information has been brought online. Pharmacists can now check the validity of a customer's health card, his pay and the latest medicine prices. Not only are health expenses being reduced – the $8m that the system cost to set up should be recouped in its first full year of operation – but a medical communication network is being built between pharmacies, hospitals and doctors. Repayment to pharmacists now takes less than a week, and customers can use pharmacies anywhere in Turkey.

Electronic procurement is one of the fastest-growing areas of e-business because it can save time and money. The same web-based technologies that are saving firms such as GE and Ford hundreds of millions of dollars could have an even more dramatic effect in the public sector. The state government of Australia's Victoria has worked with Oracle to improve the purchasing efficiency of its Department of Natural Resources and Environment by 70%. The department has deployed a paperless system with access for 5,000 users that provides complete transparency between vendor and user. Payments are electronic, and fraud is kept down by random sampling. As well as saving money, the department is providing better value thanks to the enforcement of business rules and the accountability inherent in the system. The model is likely to be adopted in government throughout Australia.

The most popular self-service e-government applications that involve transactions of real value with members of the public tend to be web-based systems for paying fines and renewing licences and permits of various kinds. Online payment of taxes is also making rapid headway, although some legitimate concerns about security and authentication remain.

One of the pioneers in using the Internet for tax collection is the Chilean Internal Taxation Service. Its web strategy was introduced in stages but has now come fully on stream, allowing tax returns to be submitted entirely on the web. Citizens can schedule payments, check accuracy and look back over their full tax history. The use of the electronic system has dramatically reduced not only the time taken over each transaction but also the number of errors made.

If something as complicated as paying income tax can be carried out on a self-service basis, is there any limit? Sceptics think that such sites may not be able to cope with the more complex aspects of benefit and entitlement programmes. But the success of a website designed by EDS for the Pennsylvania Department of Labour to resolve disputed unemployment insurance claims suggests otherwise.

The system, known as EASE (Expert Assistance System for Examiners), is designed to allow people who have been denied benefit after being fired to lodge an appeal. It establishes the exact circumstances of an individual's dismissal and narrows the ground for lawyers to argue over. It works by carefully walking the claimant through a logical series of questions. If he continues to disagree with the department's decision, he still needs to go to court, but the system reduces the scope for legal argument and makes it quicker and cheaper for everyone to resolve their differences.

Public-service innovators all over the world are trying to work out how best to move their services to the web, starting with the easy ones, then graduating to those that will require a lot of reorganisation. Technology vendors, for their part, are creating ever more sophisticated applications to deal with the range of problems that their public-service customers are confronted with. As IBM's Todd Ramsey says: "You start with one set of ideas, but then you go further."

Island site

When it comes to e-government, there is nothing to match Singapore

FOR A TINY island state, Singapore takes its defence extremely seriously. A two-year stint in the armed forces is compulsory for every able-bodied male, and after that, every year well into his 50s, he has to report back to his unit for three weeks of military grind. It does not matter whether he is a road sweeper, a teacher or the boss of a bank: he is an "NSman" (National Service man) as well.

Nor does it end there. If you leave the country for more than 24 hours, you must get an exit permit and provide a contact number. And do not forget to submit yourself for regular fitness checks and keep abreast of the latest training requirements. But although boot camp remains painfully physical, much of what goes on between the NSman and MINDEF (Singapore's Ministry of Defence) happens on the web.

The site is both a window on the armed services for the whole Singaporean community (including parents, partners and employers) and the platform for a wide variety of applications for the NSman. He can claim his training pay and get it sent to the right bank account; send information about trips and update it from anywhere in the world; book fitness tests and training courses; shorten the length of training by upgrading his skills through an online military academy; and then tell MINDEF what he thinks of the whole thing in a feedback section.

Access to confidential information, such as medical data, is by PIN number, but families can check dates of events and the whereabouts of recruits. Some sections are already accessible by WAP (wireless application protocol) mobile phone. With such a large and active community, MINDEF's managers are wondering whether to launch into e-commerce: for example, by taking advertisements from private fitness centres and by using the purchasing power of the NS community to get good prices for products and services that may have nothing to do with defence.

Small and rich, with a well-rewarded, entrepreneurial civil service and a political leadership with a liking for big strategies, Singapore is an e-government natural. The IT2000 Masterplan, which has been largely implemented, provided a blueprint for the use of IT in nearly every government department and spawned the Singapore ONE project – a broadband infrastructure of high-capacity networks and switches throughout the "intelligent island". With that foundation, the government is now putting together the ICT21 Masterplan, which will, it gushes, "transform Singapore into a vibrant and dynamic global ICT (information communications technology) capital with a thriving and prosperous net economy by the year 2010."

What is striking about Singapore's approach is that although individual departments are left to innovate in their own way, the big picture is never lost sight of. For example, when MINDEF's IT arm, the Systems and Computer Organisation, launched the world's first Internet-based government procurement system in 1998, it was not long before the Ministry of Finance and the National Computing Board (recently rechristened the Infocomm Development Agency) turned up to see whether it might be extended to the rest of the government service.

The idea – to create a one-stop, round-the-clock centre for the government's business dealings – was christened GeBIZ. The first phase was launched in April 2000, and the site was expected to be fully operational by the end of the year. As with other online B2B trading networks, the cost benefits come in the form of more competitive bidding, easy access to suppliers round the world, time saved by online processing of orders, lower stocks and automated collection of high-quality data.

Gee-whizz GeBIZ

GeBIZ works by allowing the financial systems of ministries and agencies (the usual mix of software from SAP, Oracle and PeopleSoft) and the procurement applications to work together. Trading partners can find invitations to tender and purchase orders on the site. Suppliers can also submit invoices, check payment status, post their catalogues and bid for contracts. For the moment purchases are capped at S$30,000 (US$17,341), but the Ministry of Finance reckons that once new payment and security systems are introduced, 80% of all government procurement will transfer to GeBIZ.

One agency that stands to gain more than most is the Housing and Development Board (HDB), the body responsible for the estates where most Singaporeans live. With a budget of nearly S$18 billion a year, it orders the building of new blocks of flats, as well as building and managing over 32,000 commercial and industrial premises.

Under Alex Siow, its chief information officer, HDB has developed a sophisticated information site that gets nearly half a million hits a month. Soon he expects to be able to deal with rent and mortgage payments online, but the biggest gains will come when HDB can conduct its dealings with the construction industry through GeBIZ. "When we do that, we can expect just-in-time contract manufacturing and contract construction," says Mr Siow. The key to this, he believes, is using Singapore ONE to exchange three-dimensional drawings and plans.

But GeBIZ is only one example of Singapore's lead in e-government. In an international survey of sites offering integrated service delivery through the Internet, conducted in 1999, America's General Services Administration (GSA) concluded: "Singapore's eCitizen centre is the most developed example of integrated service delivery in the world." Launched early in 1999, eCitizen was one of a series of modernising government projects designed to ensure that the public sector operated, and was seen to operate, as a single entity. This particular project aimed to bring together useful services and to deliver them to Singaporeans in convenient and easily accessible packages.

It quickly became clear that this would best be achieved by creating a single, comprehensive government web portal. This would also help to make Singapore's population more net-minded. As the education minister, Peter Chen, puts it: "The development of electronic public services is critical to setting the pace in proliferating the use of IT and creating an IT-savvy culture in Singapore. It will enhance the ability of the public to be increasingly familiar and comfortable with IT, which has become a critical component in the knowledge economy. Our people's openness to and skill with IT can offer a distinctive competitive edge to Singapore."

According to the IDA's Tan Sue Hua, one of the eCitizen team leaders, the main reason for the successful launch of the site and its ever-increasing usefulness is the way in which the Ministry of Finance and the IDA have worked together. The IDA has been responsible for the technological side, whereas the ministry has been controlling the purse strings. That has given it the clout to secure top management support and commitment across every government agency. Ms Tan also points out that the technology alone will not get you anywhere: you also have to rethink the processes you are transferring to the web, and make sure that people's attitudes keep up with the changes.

The design of eCitizen allows each agency to keep its own website but to display the content by function, giving users an overall view of the way government works. Information and the opportunity to do business on the site are delivered in one package. To make that possible, all agencies have to adopt a common infrastructure and common modules for things like form-filling, payment and security. And to achieve a single, consistent user interface, they must accept common methodologies.

The user interface the Singaporeans have chosen adopts the metaphor of a citizen journeying through life. As he travels along that road, he can stop at various "towns". Each town groups together a number of service packages that are related to each other. For example, in the Employment Town a visitor can find packages called "Employ People" (for employers), "Look for a Job" (for employees), "Retire from the Workforce", "Upgrade your Skills" and (for foreigners) "Work in Singapore". At present there are nine towns, covering business, defence, education, employment, family, health, housing, law and order, and transport, many of which link the functions of one agency with another. In the Family Town, for example, packages on "Care for the Elderly" come from the Ministry of Health, whereas packages on "Getting Married" are from the Ministry of Community Development.

The Housing Town includes packages on buying and selling flats provided by Mr Siow's HDB. It also has a package called "Move House", complete with electronic application forms for a telephone, utilities, a television licence and parking. Through the same package, a change of address can be sent to every agency that needs to know about it, including the postal service so it can redirect mail. At the Law and Order Town, a report of an incident can be sent to the police, a petition for bankruptcy can be filed and an application made for estate administration by a public trustee.

The next step will be to personalise the transactions so that frequently used services can be reached even more easily and quickly. That does, however, raise one unsettling issue. A site of this kind, which hoovers up a huge amount of data as people transact business on it, can create a centrally held "superfile" of detailed material on every user. This information can be used to improve the level of service further, but it can also be put to less innocent use. The IDA blandly says that information will be protected "in accordance with existing policies", and that individuals must have the right to withhold certain information.

In fact, Singapore's somewhat authoritarian government is probably not too bothered by the threat to privacy or civil liberties. The point of the web is that it is a two-way street. E-governments may be more transparent and accountable than the old-fashioned kind – a risk Singapore seems willing to run – but they will also know far more about their citizens than they do now, and have much more efficient ways of putting to use what they know.

A tool for learning

W HEN HE TALKS about his favourite digital project, Chan Poh Meng, the principal of Singapore's Outram Secondary School, almost wriggles with excitement. Outram, an old neighbourhood school that recently moved to multi-storied new premises in a hilly district of the city, was chosen last year to pilot the Learning Village, a joint venture between Singapore's Ministry of Education and IBM.

Mr Chan's enthusiasm is understandable. The project goes to the heart of Singapore's education policy of using the Internet to make schools more outward-looking and collaborative. "Schools can become like cocoons," says Mr Chan, "but they can't remain isolated, they must reach out." That is the idea behind the Learning Village. It is a web-based platform combining a set of Internet applications to allow communication and collaboration both within the school and beyond it, involving parents and other interested members of the local community with the school and its activities.

Parents are given passwords for logging in to the site from their homes and offices. Once there, they can go to the events calendar to check the whereabouts of their children, or visit teachers' home pages to get information about their teaching methods, grades achieved in class, assignments, homework and lessons for independent study. Parents can also participate in online "meetings" with teachers, and follow their children's online discussions. For their part, parents can offer their ideas for the school's development and influence its policies. Mr Chan recently used the site to solicit views about the school's opening hours. He is also keen to use a "private conference" application to discuss individual children's problems online.

Teachers and pupils benefit from the online bulletin board, which cuts down on the need for time-consuming assemblies and administrative meetings. In a "Teachers' Lounge", teachers can share ideas about lessons and discuss the effectiveness of teaching strategies. There is also a suggestions box for the principal. Pupils can use the site to work together on projects, not just with each other, but with counterparts in other schools, even other countries. Official "mentors" who may have some special expertise or experience to contribute to the school can do so from anywhere in the world. Recently, a class was able to put questions to a mountaineer climbing Everest.

Mr Chan believes that what makes the Learning Village so valuable is its openness – which only the public Internet, as opposed to a special schools intranet, can deliver. It offers parents the opportunity for day-to-day involvement instead of just the odd visit to the school at some critical time in their child's career. He says this transparency is not only profoundly challenging for teachers, but also makes parents rethink their commitment to their child's education.

His main concern is to increase the number of parents using the site. Although Outram's parents tend to have lower-than-average incomes, about 40% have access to the Internet from home, but fewer than 20% have applied for a password. This being Singapore, though, they are not being let off the hook. Mr Chan is bringing them in for training sessions of two hours for groups of 30 at a time.

Other schools in Singapore are keen to follow Outram's example. In January 2000, 40 principals met to hear about Mr Chan's experience. One obstacle may be cost. IBM charges a subscription of S$2.80 per student per month to cover the cost of the software and of hosting the site, though it makes little or no money out of it. With a school the size of Outram, that works out at around S$40,000 a year. The Ministry of Education's director of education technology, Tam Yap Kwang, says it will be three to five years before the Learning Village can be extended to all the schools that want it, but he may be being too pessimistic.

Whatever the time scale, it seems likely that schools in poorer areas will get priority. The ministry has a policy of aiming the bulk of its IT spending at less able and less privileged children. Mr Tam says that, in the past, most of these children would have dropped out of school at the earliest opportunity, but give them access to a PC and the Internet and they will stay in school. If they want to upgrade their skills later, they will also be far better equipped to take advantage of the many online courses to be found in "Education Town" on the eCitizen site.

Haves and have-nots

How to overcome the digital divide

E-GOVERNMENT IS NOT just e-business on a larger scale. One of the most fundamental differences is that whereas businesses can, by and large, choose their customers, government cannot. The debate over the so-called "digital divide" is like the ghost at the e-government feast. For e-government to succeed fully, the dream of Internet access for all has to become a reality.

Governments are well aware that large and expensive e-projects will command little support if only a privileged minority benefits. As David Agnew of the Toronto-based Governance in the Digital Economy Project, which is supported by eight big IT firms and 20 national and local governments, argues: "If putting government online is just a way of reinforcing access for people who probably already have more opportunity to access government and decision-makers, then it hasn't really been much of an advance after all."

When Arizona's Democrats held their state presidential primary online in March, it nearly did not happen – not because of security and authentication problems (although there were plenty of those), but because a pressure group called the Voting Integrity Project tried to have it banned. It almost persuaded a court that the vote would disenfranchise the state's minorities, so should be ruled illegal.

Raise the spectre of the digital divide with the technology vendors and e-government champions within the public sector, and their brows furrow with concern – but not for long. They are, after all, professional technology optimists. But they also genuinely believe that many of the barriers to near-universal Internet access are falling, at least in economically advanced countries (though it is worth remembering that half the world has never even made a telephone call). Survey after survey has found that the main barriers to access are the fear that it is too expensive, that computers are too complicated and that somehow the whole thing is not really relevant or useful. Those optimists argue that, one by one, each of those perfectly legitimate anxieties is being overcome.

What's the problem?

Too expensive? Internet-ready PCs can be bought for little more than

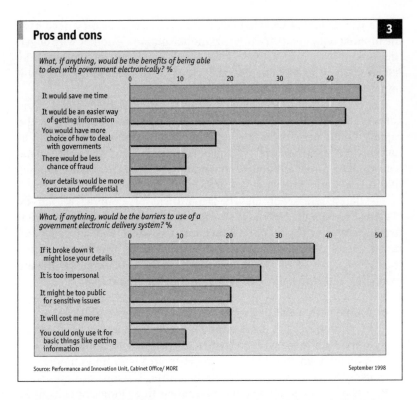

Pros and cons 3

What, if anything, would be the benefits of being able to deal with government electronically? %

- It would save me time
- It would be an easier way of getting information
- You would have more choice of how to deal with governments
- There would be less chance of fraud
- Your details would be more secure and confidential

What, if anything, would be the barriers to use of a government electronic delivery system? %

- If it broke down it might lose your details
- It is too impersonal
- It might be too public for sensitive issues
- It will cost me more
- You could only use it for basic things like getting information

Source: Performance and Innovation Unit, Cabinet Office/ MORI   September 1998

$300 – less than the price of most televisions, a device that has found its way into 99% of American homes. Some Internet service providers (ISPs) are even giving PCs away in return for two or three years' subscription, and other firms offer free PCs to users who agree to be bombarded by advertisements while online. True, access fees and telephone call charges remain high in some countries, but unmetered local calls are spreading from America to Europe, and free ISPs are evolving a range of different business models.

Too complicated and unreliable? That will soon be fixed by the proliferation of non-PC devices which provide access to the web. Among them will be simple terminals that do nothing more than run a browser and take all their applications from the web. These will be found in places such as schools, community centres, libraries and anywhere else that needs a robust machine and has an "always on" connection. An even simpler version is the kind of web kiosk with a touch-screen that is springing up in cities such as Singapore and Toronto.

Many people will be able to do as much business as they need over the Internet with inexpensive smart mobile phones. Another way of getting online is by interactive digital television. Early services, such as BSkyB's Open in Britain, are still clunky, but the technology will improve, and the set-top box decoders will often come free.

The more extreme technology optimists, such as Adam Thierer of the conservative Heritage Foundation in Washington DC, say that the rapidly falling price of both computing power and bandwidth will in fact create a "digital deluge", so any policies aimed at giving access to the "information-poor" are quite unnecessary and may be counter-productive. It is a comforting view, but probably quite wrong.

The latest release of the US Department of Commerce's survey "Falling Through the Net" paints a disturbing picture in which the digital divide between rich and poor, white and non-white, well-educated and under-schooled seems, if anything, to have widened significantly during the five years in which this information has been collected. Among the examples of the digital divide today, the survey found that:

- People with a college degree are eight times more likely to have a PC at home and 16 times more likely to have Internet access at home than those with an elementary school education.
- A high-income household in an urban area is 20 times more likely to have Internet access than a rural, low-income household.
- A child in a low-income white family is three times more likely to have Internet access than a child in a comparable black family, and four times more likely than if he were Hispanic.
- A wealthy household of Asian descent is 34 times more likely to have Internet access than a poor black household.
- A child in a two-parent white household is twice as likely to have Internet access as a child in a single-parent household. If the child is black, he is four times more likely to have Internet access than his single-parent counterpart.
- Disabled people are nearly three times less likely to have home access to the Internet than people without disabilities.

In other words, although Internet penetration has risen across all demographic groups, the digital divide remains only too real. It has also become a poignant proxy for almost every other kind of disadvantage and inequality in society.

Chalk and cheese

It would be hard to find a better real-life symbol for the digital divide than the gulf between Silicon Valley's leafy Palo Alto, home to dotcom millionaires, where the average house sells for nearly $700,000, and East Palo Alto, the desperate little town on the other side of Highway 101 that not long ago claimed America's highest murder rate. Palo Alto's website has 251 sections and is a paragon of e-government. Among many other things, it allows users to send forms to the planning department and search the city's library catalogue. During storms, it even provides live video footage of flood-prone San Francisquito Creek. East Palo Alto's site, by contrast, has only three pages, containing little more than outdated population figures and the address of City Hall.

The digital divide is not so much a question of access but of education. As Esther Dyson, an Internet pundit, puts it: "You can put computers in community centres, but only the literate people are likely to go use them." Simpler, cheaper ways of getting on to the web will help, as will content that seems relevant to those who shun the Internet today – after all, the mobile phone has conquered all social classes, thanks to its sheer usefulness and simplicity. But even with enlightened policies such as America's "e-rate", which gives cut-price web access to schools and libraries, and the growing number of private–public partnerships to spread both technology and training in its use, there is a danger that the "digital deluge" may reach only those parts where the grass is already green. The same people who have wired PCs today will collect all the fancy new web gadgets that are coming in, and the rest will continue to go without.

So what does this mean for e-government architects? First, as IBM's Todd Ramsey points out, they have to accept that some people, especially the elderly, will never want to deal with government – or indeed anyone else – online. That means some offline channels will almost certainly have to be kept open for years after everything has moved on to the web. Second, they must find ways to allow even those diehards to benefit from the e-government transformation by improving the quality of the offline channels and targeting them better. Not all the savings from electronic service delivery will be bankable.

Third, they need to think up incentives for those on the wrong side of the digital divide to take the leap. Government may be able to act as a catalyst in a way that the private sector cannot. What persuades most people to try the Internet is the promise that they will find something relevant to them. If the most convenient way of getting welfare benefits is online, a lot of people who had never thought of using the web will have a go.

Sign on the dotcom line

Paying fines or fees online is a lot more convenient

KALEIL TUZMAN says it was his discovery of a two-year-old unpaid parking ticket while moving house in 1998 that prompted him to start govWorks. How convenient it would be, thought Mr Tuzman, if he could pay the fine online. The end result of that reflection was a firm which uses the Internet to let people make all kinds of routine payments to local government – not just in one municipality, but potentially across America.

The business logic was simple. The United States alone has more than 80,000 local authorities which between them collect around $450 billion a year in fines and fees for things like building permits and business licences. If taxes are included, the figure rises to about $4 trillion.

Most of them post some information on the web, but very few as yet offer facilities for the public to do business with them that way. The standard method is still by mail or, if you can ever get through to the right person, the telephone. With limited funding for new IT initiatives and the skilled staff that would be needed for them, Mr Tuzman reckoned there was an opportunity for a web-based "infomediary" to pull all the pieces together. GovWorks would get its money partly from a convenience fee it would charge users for its service, and partly from a share of the savings made by the local authorities.

GovWorks, launched at the beginning of 2000 with $19m of venture capital, claims to be able to process credit-card payments of fines, local taxes, permits and utility bills for 3,600 municipalities. Its site is a portal designed to provide a friendly interface between citizens and bureaucrats. Its features include discounts on payments through its "govRewards" loyalty scheme, information about jobs, a who's who of officials, auction listings, bulletin boards and chat forums to encourage civic debate, among other things. Job advertising and hosting online government auctions are expected to provide additional revenue in future. To add a feelgood element, govWorks has promised to reinvest a portion of all these earnings in free PCs and pre-paid Internet access for schools, libraries and community centres. "It's all part of our effort to help bridge the digital divide," it claims with a hint of smugness.

Great minds think alike

It would be surprising if govWorks had such a potentially lucrative field to itself, and indeed several other well-financed competitors have emerged at about the same time. Some of them, such as PayTheTicket.com and Official Payments Corp, simply offer a payments service. Others concentrate on just one aspect of the government-to-business relationship. For example, Accela.com issues building permits, in the hope that this will bring traffic to its construction e-marketplace. Another firm, the National Information Consortium, offers a turnkey solution to governments, designing, building and operating Internet portals for 11 states, from Hawaii to Maine. NIC's proprietary technology provides access to many previously unconnected databases, saving its customers' time. Most of its earnings come from fee-sharing arrangements. Govhost.com is a systems integrator for local-government services. Giants such as IBM, EDS, Oracle and the big five consulting firms either collaborate or compete with e-government specialists in building and hosting sites.

GovWorks' closest rival is ezgov.com, an Atlanta-based start-up that launched its website a month before govWorks. It boasts Mario Cuomo, a former governor of New York, and Jack Kemp, a former congressman, among its directors. In March it tied up $28m of second-round funding. The main difference between the two is that ezgov.com aims to establish a deeper partnership more slowly by partly integrating its activities with the back offices of the government department concerned.

Each of the two firms has launched spirited attacks against the other. Bryan Mundy, the chairman and co-founder of ezgov.com, is particularly resentful of govWorks' claim to have been first in the field. He suggests that his rivals have hijacked his business model, and accuses them of being "a bunch of investment bankers" who lack the passion to make a difference. But this will not be a winner-takes-all market. Joe Landy of Warburg Pincus says that because government in America is so decentralised, "there's plenty of space here for a lot of winners". He is probably right. Nevertheless, a recent report by Forrester Research gives warning that this market will take time to develop. The problem, says one of the report's authors, Jeremy Sharrard, is that the contestants' "success is linked to outgunned, undermanned local governments, and as a result implementation will prove more difficult than expected". In order to move online quickly, he argues, governments often stop short of full integration, which makes their sites less convenient to use.

GovWorks, for example, does not connect to most of the governments it does business with, so payments made through the site can take up to seven days to reach their destination – about the same time as using the mail. Even ezgov.com, which does link into governments, usually processes information in batches only once a day – adequate for most things, but a far cry from the real-time efficiency of the best commercial e-tailing sites. Forrester also doubts that revenue will grow fast enough to finance traditional dotcom advertising sprees, so awareness of the sites may grow only slowly.

That said, Forrester believes the very existence of these sites will create consumer demand and put further pressure on governments to invest in improving them. Technology, especially data standards such as XML, will evolve to make the job easier, allowing the way services are delivered to be standardised, wherever they are based.

When that point is reached, government's present way of doing things will be challenged. Within America, the range of services that local government delivers is much the same, so as they move to the web, those services are bound to be commoditised. Firms such as ezgov.com will be able to do everything that today's bureaucracy can do. Unless it can respond creatively, local government in America, and many other parts of the world, may lose at least some of its *raison d'être*. Who said governments can't be "Amazoned"?

A local site for local people

THREE YEARS AGO, the people of Spain's Valencia region were rather behind with their computing skills and Internet experience. It was one of the things that put off would-be investors in the area. But today, thanks to a project called "Infoville", Valencia is on the way to becoming one of Europe's first "smart communities".

The government of Valencia, working with Oracle, a software and consulting firm, conceived Infoville not just as a local government website, but as a portal that would combine a broad range of services from both the public and the private sector. Juan Rada, who heads Oracle's service-industries practice in Europe, says that Infoville is a kind of local information utility which integrates e-commerce, e-government, online learning and virtual governance.

As well as dealing with government departments such as housing and tax-collection agencies, the portal also provides access to utilities, local bank accounts, schools, doctors' surgeries, garages, restaurants and retailers. With more than 260 services now available through the site, it is a little like an interactive version of the yellow pages. New services can easily be added, and will benefit from its centralised arrangements for administration and billing at virtually no cost to themselves.

Infoville was designed to be relevant to its users' daily lives, and to be simple enough for even the most technophobic to handle. To encourage its use, it was made accessible in a variety of ways: through not only PCs, but also kiosks in public places, as well as digital interactive television. The 100,000 citizens who are involved in the pilot were chosen as a representative sample of the population at large.

To disarm critics, Mr Rada makes a simple point: most of the activities people engage in take place within their local community. For example, 90% of telephone calls are local. He calculates that up to 80% of the information carried on a site such as Infoville is unique to the region. The pilot has been so successful that similar services will soon be launched in 35 more Spanish cities.

Another European city that set out to become a "smart community" was Naestved in Denmark. It was faced with a decline in traditional industries such as paper, steel and timber and had trouble attracting new ones because it is by-passed by main road and rail links, even though it is only an hour's drive from Copenhagen. To overcome that disadvantage, it built itself a world-class IT infrastructure based on high-speed cable and set about integrating Internet technologies into every aspect of local society – private, public and commercial.

One result is NaestvedNet, an Internet database that provides access to all regional services from a single site. Users can choose from, among other things, council services, traffic information, an interactive local phone book, a complete local business directory, tele-education, an electronic map, banking and online local shopping. A digital signature can be obtained by anyone who wants it, and so far about 20% of the population have got one. Herman Weidermann, the municipal director, says that digital signatures give citizens access to their own files and allow them to perform legal transactions over the net, despite Denmark's draconian data-transfer laws.

The mayor of Naestved, Henning Jensen, thinks that concerns about the digital divide are overdone. But to make doubly sure, six open data centres have been set up across the city to provide free PC and Internet access, as well as a distributed learning network for training and computing skills. Working with IBM, Naestved has set up a single basic system to meet the needs of government employees, citizens and private firms alike.

Neither Naestved nor Valencia have found it cheap to build their smart communities, but Oracle's Mr Rada says the costs have to be put into perspective. The cost of launching Infoville, he points out, has been about the same as building a single kilometre of motorway. If that seems too expensive, he says, Oracle offers to put up the money in return for a share of advertising or transaction revenue. "Don't invest – if you give us the site traffic, we will build a local information utility for you."

Digital democracy

Stand by for online voting, and more

IN THE 48 hours after winning the Republican primary in New Hampshire, John McCain raised $1m in campaign funding through his official website. The donations were made by credit card and averaged $110. He subsequently amassed a further $6m online, but even that was not enough to stave off defeat at the hands of a well-dug-in party organisation that had already committed itself to George W. Bush. Still, Mr McCain's use of the web to raise funds as well as to organise grassroots volunteers was not lost on the political establishment.

To date, fewer than 14m Americans have ever used their credit cards for an online transaction, but as that number surges over the next few years, candidates will be able to use the web to go straight to the voters' wallets. The traditional party machines will increasingly find themselves disintermediated.

Inevitably, much of e-government is about the delivery of services and the government's dealings with the private sector. But there is another dimension to it. This is how David Agnew, the executive director of the Governance in the Digital Economy programme in Toronto, describes it: "You can't build a fence around the citizen as simply a consumer or customer of government services. That same citizen is also an owner or shareholder of government itself. In the digital age, people have an ability to communicate, to participate and add value."

Mr Agnew believes that just as the Internet has helped to empower a new generation of well-informed and demanding consumers, it will challenge the essentially passive relationship that the majority of people have with government and politics. Just as consumers at first used the web to gather information and only later took the plunge by buying things online, in due course citizens will move from using the web to communicate with government to expecting to be able to cast online votes in a national election.

President Clinton acknowledged as much when he launched a number of initiatives under the broad heading of "e-society" – including a feasibility study into the introduction of online voting described by White House insiders as "very serious". There is support for online voting in Europe too. An e-mail survey of 500 elected officials in 14 countries in 1999 showed that more than half would back the introduction of online voting as long as traditional channels remained in place (see Chart 4).

How soon online voting will become routine is hard to say. Experiments at the local level, such as the Arizona Democratic primary in March 2000, when 40,000 people voted via the web (a 600% increase in turnout over the election of 1996), will proliferate in both America and Europe as governments learn how to run cyberpolls.

When Arizona's Democrats held their now famous election, state officials from all over America went to Phoenix to learn from the experiment and to quiz the firm that conducted the poll, election.com. Although those involved declared it a huge success, there were enough hitches to give pause for thought. The voting site went down for an hour on the first day of voting; some voters lost the PIN number that had earlier been e-mailed to them; and the helpline could not handle the volume of calls from all the people having problems. Users of Apple Macintosh computers encountered particular difficulties.

Thomas Wilkey of the New York State Board of Elections was unimpressed: "I call it chaos." Alfie Charles, the assistant secretary of state in California, agreed that breakdowns of any kind were unacceptable during an election, and added that before Internet voting systems could be certified, state authorities would have to be confident about the strength of encryption systems and the overall reliability of the software. "It just shows the need to take a cautious, incremental approach," he said.

Theo Dolan of Forrester Research echoes that sentiment, arguing that essential requirements are likely to include mass voter authentication, online databases with accurate, up-to-date electoral rolls, and voting sites that can withstand concerted hacker attacks.

How universal?

But the biggest uncertainty surrounds the digital divide. Some argue that as long as Internet access is so heavily weighted towards the better off, online voting will inevitably weaken the voice of the poor and of ethnic minorities still further. Others, such as Reed Hundt, the former chairman of the Federal Communications Commission, think that it will encourage voter registration, especially among the young. Opponents may find it hard to argue against something that could help to make America's voters less apathetic (only 49% bothered to vote in the 1996 presidential election).

Still, there is more to e-politics and digital democracy than online voting, says Janet Caldow, the director of the IBM-funded Institute for Electronic Government in Washington, DC. Ms Caldow defines digital democracy as any electronic exchange in the democratic process. As well as online voting, digital democracy includes a number of things that are already gaining ground: campaigning and fund-raising, as demonstrated by Mr McCain; voter registration; opinion polling; communication between representatives and voters; wired legislative bodies (such as Wisconsin's state assembly and Belgium's parliament); and feedback from the public on legislative drafts.

If some governments are tempted to drag their feet over digital democracy, they will find that a plethora of commercial e-politics sites has stirred interest among both candidates and voters. PoliticsOnline provides news and webcasts of campaigns, but also sells Internet tools and software to candidates and campaign managers who want to build a site for fund-raising. One of the PoliticsOnline offerings is "Instant Online Fundraiser v2.0 – your Internet fundraising solution".

Virtual votes 4

A survey of elected officials from 14 European countries, 1999, % of total*

Would you support the introduction of online voting?

Yes 50.4
No 30.6
Don't know 19

Do you believe that information technology can enhance democracy?

Yes 74.2
No 15
Don't know 10.7

Source: Institute for Electronic Government

*436 responses

New York-based election.com, the company that ran and hosted the Arizona primary, claims to have conducted hundreds of elections in both the public and the private sector around the world. It promises a seamless migration path from the traditional, paper-based election process to the Internet". Its latest project is the Youth-e-Vote, a nationwide exercise in online voting for students.

The brilliantly named E The People describes itself as "America's Interactive Town Hall". It offers a service processing traffic-violation fines, but its main activity is running online petitions on almost any subject, from "Stop Kent State Primate Research" to "Equal Rights for Children of Second Families" and "Impeach Mayor Penelas". Other participatory sites include Voter.com and Grassroots.com. Voters who find it hard to make up their minds can try CandidateCompare.com and SelectSmart, which will identify the candidates whose policies and prejudices most closely match those of the inquirer.

Some hope that this could one day lead to the kind of electronic direct democracy once advocated by Ross Perrot, the founder of EDS and maverick presidential candidate. But the unbridled populism that might be released is something that makes most e-government enthusiasts recoil. The aim, they say, is to work with the grain of the representative institutions we already have, not replace them with something so scarily unpredictable. They prefer to think of the rekindling of Jeffersonian democracy by electronic means. As David Agnew says: "In many countries, democracy is showing signs of fraying round the edges. We've seen not just lower participation rates in elections, but also a lower commitment to the political process and lower trust in government. Thoughtful governments are looking at the Internet not as a threat, but as a positive potential tool to re-engage the citizenry in the business of governing."

Perhaps. But Jeffersonian democracy was, in practice, democracy by and for an elite. Unless great care is taken, digital democracy could turn out to be something similar – providing simply a better way for political activists to be politically active.

Handle with care

E-government is mostly a good thing, but it needs watching

WHEN INTEL'S ANDY GROVE declared in 1999 that all businesses would have to become e-businesses soon if they wanted to survive, few people disagreed. The fear of new Internet-borne competitors and the sudden transformation of old rivals was almost palpable. Many businessmen were also excited about innovative business models that seemed to dazzle investors. E-government is not driven by such extremes of fear and greed as e-business, but the rewards for success and the penalties of failure are nevertheless real. So what should governments do to prepare?

The first thing they need to do is hurry up. If they are not already well on the road to e-government, they should start at once, or other countries and regions will make them look like laggards. Governments form such a crucial part of the business environment that, when all else is equal, firms and skilled people may prefer to go where the benefits of e-government are available. Not only are e-governments far more efficient than the old-fashioned variety, they are also easier to deal with and better liked by their citizens.

At a minimum, e-government might consist of nothing more than filling in a few forms on the web. But to move from there to a fully fledged e-government requires profound changes in the culture, the processes and the relationships that define government as an entity. Would-be e-governments need to adopt a four-pronged strategy.

- The first prong is the establishment of a secure government intranet and central database that reaches across all departments and enables them to work together. The intranet must also have resilient, high-capacity interfaces with public-sector agencies and local and regional government bodies.
- The second is the delivery of services that are tailored to the needs of citizens, and are accessible via the web in a convenient and secure form. The aim should be to provide a one-stop portal that is always open and caters for all aspects of the relationship between government and citizen.
- The third is the creation of a government e-marketplace where

departments can advertise their requirements, authorised suppliers can bid and post tenders for high-value purchases and public servants can purchase low-value goods quickly, efficiently and at centrally negotiated prices.

- The final prong is digital democracy. Governments and politicians must use the web to make themselves more transparent and accountable to the voters, evolve new methods of consultation and eventually offer online voting.

Implementing e-government is complex and will require not only vision, but also strong political leadership at the highest level. Prime ministers and presidents do not need to know how to write Java code, but they must have the basic familiarity with technology that comes from using computers and the Internet in their daily lives. Bosses whose idea of using e-mail is to get their assistants to provide printouts are ill-equipped to lead the transformation of their firms into e-businesses. They do not know the right questions to ask, and lack credibility with employees. The same applies to senior politicians.

Leadership also means that a team has to be put in place within government that has the political clout and the funding to knock departmental heads together, rethink current working practices and impose technology standards. Britain has made a start by appointing an impressive-sounding "e-envoy", Alex Allan. He leads a group of 35 senior officials representing the departments and agencies involved, who optimistically call themselves "The Information Age Government Champions". But the minister responsible, Ian McCartney, attached to an all-purpose department called the Cabinet Office, is rather junior, which raises doubts about the outfit's effectiveness. Singapore, by contrast, has given the job to its Ministry of Finance and the IDA, the agency responsible for IT and communications policy, both of which carry plenty of clout.

Keep it simple

So where should governments start? An e-government portal can operate on three levels of complexity: first, publishing information and providing links to existing departmental websites; second, establishing a two-way channel for dealing with individual agencies; and third, doing business involving multiple agencies. The best starting point may be to get a fairly simple portal up and running and then add more functions in stages as progress allows. An over-ambitious site that loads pages slowly and fails to deliver on its promise to work across departments and agencies will put users off.

Governments would also be well advised not to get carried away with the technology, but to stick to established Internet standards and protocols and to use tried-and-tested packaged applications with as little customisation as possible. Much of what they should aim for sounds obvious: a flexible technology infrastructure that can easily adapt to different amounts of traffic (because successful central government portals are bound to be busy, with unpredictable peaks); a single interface that provides a consistent look and feel for its users; straightforward search and navigation; the widest possible range of getting access to the system, from PC browsers to WAP phones and digital television; and the provision of links between online services and telephone call centres, because many people will still prefer to use the phone to deal with government departments and agencies.

E-government projects offer great opportunities for establishing new partnerships with the private sector, for example by developing new funding models and sharing risk with technology vendors. Each government will have to decide for itself how far it wants to go in allowing the private sector to supply public services and to package these with commercial services. But, leaving funding to one side, arrangements of this kind will be essential to bridge a growing skills gap, because the public sector is finding it ever more difficult to attract scarce IT professionals. Careful thought must be given to the regulatory framework for such partnerships. Typically, problems will arise about the way licences are awarded, and over what periods; fair treatment for governments' other commercial partners; and the safeguarding of citizens' privacy.

Privacy and security is a hugely important aspect of successful e-government. By giving every citizen a digital signature and maintaining the highest standards of data and privacy protection in its electronic transactions, e-governments will not only increase confidence in the Internet delivery of their own services, but also provide a stimulus to e-commerce throughout the economy. If they fall down on the job, they risk a damaging erosion of trust. As more and more detailed information about individuals is gathered electronically and passed between agencies, e-governments will also have to allay citizens' understandable fears that Big Brother is snooping on them in cyberspace.

Another serious concern is the digital divide, which carries the threat that the least well-off may not have access to the system. At first sight, this suggests that governments should move more slowly, but an even better case can be made for the opposite policy. Because all citizens have to deal with government, whether they like it or not, e-governments can provide incentives for them to make the web their preferred channel for such transactions, thus spurring the adoption of the Internet. In the same way, governments can encourage small businesses that want to supply them to get on to the web.

Fully fledged e-government will be neither quick nor easy to achieve – despite the sometimes extravagant promises of cyber zealots and politicians alike. In that respect, the rules for e-government are different from those for e-business, where it may sometimes be better to be fast than right. Governments have to be more cautious, they must take more care to take people with them, they are more accountable for the money they spend, and the sheer size of their operations dwarfs all but the biggest global companies. That said, for the first time since the establishment of the modern welfare state, there is now a real chance to "reinvent" government – and make it a great deal better.

For tomorrow's e-citizens and e-businesses, the coming e-government revolution is almost wholly good news. It offers the potential for services that are designed for citizens' needs, and available when and how they want them; lower taxes, as increased efficiency cuts the cost of government; more transparent ways of doing business with the different arms of government; a two-way street of consultation and collaboration; a new level of accountability for both elected and unelected officials; and more open and responsive politics.

The one important reservation is that vastly more efficient governments will also know vastly more about each and every one of their citizens. The exponential increase in the ability of e-governments to gather, store and mine data about people will raise well-founded worries about privacy and civil liberties. The price of happy e-citizenship will be eternal vigilance.

The material on pages 69–106 first appeared in a survey written by Matthew Symonds in *The Economist* in June 2000.

POSTSCRIPT

SINCE THIS SURVEY was written, there has been a huge shift towards the electronic delivery of government services around the world. For example, whereas in Britain 18 months ago the government had some ambitious ideas about what it wanted to achieve online but little to show for it, now it has a rather effective portal – ukonline.gov.uk – that enables people to apply for passports, pay their taxes, notify a change of address or buy a TV licence. Like Singapore's pioneering citizens' portal, it also organises information and services around so-called "life events", such as moving house, getting married or dealing with the death of a relative. In the Netherlands, nearly 40% of Internet users already deal directly with government websites by submitting forms and making payments online. Governments are increasing their Internet investments, unlike many businesses which have cut back on theirs over the past year or so. Dotcoms may have faded, but dotgov is just getting into its stride .

III

E-COMMERCE

Define and sell

Where e-commerce wins hands down, and where it doesn't

IT IS TEMPTINGLY easy to use the phrase "electronic commerce" without troubling to define it, but there is something to be said for being a little more rigorous. E-commerce refers to trade that actually takes place over the Internet, usually through a buyer visiting a seller's website and making a transaction there. Clearly the influence of the web – for example, as a source of information – stretches much wider than this. One estimate suggests that, although only 2.7% of new-car sales in America in 1999 took place over the Internet, as many as 40% involved the net at some point, with consumers using it to compare prices or to look at the latest models.

Within this broad definition of e-commerce, it also helps to be clear about the main actors, ie, businesses and consumers. At present the biggest volume of trade by far is business-to-business, typically for suppliers to such large companies as General Electric. Several technology companies, including Cisco and Oracle, have transferred almost all their purchasing (and indeed most of their sales) to the web. Web-based business-to-business exchanges have mushroomed. Even such long-established businesses as America's two biggest car makers, Ford and GM, say they are transferring all their purchasing to the web within the next few years.

This survey, however, is concerned with the remaining three segments of e-commerce: business-to-consumer, consumer-to-business and consumer-to-consumer. The first embraces normal retail activity on the web, such as bookselling by Amazon.com or online stockbroking by Charles Schwab. The second, as yet smaller, takes advantage of the Internet's power to drive transactions the other way round: would-be passengers bidding for airline tickets on Priceline.com, for example, leaving the airlines to decide whether to accept these offers. The third covers the new fashion for consumers' auctions, epitomised by the auction site eBay.com. The shorthand for these segments is B2C, C2B and C2C (see chart 1).

What has the Internet, or more specifically the World Wide Web, got to offer in all these areas? John Hagel, a consultant at McKinsey in Palo Alto, points to ease of price comparison and greater choice as its two

The e-commerce matrix   **1**

	Business	Consumer
Business	**B2B** GM/Ford EDI networks	**B2C** Amazon Dell
Consumer	**C2B** Priceline Accompany	**C2C** EBay QXL

Source: *The Economist*

biggest plus points compared with the physical world. The web's reach is global – no limits of shelf space or warehousing – so it ought in some respects to be able to outdo the real world. Amazon quickly proclaimed itself the world's biggest bookseller, even though its own physical stock of books was tiny. EBay aggregates bidders around the world, giving it a huge advantage over the local flea market.

On the other hand, websites are not much good for replicating the social function of shopping, or for browsing around, or for producing the serendipity and impulse purchases that come from visits to a shopping centre. Nor, because it usually depends on separate delivery, can e-commerce offer the instant gratification that today's consumers have come to expect. The Internet may work better for replacement buys than it does for new purchases. And fulfilment must always be brought into the equation: goods that are heavy or bulky will be harder and more expensive to sell online than light, easily transportable ones.

Then there is the question of what kinds of goods and services sell well electronically. The simplest distinction here is between "high touch" and "low touch". The first category includes clothes and shoes, as well as many groceries, which consumers often prefer to be able to see and poke before they buy. The second takes in such products as computers, books and CDs. Pornography, the earliest big seller on the web, may rate a special category of its own. In general, it is low-touch items that have sold best on the web so far (see chart 3). Yet the distinctions are blurring: catalogue experience suggests that consumers will buy clothes without trying them on, and the Internet may yet offer "virtual" fitting.

Chart 2 also shows some goods and services that are tailor-made for the web: those that can actually be delivered over the Internet. Computer software is the most obvious example, but increasingly such things as airline tickets, stockbroking services, banking and insurance, and even books and newspapers are being delivered electronically. Anything with content that can take digital form, which includes most recorded music and film, is especially suitable for the web – indeed, the economics of the web could easily kill off the traditional distribution channels for such products. The same may apply to many financial services.

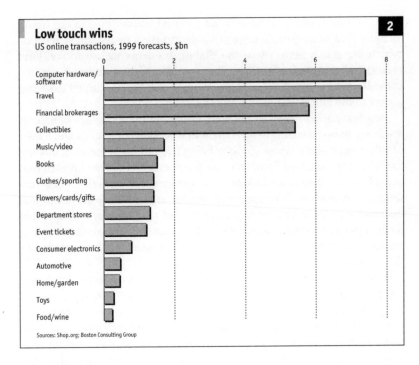

Low touch wins
US online transactions, 1999 forecasts, $bn

2

Computer hardware/software
Travel
Financial brokerages
Collectibles
Music/video
Books
Clothes/sporting
Flowers/cards/gifts
Department stores
Event tickets
Consumer electronics
Automotive
Home/garden
Toys
Food/wine

Sources: Shop.org; Boston Consulting Group

Cash and carry

What works well in e-commerce is not predetermined for ever. Technological change is increasing the possibilities all the time. Development of more secure payment systems could help. Consumers are often advised against giving their credit-card numbers freely over the Internet, and this remains one of the most-cited reasons for not buying things online. Smart cards and digital cash would get round this, as well as making it possible to use the web for small purchases. Yet it should be said that the great credit-card fear has not, so far, proved well-founded: there have been very few instances of theft of credit-card numbers, and, even if there were any, the credit-card firms would shoulder most of the liability. Using a credit card to pay over the telephone is riskier. The increasing acceptance of the credit card for payment over the web may, indeed, be hampering the development of smart cards and digicash.

Technological change to improve speed and capacity is probably more important. The biggest boost to e-commerce over the next few years will come not from snazzier websites or snappier marketing, but from the proliferation of broadband Internet connections to the home as more and more people acquire cable modems or DSL lines, both of which are much faster than the dial-up modems in use today. One reason why America Online, the world's biggest Internet service provider, was so eager to buy Time Warner was that it wanted to secure access to the entertainment company's cables.

Besides quicker access to the web, two other huge changes loom that are likely to boost e-commerce. One is that mobile telephones and a host of other electronic devices are now being hooked up to the web, ousting the personal computer from its monopoly position in providing Internet access. The second is the linking of websites to call centres, so that customers can ask for direct personal advice as they surf; the telephone operator may even be able to take control of the browser to help guide a customer. A catalogue operation such as Land's End is already exploiting the link between its website and its call centre to make better use of the Internet.

Indeed, the experience of such catalogue sellers as Eddie Bauer and L.L. Bean suggest that the scope for web retailing may turn out to be a lot bigger than it seems at first sight. Clothes and shoes have both won a place in the catalogues, even though they are high-touch goods. An even more striking example is raw steaks: Omaha Steaks has a strong catalogue business selling steaks by post that it is transferring to the web. Many adventurous online merchants now reckon there may be no such thing as a product that is unsuitable for being sold over the web.

A walk through webonomics

That is the more striking, given the economics of the web, which look powerful. The most obvious gains come from cutting out shopfront costs and intermediaries, and from cheaper distribution. Book retailing on the web, for instance, means dispensing with big shops full of slow-moving stock: orders at Amazon.com go straight to the wholesaler (the reason the firm is based in Seattle is that America's biggest book wholesaler is there). Not needing much working capital is also a bonus: an online retailer usually gets paid before he has to pay his distributor, whereas in the physical world it is the other way around.

These gains should not, however, be exaggerated, if only because there are big offsetting costs. Running and servicing a website to ensure that it is 99.999% reliable (the ideal "five nines" goal) is not cheap. Logistics and distribution, too, are so critical to an e-commerce venture's success that it often has to spend heavily on them. And there are big marketing expenses (optimistically known as customer-acquisition costs) as well. Although many online retailers have managed to undercut their offline rivals on price, they have also managed to lose huge amounts of money in doing so.

Then there are the economic gains from the things that are possible on the Internet but not in the real world. Paul Saffo, of the Institute for the Future in Palo Alto, suggests that the biggest impact of the Internet may come when, like earlier technological revolutions, it gives rise to entirely new products. For existing ones, price comparisons become far easier (and can be made automatic) when it is a matter of clicking, not walking. That suggests the Internet could be a strong price-deflation mechanism: raising your prices is harder when your customers instantly compare them with everyone else's.

Economies of scale and scope are also easier to obtain online than offline. A single website can be used to cover the globe: once it is set up, and subject always to fulfilment problems – of which more later – it is eminently scalable (the current buzzword for the ability to get bigger without a big rise in costs). Similarly, it is far easier for a website that is successful at selling one product to branch into others. And the web allows things like customer aggregation and auctions to be done in ways that are impossible in the physical world.

Data can also be exploited far more readily on the Internet than in the real world. Everything can be recorded: not just every transaction, but which web pages a customer visits, how long he spends there and what banner ads he clicks on. This can produce a formidable array of data that makes possible both one-to-one marketing – directing sales pitches at particular individuals – and "mass customisation" – changing product specifications, for instance for jeans or computers, to match individual orders to the individual customer's preferences.

The Internet could, in short, overturn much of the traditional economics of retailing. Philip Evans and Thomas Wurster, consultants with the Boston Consulting Group, have proposed one methodology for assessing the web's effects on the industry. They consider three factors: reach, richness and affiliation. Reach refers to the size of your audience, richness to the intricacy and customisation of the services you can offer, and affiliation to the extent to which you are seen to respond to your customer's interests.

In the physical world, these often have to be traded off against each other: go for a big audience and you lose familiarity with individual preferences, for example. The prize offered by the Internet is that you

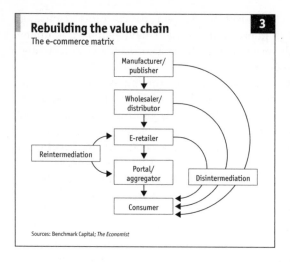

Rebuilding the value chain
The e-commerce matrix

Sources: Benchmark Capital; *The Economist*

can do them all together. Reach can be huge because you are no longer constrained by physical space; but richness need not be sacrificed, since online – and assuming the right plant design and relationship with suppliers – a bespoke service such as Dell Computer's can be as easy and cheap to offer as an off-the-peg one.

Another approach is to re-examine the normal value chain for retailing. The web makes it possible to dispense with much of this chain altogether, through direct sales by manufacturers to consumers. But as Bill Gurley of Benchmark Capital, a venture-capital firm in Silicon Valley, notes, it can also create new points on the value chain, such as Internet portals that act as shopping malls, or aggregators that offer a new way of amassing buying power (see chart 3). Pierre Omidyar, who founded eBay, stresses the power of portals: many web-surfers arrive first at Yahoo! or AOL, giving such firms huge leverage over retailers that need to be on their sites.

Examples now abound of how this works in practice. Dell, famously, has carved out its profitable niche by selling direct, first by telephone and now via the Internet. Michael Dell, the firm's founder, suggests that the car industry should adopt his model; and, indeed, both GM and Ford have been buying out dealers with a view to moving to direct distribution. (On the other hand, Levi's has stopped selling its jeans direct over the web.) Travel agents are being disintermediated all over the place; so are the record shops. Electronic goods are being sold to buying clubs or through auctions. Books are being delivered in packages of one by wholesalers, or even published direct on the web, cutting out not just booksellers but publishers and wholesalers as well.

Its early apostles thought the web was a good thing because it would disintermediate in these ways, bringing savings to both manufacturers and consumers at the expense of those in-between. Yet what is emerging as often is that the Internet is changing the role and function of intermediaries, not eliminating them. Mr Evans talks of "navigators" that will spring up to represent customers seeking to get the best out of the web. McKinsey's Mr Hagel has coined the word "infomediary", a new beast that will look after individuals' concerns over privacy or payment security.

Whichever fancy-sounding model comes to pass (disintermediation, navigation, infomediation, whatever), they all share one common feature. They threaten the cosy world of the traditional intermediaries in commerce, many of whom have for years lived off mark-ups while adding little value. The fiercest battles between old and new intermediaries are being fought in retailing.

Something old, something new

There may be room online both for established firms and for virtual newcomers. But which ones?

THE CONVENTIONAL WISDOM about traditional retail firms (often known as legacy businesses) and the web has been redefined several times in recent years. In the early phase, from about 1995, it was said that the legacy folk just did not "get it" – leaving the field clear for electronic traders such as Amazon, CDNow and E*Trade, an online stockbroker, which are known as "pure plays". Even when the legacy businesses started to make their own forays into trading on the web, their efforts were often derided in Silicon Valley and Seattle. A new verb entered the language: to be "amazoned", meaning to lose a big chunk of your business to a web upstart.

During 1998, however, this view started to be modified. The relative success on the Internet of Barnesandnoble.com (now bn.com), a joint venture between the real-world bookseller and Bertelsmann, a German publishing giant, and of such catalogue businesses as Land's End, began to make the pure plays sit up. Even Toys "R" Us, a troubled toy retailer that seemed to have been amazoned by eToys, a new pure play, started to recover some lost ground, despite serious problems with its Toysrus.com website and with order fulfilment.

The conventional wisdom has now shifted once again, to almost the opposite of where it started. A popular line is that the pure plays have had their years of fun, but now that the big boys, such as Home Depot, Merrill Lynch, Kmart and Wal-Mart, are seriously moving on to the web, they are likely to demolish their virtual rivals: Amazon.toast, in the vernacular. All figures about commerce on the web are murky, but one bit of evidence to support this new view is the online retailers' stunning lack of profits – and their falling share prices. Yet of the top 20 retail websites ranked by visitors during the holiday period at the end of 1999, only five were offshoots of legacy businesses – though these have generally been the fastest-growing sites (see chart 4).

On the face of it, the legacy businesses ought to have some significant advantages over the pure plays. They have an established brand name, a huge customer base, and strong links with suppliers that enable them (especially if they are big enough) to demand deep discounts. They also

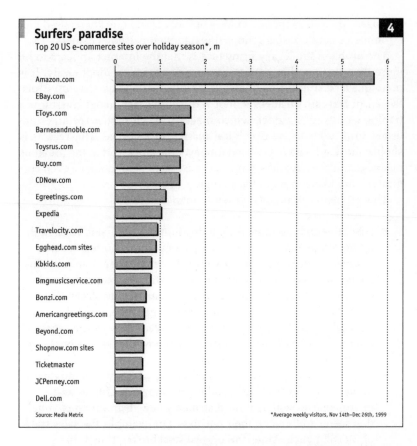

Surfers' paradise
Top 20 US e-commerce sites over holiday season*, m

4

Amazon.com	
EBay.com	
EToys.com	
Barnesandnoble.com	
Toysrus.com	
Buy.com	
CDNow.com	
Egreetings.com	
Expedia	
Travelocity.com	
Egghead.com sites	
Kbkids.com	
Bmgmusicservice.com	
Bonzi.com	
Americangreetings.com	
Beyond.com	
Shopnow.com sites	
Ticketmaster	
JCPenney.com	
Dell.com	

Source: Media Metrix   *Average weekly visitors, Nov 14th–Dec 26th, 1999

have a range of shops that should be of some value for fulfilment and distribution, as well as making life easier for customers who like to combine online and offline shopping, for instance by returning products they have bought over the web to a physical shop. For those who look on the Internet as essentially just another distribution channel, along with shops, telephone selling and catalogues, it seems natural that the legacy firms should come to dominate it.

A legacy too far
Maybe they will. But there are several large obstacles in the way, and it is these obstacles that provide the answer to another puzzling question: why have almost all the legacy businesses, whatever field they are in, been so slow to respond to the arrival of the web? It is not as if all their

managers are technophobes, knaves or fools. Many quickly realised that e-commerce could pose a serious threat to their existing business.

The problem the legacy firms face is that the Internet appears to be more than just another new technology for them to adopt easily. It is more like a "disruptive technology", a phrase coined by Clayton Christensen, of Harvard Business School. A disruptive technology overturns a traditional business model, which makes it much harder for an established firm, with its own cultural inertia, to embrace. The reason is simple: those who have invested money, time and effort in the previous business model – suppliers, employees, bankers, shareholders, even customers – are likely to resist the change.

This problem can manifest itself in many ways:

- ◪ Fear of cannibalisation. A new distribution channel is vulnerable to the charge that it is not creating a new market or extra sales but merely siphoning off existing sales from other channels. This charge will be especially pointed if, as with the Internet, the new channel tends to offer lower prices: if it merely displaces sales, the net effect on the bottom line will be negative. Several booksellers, including Barnes & Noble, were slow to embrace Internet retailing partly because they feared that selling books at lower prices would dent their profits.
- ◪ Partly because of fear of cannibalisation, a legacy firm moving on to the web risks "channel conflict". Existing sales forces and intermediaries will fight hard against a new distribution channel that seems to threaten their business. For example, the sales staff of Merrill Lynch, America's biggest stockbroker, fought hard against starting an online stockbroking operation; at one point Launie Steffens, boss of the firm's retail operation, reassured them with an embarrassing attack on the Internet, which his firm then embraced soon afterwards. Levi's, which made great play of direct sales of jeans over the web when they began in 1998, has been forced to stop because its other retail channels objected vociferously to being sidestepped.
- ◪ Firms that have tried to branch into e-commerce have found the capital markets particularly awkward to deal with. For most of 1997–2000, the markets rewarded pure plays with stratospheric share prices, making capital virtually free, even as most of them were running up huge losses. Yet legacy firms have continued to be judged on normal earnings criteria, so when they have

plunged into a web operation that has lost money, their shares have often been punished, not rewarded. In effect, the capital markets have, at least until recently, been sharply tilting the playing field in favour of the pure plays (see chart 5).

▪ The high valuations placed on Internet shares have also caused problems over both pay and acquisitions. Offer a talented young website-designer a choice between joining a company like Amazon, which can reward him with stock options that hold out the promise of untold wealth, or a boring old retailer like Macy's, and which will he take? Equally, frothy shares have given pure plays a new currency to use to buy up promising competitors, which legacy firms have been unable to afford.

▪ Offline companies might be expected to have an advantage in distribution, but their distribution systems tend to be of the wrong sort, geared to shifting pallets of goods from large warehouses to store shelves. Selling over the web requires a completely different warehouse-cum-truck system that can deal with delivery of a single package to an individual household.

▪ Thanks to their brands and their customer base, established firms may indeed be able to spend less than pure plays on marketing. But their information about customers also turns out to be of the wrong sort. A big discount retailer will know exactly what sells and in what quantity, allowing the firm to fine-tune its stock and shelving plans. But it often has little idea who its individual customers are; nor are retailers in general much good at mining the data they do collect, eg, from credit-card slips. The web, however, is perfect for collecting, storing and exploiting individual data – recommending similar book titles, offering information on what a customer has bought before and so on.

▪ Lastly, the economics of the Internet seems to offer a powerful first-mover advantage. An e-commerce operation on the web can be scaled up at low cost in a way that its physical equivalent cannot. This does not mean that the first mover will automatically win: Amazon was not the first bookseller on the web, nor was eBay the first auction site. But if an early mover gets everything right – its website, its order fulfilment, its distribution – a newcomer might find it much harder to knock it off its perch than it would in the physical world.

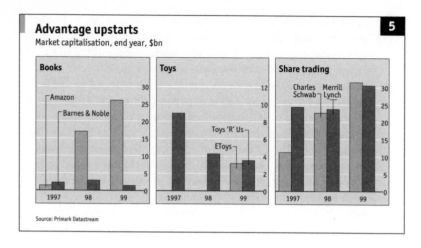

Advantage upstarts
Market capitalisation, end year, $bn

5

Books — Amazon, Barnes & Noble, 1997, 98, 99

Toys — Toys 'R' Us, EToys, 1997, 98, 99

Share trading — Charles Schwab, Merrill Lynch, 1997, 98, 99

Source: Primark Datastream

There are plenty of recent examples to show how hard it can be for established firms to adapt in time to fend off new challengers. Look at the computer industry, for example: ranging from the capture of a large chunk of the PC-making business by Compaq (thanks to its low-price strategy) to the recent squeeze that Dell and Gateway have, in turn, put on Compaq. Or, in retailing, look at the way that Sears Roebuck knocked out a whole bunch of small hardware stores when it introduced its catalogue early last century, only to be ousted in turn from its position as America's largest retailer by Wal-Mart's giant out-of-town superstores in the late 1960s.

Blessed are the pure

It would be a mistake, then, to regard the newcomers as just so much cannon fodder for their experienced rivals. But it would be equally wrong to minimise the problems that the pure plays themselves now face on the web. There is so much noise on the Internet that getting sufficiently well known to win customers can be quite hard. In the physical retail world, you can simply put a snazzy new store on a good site and wait for passers-by to come in. On the web you need to do far more to attract attention. Hence the wave of dotcom advertising in 1999, which in the end threatened to swamp many of the advertisers – some of whom spent as much as three-quarters of all the cash that they had raised from venture capitalists on marketing.

The capital markets, too, seem, slowly and belatedly, to be overcoming their infatuation with dotcoms. New companies of this kind now

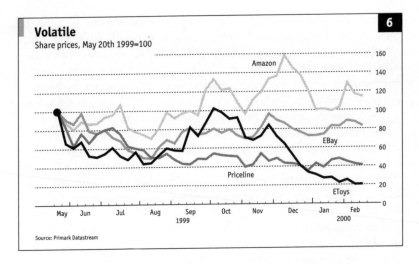

Volatile `6`
Share prices, May 20th 1999=100

Source: Primark Datastream

often find it harder than they expected to raise venture capital, unless they can show that they have a genuinely new business model; if they are following in somebody else's footsteps, their lack of originality will be held against them. And lack of profits could yet turn out to be a big problem: the shares of several e-commerce firms have fallen back sharply from their peaks (see chart 6). It is notable that nearly three-quarters of e-commerce-related IPOs since mid-1995 are now trading below their issue price. The markets are beginning to fear that a dearth of profits may be something other than just a reflection of high customer-acquisition costs. Instead, it may be a consequence of the increased competition that the Internet has created, and of high fulfilment and distribution costs.

Moreover, as established firms begin to grasp why they have found it so hard to succeed on the web, that knowledge could help them to overcome the obstacles. Consider the example of Charles Schwab, now America's biggest online retail stockbroker. Until 1996, it operated largely as a discount telephone broker, though it had a few physical offices as well. Then it set up an online operation called eSchwab which charged lower commissions, but soon became engulfed in the usual problems of cannibalisation and channel conflict. So in January 1998 the firm restructured to turn itself into a single, mainly online business, subsuming the former discount-broking business. The changeover was both painful and expensive, but Schwab has emerged with a capitalisation that rivals Merrill Lynch's.

There are other signs that established firms are starting to "get" the web. To avoid overt channel conflict and also to benefit from high Internet share valuations, several of them have set up separate online subsidiaries, such as bn.com or Toysrus.com. Others have formed partnerships. Kmart, for instance, has linked with Yahoo!, an Internet portal, to market its Bluelight.com site on Yahoo's shopping mall. Bestbuy has teamed up with Microsoft's online service. And in the pharmacy business, several of the physical chains now have links with online operations. Peter Neupert, chief executive of the biggest online drug retailer, drugstore.com, has cemented an alliance with the Rite Aid chain, in part to make it easier for drugstore.com to capture some prescription trade from the offline "pharmacy benefit managers" that control it.

But the most interesting recent deals have been those of Wal-Mart, the world's biggest retailer. Although feared by its competitors everywhere, the firm has nevertheless suffered many of the same problems on the Internet as other legacy firms: its website has been poorly designed, choice has been minimal and delivery uncertain. Yet it seems to be learning. It teamed up with America Online, the world's biggest Internet service provider, to promote a relaunched and vastly expanded website that opened for business in January 2000. It also formed a joint venture with Accel Partners, a Silicon Valley venture-capital firm, to set up a separate Wal-Mart.com subsidiary that it planned to float off.

Jim Breyer, of Accel Partners, who will sit on the board of Wal-Mart.com, declares that the future lies with firms that seamlessly integrate the online and offline worlds. He cites the AOL/Time Warner merger as an example of what is to come. Within a few years, he predicts, the distinction between online and offline will become blurred anyway. His aim is to use Wal-Mart's brand, financial strength and offline logistical strength to build Wal-Mart.com into the best in the online world as well.

Will mixing the best of online and offline, which has been variously labelled as "clicks'n'mortar", "clicks and bricks" or the more prosaic "multichannel", produce the e-commerce winners of the future? Maybe; and it is noteworthy that several pure plays are building physical shops. Michael Dell argues that offline shops will survive by becoming showrooms or places for enjoyment, with most buying taking place on the Internet. But the mixed approach has not always worked so far. For example, Barnesandnoble.com, either despite or because of being set up as a different company, has had several clashes with its parent, and recently shed its chief executive. Toys "R" Us tried a joint venture with Benchmark Capital that fell apart over policy differences after a few months. It too has lost several top managers. It has also failed, despite determined efforts, to oust eToys as the top online toy retailer.

So who, if not the Arkansan firm itself, will emerge as the Wal-Mart of the web? As it happens, there is at least one strong and well-established pretender to the title: Amazon. For Amazon has chosen to test the inherent "scalability" of the Internet by expanding well beyond its starting point of bookselling. Indeed Jeff Bezos, Amazon's cheery founder and boss, has said at various times that he is willing to sell anything except cement, or simply that he is willing to sell anything. But in saying this, he raises a big question: does web retailing favour generalists or specialists?

Webmaster or web servant

In favour of the generalists – the Wal-Marts of this world – is the web's scalability. Once you have invested enough in customer acquisition, by building a brand name and a reliable logistics and distribution system, the marginal cost of adding extra lines is tiny, and certainly much smaller than in the physical world. So in order to maximise revenues from those customers, you might as well sell them whatever you can. This, in essence, is what Amazon is now trying to do. If this line of thought were carried to its logical extreme, e-commerce would end up with a single retailer: Amazon as not so much the Wal-Mart as the Microsoft of the web.

Yet even though economies of scale and scope are greater on the Internet than in the physical world, this still seems highly implausible. Some features of the web point in the opposite direction: in favour of the specialist, or "category killer". If you can create the best website for selling Pokemons, for instance, the Internet's scalability could allow you to capture this niche all over the world. Numerous specialist websites have indeed established powerful niches. Ehobbies.com, based in Los Angeles, has established itself in the niche of collectibles such as model trains or cars. Garden.com has grabbed much of the gardening space. EToys regards the children's market as its oyster.

So which model will win out? There is probably room both for some specialist category killers that identify and conquer specific markets, and for a few large generalists that cover a broader front. But the pickings for anybody positioned in-between, on the model of a classic department store such as Macy's or Bloomingdale's, are likely to be thin. And the web's economics also suggest that, within each defined market segment – whether for beanie babies, books or beauty products – there may eventually be room for just one or two e-commerce firms.

Over the coming years, therefore, web retailers whose business models are insufficiently different from better-established rivals will be winnowed out. The process has started as the industry concentrates. As much as 75% of business-to-consumer e-commerce is now done through five sites: Amazon, eBay, AOL, Yahoo! and Buy.com. Two web retailers, Beyond.com and ValueAmerica.com, have sharply cut back operations. Some new entrants to the game, who failed to make enough impact in the holiday season, may pull out altogether. Even established firms such as eToys have seen their shares tumble precipitously.

The capital markets may not be voting for the success of traditional retailers in e-commerce; but they are raising doubts over the survival prospects of many business-to-consumer pure plays. And it is not just the obvious things such as price, choice, website design or marketing that such firms need to get right. At least as important, e-commerce ventures must pay more attention to what has often proved their weakest spot: fulfilment and distribution.

Distribution dilemmas

Can the web merchants deliver the goods?

W HAT MIGHT A Martian looking at America pick out as the clearest physical indicators of an e-commerce revolution? Those millions of PC users clicking away at their desks might well be buying things over the Internet, but then again they might be sending e-mails to their best friend, or looking at rude pictures. No, the most tell-tale signs that e-commerce has taken off are hundreds of huge new automated warehouses and thousands of vans delivering little packets to households: in short, the fulfilment and distribution end of the Internet revolution. And, as so often with an unglamorous back-office business, it is this end of e-commerce that has proved the most troublesome.

Indeed, the late-1999 holiday season will probably be remembered not so much for the huge explosion in Internet retail sales as for its terrible tales of delivery snarl-ups. Toys "R" Us and Wal-Mart announced as early as the second week of December that they could no longer guarantee delivery of website orders by Christmas. Consumers started to scream when they found they were unable to cancel or amend orders; many made it clear that they would have liked some real live human contact to sort things out. Some observers thought things were so bad that frustrated customers might abandon their efforts to shop on the web, and that e-commerce, instead of growing by leaps and bounds, might start sliding downhill. They may have been overreacting, but it seems certain that one thing that will count a lot in the coming e-commerce shake-out will be fulfilment, not price.

The early e-commerce pioneers concentrated on the end of the action that they reckoned to understand: website design and snazzy marketing. Many outsourced the whole tiresome business of order checking and distribution. In its early days, even Amazon relied wholly on Ingram's book-wholesaling operation. Consumer-electronics sites left the business to Micro, another big wholesaler and distributor. Everybody used United Parcel Services (UPS), Federal Express or the post for delivery.

Yet two things soon became clear. One was that shipping costs were (and remain) one of the biggest deterrents for consumers considering online purchases of physical products. The second was that traditional warehouse and distribution centres were not well suited to the business of e-commerce fulfilment: if it is to work properly, it needs newly designed systems. Both these things have combined to undermine some of the economic advantages of online shopping.

Perhaps this should not have come as a surprise. Physical shoppers, after all, handle their own order fulfilment, by choosing the goods and paying for them at the check-out, as well as their own delivery, by personally taking them home. And they do all this at their own expense, in both time and money. Merely to replicate this system efficiently, down to the individual consumer, is demanding enough; financing it, whether by absorbing the cost or by adding it to the bill, makes it even harder. It might have been better had e-commerce firms given more attention to this end of their business first.

Ironically, the delivery problems encountered by pure plays were one of the things that led many traditional retailers to assume that they could do better. Ironically because, here as elsewhere, many quickly found that their own distribution systems, geared to moving goods on pallets from warehouses to shops, proved a disadvantage, not a benefit. Wal-Mart, for example, has the most highly praised distribution system in the world: even the tyre pressures of its lorries are calibrated so that, when fully laden with pallets, the vehicles will be at exactly the right height for the unloading docks at Wal-Mart stores. But such a system is unable to cope with individual orders that have to be delivered to people's homes. So Wal-Mart has had to outsource its website distribution to two rivals: Fingerhut, a distributor that now belongs to Federated Department Stores, and Books-a-million. EToys, the biggest online toy retailer, also used Fingerhut.

Catalogue retailers, such as Land's End and J. Crew, are a different matter. Their warehouses were already aimed at delivering individual orders to people's homes. Adapting their order-taking to the web has not been simple, but distribution has certainly been far easier than building new warehouses from scratch. This remains the biggest reason for expecting the catalogue businesses to emerge as a success story of the web.

For the rest, is outsourcing the answer? Sometimes it can be. It certainly seems to work for long-distance deliveries by road. UPS has been one of the biggest beneficiaries of the e-commerce boom: it reckons to handle around two-thirds of all goods ordered online. It has also developed a sophisticated (and very popular) website-cum-tracking service that allows consumers to check exactly where their order is at any time of the day or night. It has edged out its biggest rival, Fedex, which realised too late that its main business, document delivery, was being disintermediated by e-mail.

But although actual delivery can be outsourced, many e-commerce firms are finding that it is risky to do the same with picking and packing, because a contractor working for many web merchants will never be able to give all of them priority, especially when the pressure is on during the holiday season. So more and more e-merchants have decided to follow the example of Amazon and build their own giant automated warehouses around the country.

Yet it will take time to establish whether online shops really can replace the supermarket trip, or merely supplement it. And even if they can provide the wired and short-of-time with everything they need, online shops are never likely to take more than a smallish share of the market.

Digital delivery

That is not, however, likely to be true for those goods and services that have the best possible answer to the nightmarish logistics of e-commerce: deliver over the Internet itself. For if you do that, you can eliminate most of your physical rivals altogether. A lot of computer software is now digitally downloaded by the user. The next businesses to head this way are music and films – and perhaps newspapers and books. Already you can buy the "Rocket e-book" that allows you to download new books at the click of a button; improvements in home printing and binding may one day make electronic distribution the preferred way to sell books.

The music business is already heading towards e-distribution, and is in turmoil as a result. In 1999 a new release from Public Enemy became the first commercial piece of music to be distributed over the Internet before appearing in the record shops. The spread of the MP3 digital music format has put fear into the hearts of record-industry executives, because it seems to hold out the possibility of people making an infinite number of perfect (albeit illegal) copies of any music that is ever put out on the web. They have responded with floods of lawsuits, and have launched a "secure digital music initiative" (SDMI) to stop the pirating; but so far there has been little sign of agreement. Broadly speaking, most people in the industry have treated the Internet as a threat, not an opportunity.

Al Teller, president of Los Angeles-based Atomicpop.com, which released the Public Enemy number, thinks this is misconceived. He has little sympathy for the traditionalists, even though he himself, as president of CBS records, used to be one. He recognised the scope for digital downloading several years ago, and tried to persuade the industry to embrace it. Having failed, he set up Atomicpop instead. His vision of the future is radical: "The only permanent bits of this industry are the people who make the music, and the people who buy it." But it is also expansive, for he sees the web as a way greatly to enlarge the market for music.

Does that mean there is no future for the record labels, and that artists will release music straight over the Internet? No, says Mr Teller, because the web needs filtering, editing, marketing and promoting even more than the physical world. But he thinks the labels need to restructure themselves from top to bottom: it is not enough to treat the Internet as just one more distribution channel. If they fail to adapt, labels like his stand ready to take over, along with a new infrastructure of portable music players, listener-driven music reviews – and artists keen to escape the record companies' clutches.

What about copyright? Mr Teller strongly supports it. In a digital world, he says, it should not be too hard to find a way of protecting copyright, even if the SDMI proves not to be the way to do it. One firm that may offer a better solution to digital-rights management on the web is itself a member of the SDMI: InterTrust, based in Santa Clara, California. It has devised what Victor Shear, its chairman, calls a "meta-utility": a platform for the conduct of digital e-commerce that protects copyright and deals with payments at the same time.

To use the InterTrust model, a record company (or, indeed, anybody who has digital content) gets it packaged into an encrypted file known as a "Digibox", which comes complete with rules about use, access and payment methods. These can be as flexible as the content provider wants to make them: three free plays followed by a charge, say, or a fixed price for the whole thing, or a system that charges a small amount for every replay.

Mr Shear's outfit is not the only one trying to make such a thing: Liquid Audio, Xerox, even Microsoft are devising their own digital-rights management systems. But InterTrust has the most advanced and capable technology, protected by at least 12 patents. It has the backing of Universal Music and Bertelsmann; PricewaterhouseCoopers, the world's biggest consulting firm, has signed on to use the technology for its clients. InterTrust's may indeed become the main operating system for the new market, and could yet come to govern the distribution of all digital content over the web, whether it be films, news or books. So has Mr Shear heard from the Justice Department? Not yet, he responds evenly.

In the great web bazaar

Everything has its price. The web can help you find out what it is

THE REAL REVOLUTION in e-commerce may lie in something not immediately obvious. Never mind the new retailers, the lower prices, or huge changes in logistics and delivery. What is truly new about the Internet is its ability to generate different pricing mechanisms, and in particular to allow price and product comparisons to be made and various kinds of auctions and exchanges to take place.

Two things are making these possible. One is that the Internet provides a perfect medium for aggregating buyers and sellers from all around the world. If you are a collector of beanie babies, in the physical world you would have to spend a lot of time and effort to draw together all your fellow enthusiasts; online you can do it effortlessly. The second is that the Internet offers an excellent way of comparing prices and collecting information, for example on new products, or on recent bids. Once again, to replicate this offline would be costly and time-consuming.

Price and product comparisons have been made easier by the development of "shopping bots". Websites such as Mysimon.com and DealPilot.com enable buyers quickly to compare products, prices and availability. For cheap items such as toothpaste or lettuces, most web users may not want to bother, any more than real-world shoppers do. It is also noticeable that, at present, the cheapest providers on the web are not always the most popular; and that, conversely, higher prices do not necessarily put customers off. But as e-commerce grows, and as people start to use it more to buy such bigger-ticket items as cars and white goods, the role of comparison-shopping agents could become far more significant.

Price it and see

The technology is improving fast, and so is the choice of agents. Both DealPilot.com, a German company, and RuSure.com, based in Israel, equip users with a piece of software that automatically informs them of the best available prices on the products they are looking for. If you have the software and are about to buy a book from Amazon, say, a box will pop up to tell you that you could get it cheaper elsewhere. As consumers get used to this idea, price competition on the web, which has so far been less energetic than expected, may intensify.

Shopping agents also judge how well a product does its job. Frictionless.com, of Cambridge, Massachusetts, offers the electronic equivalent of a well-informed shop assistant. Buyers of electrical goods, for instance, can use the technology to rank products by performance on criteria of their choice – weighing ease of use, say, against quality of service. Only once the product has been chosen do criteria such as price and availability come into play. This model can be applied to any number of products: recently, Frictionless teamed up with Wingspanbank.com, an online banking subsidiary of Bank One, to help consumers choose mutual funds.

A plethora of other agents that help consumers, all offering slightly different services, is now vying for attention. From New York, Brandwise.com offers product reviews and price information about white goods and electrical appliances. Epinions.com, based in Silicon Valley, has chosen what its boss, Naval Ravikant, calls an "open-source" style of review for products of all kinds, with visitors providing all the material. The company makes no attempt to edit any of this, but once posted, the reviews themselves are rated by other users. Listen.com, also in San Francisco, is using a similar idea to build a directory of music on the web, complete with reviews written initially by its own reviewers, but incorporating the comments of visitors to its website as well.

Then there are the intermediaries that specialise in ranking other websites. Gomez Advisors, in Massachusetts, assesses the performance of e-commerce websites in areas such as personal finance, travel, shopping, car buying and auctions. For each of them it offers a table of rankings based on marks out of ten. Rivals include Yahoo!, the biggest Internet portal, which offers its own frequently updated guides to the best websites around. AOL, the biggest Internet service provider, also helps to guide customers around the shops.

If it seems hard for anyone to make himself heard amid this cacophony, that is largely because e-commerce is at such an early stage of development. The shopping agents and rating sites will themselves no doubt be ranked in due course, and their numbers are bound to come down. But one big question they will all have to answer is: which side of the transaction are they on?

The same question applies to another new band of intermediaries: those that are running auctions and other price-discovery mechanisms. The arrival of the Internet auction is one of the biggest things that has happened in e-commerce – as some of the crazier stockmarket valuations attest. EBay, the biggest single consumer-to-consumer auction-site, is worth 20 times as much as Sotheby's (in which Amazon, which runs its own auctions, recently bought a stake). By market capitalisation, Price-line.com, which runs reverse auctions for airline tickets and a few other things (including groceries), is worth as much as American Airlines.

The power of auction sites, says eBay's Pierre Omidyar, lies in their strong "network effects": the more people that use them, the more useful they are. That also confers a potentially bigger first-mover advantage on auction sites than on electronic retailers – an advantage eBay is so bent on preserving that it has triggered antitrust inquiries into its efforts to stop rivals logging into its site for price and bidding data. EBay has nearly 10m registered users bidding for some 3m items. Each user spends nearly two hours a month on the site, a lot more than Amazon's equally loyal clientele. And eBay's charges are typically no more than 7.5%, compared with over 25% for offline auctioneers. Mr Omidyar also takes pleasure in pointing out that eBay is almost the only pure-play e-commerce firm that has been consistently profitable from the beginning. No wonder auction-houses and flea-market operators fret that they may soon lose much of their business.

Monetising the audience

Nor is eBay the only threat. Amazon and Yahoo! run their own auction businesses. Jerry Kaplan, who started Onsale, the first B2C auction site (before eBay), which dealt mainly in surplus computers, has subsumed it into Egghead, a big retail site for computer hardware and software that also does auctions. He claims the most loyal consumers on the web. And it is not just a matter of sales; as he points out, there are "multiple ways of monetising the audience besides selling stuff", a claim confirmed by Egghead's high number of repeat customers.

Even Priceline is only the best-known of several reverse-auction sites. Under its (patented) model, bidders name their prices for a flight and leave them on the table for airlines to accept or reject. Jay Walker, the firm's founder, claims the model has been so successful that he is now the largest seller of leisure air tickets in America, with 3m customers. He dismisses critics who say that few of their bids (under 5%, according to some) are fulfilled, and that many consumers dislike the rules preventing them from withdrawing their bid or from putting in a fresh bid. He says these grouses come "from university professors, not from our customers." Priceline is now expanding its model from air tickets and hotels to cars and groceries, for which it claims to be able to deliver savings of as much as 50%.

Yet Priceline falls far short of a perfect consumer-auction model. Many people wonder if it is really working mainly for the benefit of the consumer, or mainly for the airlines; its refusal to publish price information certainly leaves some doubt. A purer model is offered by NexTag.com, an auction site that at present deals primarily in new computers and consumer electronics but has plans for much bigger things. Purnendu Ojha, its co-founder and boss, describes NexTag's goal as nothing less than eliminating the "deadweight loss in the supply/demand curves caused by fixed pricing".

The model that Mr Ojha cites is the stockmarket, in which supply and demand are continously adjusting to each other throughout the day. Buyers on NexTag submit bids, but can then change them; sellers can accept or reject them, or even counterbid. This is not so much a reverse auction as a continuous buy-and-sell auction that avoids the delays built into eBay and Onsale and the inflexibility and secrecy of Priceline. The effect is to turn the market for, say, Palm Pilots into the equivalent of the Nasdaq stock exchange. Now another new firm, Perfect.com, is trying to go one better by adding such dimensions as delivery and customer service to the continuous auction.

Yet another auction-style model is employed by aggregators such as Accompany.com or Mercata.com. These websites seek to gather together the buying power of consumers to achieve the best possible prices. As more bidders join in, the price of whatever they are after will tend to drop. Jim Rose, chief executive of Accompany, describes his site as offering "many-to-many" trading (as opposed to eBay's one-to-one or Nextag's one-to-many). It is especially useful for people who have time to wait for an item but are highly price-sensitive.

All the world's an exchange

Any believer in Adam Smith's ideas must welcome this kaleidoscope of different trading models. In some ways, the notion of fixed prices is itself quite recent. Until, say, 200 years ago, most trading took place in open marketplaces or bazaars, in which prices moved continuously and it was easy to check up on competitors. Now the Internet seems to be creating the possibility of a permanent worldwide bazaar in which no prices are ever fixed for long, all information is instantly available, and buyers and sellers spend their lives haggling to try to get the best deals.

Of course it will not happen like this. The popularity of Internet auctions and other buying techniques such as aggregation owes a lot to their novelty effect: they are not so much a new marketplace as a new form of entertainment, the appeal of which could easily wear off. Auctions are better suited to some items than to others. Fixed prices did not develop by chance; for many products they suit both buyers and sellers. A shopper anxious to get through the weekly grocery list will have little appetite for bargaining over every item.

Yet despite these reservations, the new pricing models will not disappear. More likely, they will be used for goods and services where they work well – new computers, cars much in demand, anything second-hand, airline seats and other travel services. And as consumers get used to auctions as a way of buying travel or computers, say, they may be readier to try them out for other goods and services, such as buying and selling property or catering for a party.

But the intermediaries will still have to answer the same old question: whose side are they on? John Hagel's infomediaries and Philip Evans's navigators around the web act as agents for the consumer. The best Internet portals, aggregators and auction-sites will have to do this too. In the physical world similar questions can be fudged: an auctioneer works for both seller and buyer, a travel agent tries to get the best deal for both sides. But in the online world, where any transaction can be unbundled and any intermediary disintermediated, such ambiguity may not be sustainable. The best indicator will often be payment or ownership: that Brandwise, the reviewer of white goods, was set up by Whirlpool, a manufacturer, arouses immediate suspicions. Once profits start to be made on the web, it will always be worth asking an intermediary: who pays you?

It is worth noting, too, that auctions and aggregators, like the Internet itself, are inherently global in scale. Indeed, to make the most of network effects, they need to operate worldwide. However, this raises an unexpected difficulty, given America's lead in the Internet: in Europe, at least, rivals are already at work. Hence the next big question in e-commerce: will it spawn global merchants, most of which are likely to be American?

First America, then the world

But before e-commerce can go global, it needs to overcome a few tiresome obstacles

SOMETHING LIKE THREE-QUARTERS of all e-commerce currently takes place in the United States. The country also accounts for 90% of commercial websites. Given that the Internet is, by its very nature, global in reach, these two facts raise a vital question about e-commerce for the rest of the world: is America in general, and are American websites in particular, inevitably going to dominate it?

The answer is not immediately obvious, for several reasons. For a start, it is surprisingly hard to cross borders in the retail world. In industries such as drink manufacturing, car making or investment banking, it is comparatively easy for a Coca-Cola, a Ford or a Goldman Sachs to establish itself around the globe. But in shopping or retail banking, cultural, linguistic and regulatory barriers often get in the way. Intriguingly, the physical world may be about to see some of these barriers surmounted for the first time: Citibank is on the way to becoming a global retail bank, and Wal-Mart is doing its best to turn itself into a global retail giant.

Surely the Internet will give such aspirants a further boost? It may well do so; but that does not mean going global will suddenly become straightforward. One big reason is the crucial importance of fulfilment and delivery. Being good at these inside the United States is clearly a plus, but, because they are by definition local, that does not guarantee success in other markets. Several American websites no longer take orders from outside the country for this reason (although Amazon reckons that as many as a quarter of visitors to Amazon.com live abroad). And in Europe, the biggest marketplace for e-commerce after America, a clutch of perceived or real government, tax and regulatory obstacles are deterring would-be American electronic vendors.

The third reason that American global domination should not be taken for granted is that many American web retailers may have left it too late: a surprising amount of e-commerce is starting to take place in other countries. In most Scandinavian countries Internet penetration is now higher than in the United States, and in Britain and Germany it is catching up. According to a study by the Boston Consulting Group, retail sales on the web in Europe in 1999 were worth $3.6 billion, and were predicted to rise to $9 billion in 2000. Asia too is taking to the web: BCG predicted sales of $6 billion there in 2000, with Japan, Australia and South Korea in the lead. Japan's huge Seven-Eleven retail chain unveiled ambitious plans for a leap into e-commerce, with consumers placing their orders over the web and collecting them at their local store on the way home.

To be sure, in some ways Europe and Asia are at a disadvantage compared with America. Credit cards are far less widely used, which is one reason for the success of mixing online (to place orders) and offline (to pay and pick them up) business. Another problem is telecoms costs. In Europe these can be five times as high as in America, where local calls are often free. Despite the success of "free" Internet service providers in Europe, BCG reckons it still costs twice as much there as in America to surf the web.

Yet in telecoms both Europe and Japan could now find themselves at an advantage, thanks to their lead over America in mobile telephony. Those with a stake in the PC business, such as Michael Dell, like to argue that nobody wants to surf the web on a one-inch screen. But it seems more likely that mobile telephones and other handheld devices are about to become the instrument of choice for Internet access. Brent Hoberman, one of the founders of Lastminute.com, a London-based website dealing in travel, entertainment and gifts, argues this will mean that "Europe can join at the next level up".

Europe gets wired

There are also plenty of non-American, and especially European, websites around (see chart 7). Mr Hoberman's firm is just one of the hundreds of e-commerce ventures that started up all over Europe in 1999. Stockholm and London now seem to be in the grip of the sort of Internet fever that hit California a few years ago, with a slew of billboard and bus advertisements, a bubbling new venture-capital industry, and hysterical stockmarket trading of newly floated companies. One such firm, QXL.com, has established a string of auction sites across Europe that are specifically designed to cope with different languages and currencies. When eBay first tried to launch around Europe in dollars only, QXL left it flat on its face. The proportion of eBay's customers outside America is still under 3%, but the online auctioneer is trying to boost these numbers with its recent purchase of a big German online auctioneer, Alando.de.

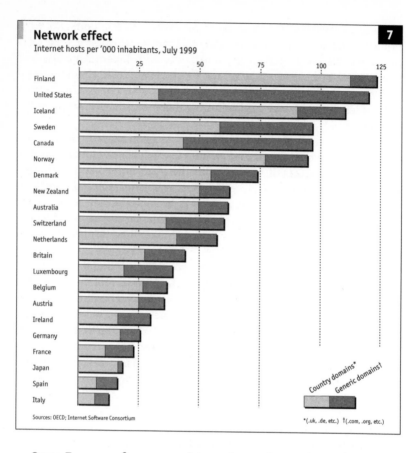

Network effect   **7**
Internet hosts per '000 inhabitants, July 1999

	0	25	50	75	100	125

Finland
United States
Iceland
Sweden
Canada
Norway
Denmark
New Zealand
Australia
Switzerland
Netherlands
Britain
Luxembourg
Belgium
Austria
Ireland
Germany
France
Japan
Spain
Italy

Country domains*
Generic domains†

Sources: OECD; Internet Software Consortium   *(.uk, .de, etc.) †(.com, .org, etc.)

Some European firms even claim to be ready to take on the Americans at home. Letsbuyit.com is an aggregator that started in Sweden but has big ambitions: not only to launch all over Europe, but also to compete against Mercata and Accompany in America. Another firm with Swedish origins, but a base in London, is Boo.com, an ambitious sports-good retail website that spent a fortune on marketing its global launch in 18 countries round the world in November 2000 before it had sold a single sweatshirt. But Boo.com has run into problems; recently it sacked some of its excess staff.

Physical European firms are also exploiting the web, though not without the usual teething troubles. MeritaNordbanken, a Finnish-Swedish bank, says that 50% of its retail customers now bank online; Sweden's SE bank claims 25%. These are far higher penetration rates than any bank has achieved in America. Tesco Direct, the online grocery service of Britain's biggest supermarket chain, is being derided for its dismal performance – but it has ambitious expansion plans. The airlines too have taken to the Internet: Easyjet and Go, two low-budget British carriers, claim to be selling 60% and 50% respectively of all their tickets online.

One European country, France, even claims to have been the global leader in e-commerce long before America ever woke up to it. In 1994, 1.2m French people bought something electronically, whereas fewer than 800,000 Americans did. The difference was that the Americans used the Internet but the French used Minitel, a proprietary network set up and operated by France Telecom.

The contrast between the two is instructive. The Internet grew up in an anarchic manner, with commerce tacked on as an afterthought, whereas the Minitel was built and run by the government on special terminals owned by France Telecom. That may have helped to boost e-commerce in France in its early stages, but it soon became clear that the Minitel was being used for buying train tickets but little else. It did not have an e-mail facility, because that would have put it in direct competition with the state-owned post office. Now Minitel use is declining as the Internet takes over. In 1999 commerce on Minitel actually shrank, but on the web it tripled (from a smaller base). France Telecom is about to relaunch the business in a form that combines the Minitel with the Internet. Even so, the French prime minister, Lionel Jospin, has conceded that the Minitel experiment may have slowed down France's move into the Internet age.

Yet despite this particular local difficulty, the scope for the Internet to shake up retailing in Europe ought to be huge. In almost every country the existing marketplace is vulnerable. Property prices, planning restrictions, regulations about advertising and rules on opening hours all conspire to make shopping in Europe less efficient than in America. So the gains from the Internet – and the profits for those American websites that can successfully make the transition – should be that much bigger if the web is allowed to work its magic.

Most American e-commerce firms say they are too busy at home to contemplate expansion in Europe yet, even though they already have about a fifth of the market there, often through their domestic operations. But in truth they may be bothered about something quite different: government. The popular view in Silicon Valley is that government should be kept out of the nascent world of e-commerce, lest it strangle

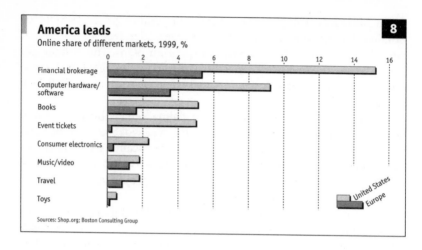

America leads   **8**

Online share of different markets, 1999, %

Financial brokerage
Computer hardware/software
Books
Event tickets
Consumer electronics
Music/video
Travel
Toys

United States
Europe

Sources: Shop.org; Boston Consulting Group

the baby at birth. And the geeks are convinced that government in Europe, which they believe regulates anything that moves, poses an even greater danger to the infant than in America.

Who needs government anyway?

Clearly this view is a huge oversimplification. The American government has in fact given a big push to e-commerce, in two ways. One is by permitting the patenting, sometimes on fairly shaky ground, of a host of business processes, so that no self-respecting e-commerce firm is now without its in-house patent lawyer. The second and even more significant one is by imposing a moratorium on new taxes on anything bought over the web. Web purchases are also generally exempt from sales taxes: so long as a website operator has no physical presence in a state, it can avoid collecting sales tax on any goods sold in that state.

However, the tax debate is coming to a head as e-commerce grows. One reason is that the American states are becoming increasingly worried about the sales-tax exemption, because that tax is often their biggest source of revenue. Ernst & Young reckon that in 1999 Internet business caused the states to lose $170m in tax revenue. Another reason is energetic lobbying by physical retailers, who reckon they are being put at a disadvantage by the pure plays' ability to escape tax. A congressional advisory committee is at work to consider the issue, but there seems no easy solution in sight. Internet consumers and websites will fiercely resist imposition of the sales tax – after all, catalogue retailers have avoided it for decades. Instead, they suggest abolishing the sales tax altogether, pointing to the jungle of America's 30,000 individual tax jurisdictions.

But the tax debate is not just about what happens inside America. Tax is also an international issue, and happens to be one of many relating to e-commerce that is being examined by the OECD, the rich countries' think-tank in Paris. If American Internet addicts hope they may be able to avoid taxes for ever, a visit to the OECD will soon persuade them otherwise. In several EU countries the rate of value-added tax (the nearest thing to American sales taxes) can be as high as 25%, and VAT accounts for an average of 40% of Europe's tax revenues. There is no prospect that, once e-commerce takes off, goods or services sold over the Internet will be allowed to avoid such an important tax. As it happens, even the Americans have accepted (at a meeting in Ottawa in 1999) the principle that tax should be levied on goods sold electronically, at the rate prevailing in the country where the consumer is based.

But it is regulation, rather than tax, that American e-commerce firms are most concerned about when they contemplate Europe. Much is made of a small number of unhelpful cases. In one, Compuserve, an online consumer network, temporarily had to close down its operation in Germany after a Bavarian court found it liable for the content of race-baiting material on a website over which Compuserve had no control. In another case, Land's End, a clothes retailer, was deemed to be breaching German consumer law by offering its normal 100% replacement guarantee for any clothing that wore out. Advertising on websites in Europe is a colossal headache: for example, Denmark bans advertising to children, France bans advertising in English and Germany bans comparative advertising.

An even fiercer argument is raging about privacy and data protection. The European Union has adopted a directive that, if implemented, would prevent the transfer of data about consumers to third countries where the level of data protection is, in the EU's terms, "inadequate". Because American data protection is non-statutory and there is no government data-protection office, it is regarded as inadequate by definition.

Yet blocking the transfer of business data to America is widely regarded to be all but unthinkable. Since 1998, the two sides have been trying to find a compromise, with the Europeans carefully holding back from sanctions but keeping the threat in reserve. Meanwhile, concerns over privacy are starting to grow in America itself, encouraging both government and business to concede some ground to the EU. Now a solution is in the air: the Americans are offering to endorse a "safe harbour" principle, whereby companies within the safe harbour undertake to apply standards of privacy protection equivalent to those in the EU. In return, they will be allowed to collect data unimpeded. The deal may be struck at the next summit between America and the EU in March 2000.

The Europeans are showing surprising flexibility not just over data protection but over e-commerce in general, because they too are anxious to secure its benefits. The Americans have followed with keen interest (and may even learn from) the legislative framework that the EU is trying to put in place to govern such matters as electronic signatures and copyright. Another example of European flexibility is forthcoming legislation to apply a convention that governs the jurisdiction for consumer disputes. At one point, this legislation had been drafted in such a way that it would have permitted a consumer to sue any website operator in the consumer's home jurisdiction, even if the website never intended to sell anything there. Under pressure from the industry, that now looks like being changed to allow a website to specify the jurisdiction in the event of any lawsuit.

Still, too many governments and regulators, especially in Europe and Asia, are frightened by the Internet, which they consider subversive, as well as too American. They see it as undermining consumer protection, jeopardising taxes and risking even more inequality. In places such as Malaysia – to say nothing of China – the Internet is deeply mistrusted because it seems to threaten government control. Yet although these fears may have some basis in reality, governments would be well advised to find ways of assuaging them that stop short of shunning the Internet altogether. They are likely to find it impossible to ban the web anyway, because it has already spread too far. Mr Dell notes that 40% of orders for his computers in China are now placed online.

Unprotected, unwired

It is true that consumer protection may be harder to enforce in a world where Internet gambling sites operate out of Gibraltar, porn rings are run out of the Caribbean and American patients are already using the web to buy drugs in Mexico that have not been approved by the Food and Drug Administration. But the web does offer both regulators and taxmen other advantages, such as a perfect audit trail. Taxation will even so be a continuing problem, especially on activities such as gambling that can take place overseas and in digital form. But for many other goods, it should be technically quite possible to find a way to collect taxes on Internet commerce.

What should perhaps be a bigger concern is inequality, or what is often called "the digital divide". Electronic commerce holds out the promise of greater efficiency and transparency, and of lower prices, all of which would be as welcome in Europe and Japan as in America (and, indeed, poorer countries). Yet Internet usage is skewed towards the upper end of the income scale, so the first to reap these benefits will be the better-off. Some way needs to be found to spread them to people on lower incomes – for instance, by putting terminals in public libraries and schools, and by teaching every schoolchild how to use a computer – and also to poorer countries. Inequality within and between countries is certainly not the fault of the Internet. And the web is by its nature democratic, open and transparent; if its benefits are spread wide enough, it might, in the long run, actually help to reduce rather than increase inequality.

The material on pages 111-43 first appeared in a survey written by John Peet in *The Economist* in February 2000.

POSTSCRIPT

IN RETROSPECT, there was always something too smooth about those curves forecasting exponential growth of electronic commerce. When this survey appeared, the Internet bubble was at its height, boo.com and Webvan were walking tall and most retailers were struggling to come to terms with their fear of the web. In the two years since, the bubble has popped spectacularly – and the vogue for e-commerce has disappeared with it.

The Achilles heel of most e-commerce firms was easy to spot, and the survey dwelt on it extensively. The bubble had meant that capital was, in effect, free for many dotcom firms, so it was bound to be wasted; and it was equally inevitable that profits would seem unimportant, and so would not materialise. Instead, the emphasis was on market share and first-mover advantage. The e-commerce market, it was believed, would be captured by these pioneers, who could then make big money as it expanded.

It now seems that this was overblown. First-mover advantage clearly existed, but setting up shop on the web proved a lot easier than it had seemed, and consumers' brand loyalty was if anything even smaller than in the physical world. But the real danger for the dotcoms was always that they would be left high and dry when market sentiment changed and new capital became almost impossible to raise.

That is exactly what happened. The survey suggested that as many as 80% of e-commerce firms might go bust. That number now looks conservative, though the two biggest, eBay and Amazon, still survive – in Amazon's case, by the skin of its teeth. The bricks-and-mortar retailers that moved more cautiously into the Internet seem to have been vindicated: Tesco has a big online grocery operation that it claims is profitable, whereas most pure-play online grocers have gone under.

And yet it would be wrong to conclude that e-commerce was a blind alley. Internet penetration rates have continued to rise and mobile-phone access to the web has only just begun. In some industries, the effects have been just as stark as this survey suggested. The bulk of airline-ticket sales in the booming low-cost airline business are now made online, cutting out travel agents; the music industry has still not resolved the problem of digital downloading and file-sharing of new music; and online banking, insurance and stockbroking continue to grow.

The Internet's secondary effects have been more important. The ease of price comparison through the web has made it far harder to raise prices. In the business-to-business field, many companies have quietly persisted with using the Internet to cut costs and simplify their supply chains, even though several B2B marketplaces have gone belly-up. And the Internet as a matchless source of information, not just about prices but about everything else, has continued to shift power from intermediaries to end-consumers. The e-commerce revolution has been slowed by the dotcom crash and recession, but it will remain all-pervasive.

Dotty about dot.commerce?

The e-commerce boom is changing business – for the better

Is IT A fleeting fancy, or a life-changing love? The question lingers over today's obsession with the Internet. There is much hype about e-commerce, but there is also a solid base in reality. And this suggests that the second answer to the lingering question is closer to the mark. The Internet is profoundly affecting almost all business and commerce. A common slogan is true: any big firm that fails to grapple with the consequences is putting its future at risk.

Numbers tell only part of the story. In 1999 global e-commerce was worth a little over $150 billion. Around 80% of those transactions were between one business and another. Yet growth of all forms of e-commerce is hectic. Business-to-business web exchanges are mushrooming. In America e-retailing revenues tripled last year; in Europe and Japan they rose faster still. The absolute level of consumer e-commerce is still small, at around 1% of American retail sales. But in some areas its share is far higher: 15% of retail stockbroking and 5% of book sales, for instance.

The influence of the Internet stretches farther than such figures suggest. It is used more as a source of information than as a place to buy. Surveys in America suggest that in 1999, when less than 3% of new car sales were made over the web, 40% involved it at some stage, usually for price comparisons. And e-commerce often needs to take only a small share of the pie to have a big impact on what is left. Travel agents' margins are so flimsy that a loss of 3% of their market to the web would be enough to push many out of business.

One impact is thus to intensify competition, producing benefits to consumers in lower prices and more choice, at the expense of producers and intermediaries. But a more profound effect is a wholesale restructuring of businesses. Thanks to the rise of e-commerce, such giant firms as General Electric, General Motors, Ford and Unilever, are embracing the Internet for many of their activities. And the giants' move means that their suppliers, as well as intermediaries through whom they sell, must take to the Internet too.

The promised land that this ripple effect holds out is one of more transparency, greater efficiency and, yes, higher productivity. Hard macroeconomic evidence of these things is admittedly still elusive. But it is at least conceivable that new technology and the Internet may be partly responsible for America's unexpectedly strong recent productivity growth, as well as its surprisingly low inflation.

Bubbling under

So is the Internet leading to economic nirvana – and does dotcom mania make sense? Not quite. The Internet may be transforming business; but it is not rewriting economics to enable the American economy to expand at its current rate without dangerously overheating. Nor does the explosion of e-commerce justify the current [February 2000] high prices of many Internet-related shares. On the contrary, all the signs are that the markets remain in the grip of a technology-related bubble.

Many equity markets look overvalued. But Internet shares are causing particular concern, for two reasons. One is simply that their overvaluation looks so huge. Because few make profits, normal earnings-related measures cannot be used. But even enthusiasts concede that as many as 80% of today's Internet companies may not survive – just as almost all the early railroads, car makers and airlines did not. Behind the hype, investors are reflecting this. Shares in profitable technology providers such as Cisco, Oracle or Sun Microsystems are doing well. But nearly three-quarters of Internet-related initial public offerings in America since mid-1995 now trade below their issue prices. And over the past year the shares of several big American e-commerce firms have crashed: eToys (down 80% from its peak), Priceline (down 69%) and E*Trade (down 66%), for example. The Internet bubble in Europe and Asia seems sure to be followed by similar collapses.

The second concern is one voiced by managers of traditional (non-tech) firms. Their share prices, they say, have been savaged just because investors are, irrationally, redirecting their capital into glamorous technology and Internet firms. Some say they are being punished despite healthy results; several talk of a misallocation of capital from profitable to loss-making activities.

Yet many of these gripes are misplaced. The Internet bubble is clearly distorting the market, and its bursting may well cause pain across the whole economy. Moreover, whenever firms can raise capital virtually free, they are likely to waste much of it, as so many dot.coms are now doing with their marketing expenditure. But there is little evidence that traditional firms' shares are undervalued, or that they are finding it hard to raise capital. And the capital going into Internet firms is not all being wasted – even if today's shareholders may not reap the benefits. This is especially true of makers of computers and the Internet's infrastructure. But even an e-retailer such as Amazon needs robust computers, software and delivery systems – all of which necessitate investment.

The bigger point is that the markets are sending traditional companies two useful signals. One is that they ignore the Internet at their peril. The second is that merely adding a website on to an existing business is not enough: the whole business needs to be redesigned around the cost-saving, communication-easing properties of the net. Most companies must become Internet firms if they are to survive. Even when the bubble deflates, that signal will be worth heeding.

The material on pages 145-7 first appeared
in *The Economist* in February 2000.

IV
E-MANAGEMENT

Inside the machine

**Companies need more than good technology to make the most of
the Internet. They need flexible and self-confident management too**

IN THE MID-1990s, the managers of established, old-economy companies concentrated on running their business well: making cars, perhaps, or selling life insurance. They had to contend with constant
change, of course, but normally of a fairly predictable kind: costs had to
be cut, new products launched, mergers and acquisitions dealt with.
Now life has become much more difficult. Change has not only become
more rapid, but also more complex and more ubiquitous. Established
companies are no longer quite sure who their competitors are, or where
their core skills lie, or whether they ought to abandon the particular
business that once served them so well. Behind this new uncertainty lies
the Internet (which in this survey is used as shorthand to include the
whole cluster of technologies that depend upon and enhance it). Since
1995, this has begun to transform managers' lives.

Why is it causing so much trouble? After all, the Internet as now
used by many companies performs familiar functions, although more
cheaply and flexibly. The e-mail is not really so different from the
memo; the electronic invoice looks much like an on-screen version of
its paper predecessor; the intranets that companies install to connect
different departments resemble the enterprise resource planning (ERP)
systems that many companies bought in the 1990s; even the networks
that link companies with their suppliers had their electronic predecessors.

But new technologies often begin by mimicking what has gone
before, and change the world later. Think how long it took companies to
realise that with electricity they did not need to cluster their machinery
around the power source, as in the days of steam. They could take the
power to the process, which could even be laid out along a production
line and set in motion. In that sense, many of today's Internet applications are still those of the steam age. Until they make the next leap, their
full potential will remain unrealised.

Yet even what the Internet has already achieved is puzzle enough for
many managers. Why should it cause more bewilderment than, say, the
arrival of the mainframe or the PC before it?

The answer lies in the Internet's chameleon qualities. It is not simply a new distribution channel, or a new way to communicate. It is many other things: a marketplace, an information system, a tool for manufacturing goods and services. It makes a difference to a whole range of things that managers do every day, from locating a new supplier to co-ordinating a project to collecting and managing customer data. Each of these, in turn, affects corporate life in many different ways. The changes that the Internet brings are simply more pervasive and varied than anything that has gone before. Even electricity did not promise so many new ways of doing things (see chart 1).

At the root of the changes is a dramatic fall in the cost of handling and transmitting information. Almost every business process involves information in some form: an instruction, a plan, an advertisement, a blueprint, a set of accounts. All this information can be handled and shared far more cheaply than before. That has its drawbacks, of course: a fall in production costs is all too likely to lead to an increase in supply, and plenty of managers now feel that they are drowning in information. But it also brings immense advantages.

In particular, the investments that companies need to make in hardware and software are small in relation to the pay-off. Gary Reiner, chief information officer of GE, one of the pioneers of the Internet, describes how the company set out to build its own electronic-auction site. "It was less expensive than we thought," he says. "We built the software for $15,000 internally." Charles Alexander, who heads GE Capital in Europe, makes the point even more forcefully. "What we have rapidly begun to understand is that the incremental investment required is extraordinarily small compared with our overall investment. A $300m investment for a company making $4 billion–5 billion is weeks of cashflow to get a payback which is months away, not years. We thought, if we are completely web-enabled in two years, that's good. Then we realised we could do it in weeks – and the productivity gain is almost instant."

Of course it is not really that easy. Companies have to do a lot more than buy terminals and write software. Erik Brynjolfsson, a professor at the Massachusetts Institute of Technology's Sloan School of Management, argues that software and hardware account for only about a tenth of true corporate investment in information technology. A far larger investment goes into new business processes, new products and the training of employees. Such spending does not show up as investment on corporate accounts. Instead, it generally appears as expenses, such as payments to consultants, and is treated that way by the taxman (unlike

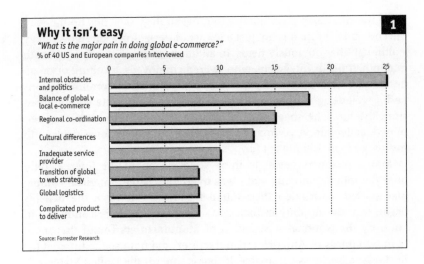

Why it isn't easy

1

"What is the major pain in doing global e-commerce?"
% of 40 US and European companies interviewed

Internal obstacles and politics
Balance of global v local e-commerce
Regional co-ordination
Cultural differences
Inadequate service provider
Transition of global to web strategy
Global logistics
Complicated product to deliver

Source: Forrester Research

investment in old-economy physical goods, which can be capitalised and depreciated). Nor does it appear on national accounts. Yet American companies, Mr Brynjolfsson calculates, have created a total of $1.5 trillion of "organisational capital" in the past decade.

A survey of 416 companies conducted by Mr Brynjolfsson, together with Lorin Hitt of the Wharton School at the University of Pennsylvania and Shinkyu Yang of the Stern School at New York University, identified such organisational capital by picking out companies in similar industries where employees tended to have a great deal of authority over the way they worked; where they were paid for performance; where they were well educated; where the emphasis was on teams; and where most information flowed freely across the company. Financial markets valued organisational capital highly, to judge by market capitalisation, but they were keenest on those firms with both high investment in information technology and a high degree of organisational change.

How's your organisational capital?

The reason for this correlation, Mr Brynjolfsson believes, is that in order to get the best out of their information technology, companies need to make a host of changes in a co-ordinated fashion. They cannot simply mimic others: if they do, the results may be disappointing or even disruptive. Organisational capital is far harder to reproduce than the more visible, marketable sort.

So the success of a company's Internet strategy depends on the way the company is run. It is not just a matter of designing a shrewd strategy – although that obviously helps. In addition, a company needs strength in depth. It needs intelligent, empowered employees, a culture of openness and a willingness to experiment, good internal communications and a well-designed pay structure. It also needs absolute commitment from the top. The sheer scale of change is impossible without determined leadership. A company whose chief executive never deals with his own e-mail will not get far.

Many managers began to think seriously about the Internet only once the millennium-bug scare was over. Most large American companies started to develop their Internet strategies in 1999; the benefits began to show up only in their 2001 accounts. A survey conducted in 2000 by the National Association of Manufacturers found that more than two-thirds of American manufacturers did not use the Internet for business-to-business commerce. If that is true for the United States, it is true in spades for the rest of the world. "Whenever I visit software companies," says Andrew McAfee, a professor at Harvard Business School, "I get them to complete the sentence, 'The business-to-business revolution is x% complete.' The biggest number I have heard is 5%. Many say 1%."

Some of the remaining 99% will be accomplished only as the network effects of the Internet feed through. For example, only half of Dell Computer's customers are "online-enabled", as Joe Marengi, one of the company's senior executives, puts it: able to use Dell's web pages to configure their orders. And only 15% of customers place the order electronically. The remaining 35% design the order online and then submit it in some way that requires Dell to take a second step to feed it into the system, such as e-mailing or faxing it. As more companies can submit electronic orders, both they and Dell will cut costs and speed the process. Many more processes involving transactions among companies bring the same double benefits, and ripple outwards as more firms share some particular capability.

The companies that have gone furthest claim to save astonishing amounts. GE plans to cut 15% from its cost base of $100 billion in both 2001 and 2002. That is five times the typical annual growth in productivity, even for this fast-moving firm, of 3–4%. In addition, the company hopes to reduce the prices of the materials it buys by making most of its purchases in electronic auctions. That should save a further $2 billion over the same period.

Other companies are discovering the same sort of magic. In the summer of 1999 Larry Ellison, boss of Oracle, announced that the company would cut $1 billion from its global corporate expenses of $7 billion. Now it expects to have cut a second billion by October 2001, and has its eye on a third. "I've been in business for more than 30 years, and I think this is by far the biggest productivity advancement I've seen in my life," says Jeff Henley, Oracle's chief financial officer.

Such figures will have a dramatic impact on the overall productivity of the economy. In America, productivity growth appears to have accelerated sharply in the past few years – although Mr Brynjolfsson would argue that the acceleration would look milder if the costly investment in organisational change could be properly measured. Even if the pioneers of the Internet are exaggerating, a long period of big productivity gains seems to lie ahead.

These gains come from several different directions. Some of them are achieved simply by transferring part of the work from the company to its customers. If you are having trouble recruiting enough staff for your help desk, and you want to spare your customers having to listen to an entire CD's worth of gruesome music on hold, then well-designed online information may actually benefit everybody. If your customers can submit electronic orders, it saves you the trouble of doing the job yourself, and also greatly diminishes time-wasting wrangles over mistakes. You just e-mail the querulous customer the original order.

Other cost savings come from being able to feed better information to suppliers all the way up the chain, and thus reduce stocks. Paul Bell, who runs Dell Computer's operations in Europe, the Middle East and Africa, describes inventory as "the physical embodiment of bad information". His company now measures its inventory in hours rather than days. Still more savings come from being able to buy from a wider market, allowing competition to drive down prices.

These changes allow measurable savings. But there are other, more far-reaching benefits. As companies install software applications to take over many of the tasks that employees now do – such as "running errands" to keep information moving – they alter the balance between the internal and external demands on a company, an issue described in a book called *Harmony: Business, Technology & Life After Paperwork* by Arno Penzias, former boss of Bell Labs. "Though to my knowledge no computer has yet managed to replicate the performance of a single office worker," he says, "the right combination of computing and communications can frequently replace whole departments."

The "errands" jobs are not only wasteful; they turn the attention of organisations inwards towards the smooth operation of internal processes, rather than outwards to the customer. Now, though, Internet-based software applications are shifting the balance, shrinking the amount of human time and effort that needs to be spent on internal co-ordination. Companies will no longer need layers of white-collar workers to manage the steps between what the customer wants and what he gets.

All sorts of things can be done differently. Manufacturers can talk directly to their customers, or to their suppliers' suppliers. Customers can click a mouse and start a production process rolling, far along the supply chain. Training sessions can be carried on a laptop. Sales staff can do presentations to customers a continent away. Companies will often find it difficult to tell whether such measures save money or whether they simply provide a better service, but overall they will increase the proportion of time that goes on keeping customers happy rather than keeping the business running.

Once a company sets off down the Internet track, what does it find? First, a change in familiar boundaries, starting with those of the firm itself. Collaborating with others becomes easier and less expensive, as does linking different operations within and between firms and buying in everything from management skills and innovation to the human-resources department. The boundaries for employees are redrawn too: those between home and work, as people work from home and shop from work, and those between the individual and the company, now that employees' knowledge and skills may be worth almost as much as the company they work for. Geographical boundaries start shifting as well: within companies, different regional divisions need to collaborate and share customers more than they did, businesses can source more products globally, and new ideas and competition can spring up anywhere on the planet.

Life in the fishbowl

Second, the Internet brings management out into the open. "You put a piece of glass into your organisation and expose all your internal strife," says Pete Martinez, who runs the worldwide consulting arm of IBM's global services. Managerial privacy dwindles. "I have to assume that every bit of information about me is broadcast back to our employees and customers. It's the fishbowl effect," says Eric Schmidt, boss of Novell, a troubled software company. If you have failed to reply to an e-mail, you can no longer hide behind "my secretary lost the message". Pricing is also more transparent, as more deals can be put to the test of an occasional auction. Companies need to allow customers and suppliers "inside the machine", in the phrase of Peter Martin, editorial director of Internet activities at ft.com. Thus customers can track the progress of their orders, and suppliers are growing used to scooping information straight out of their customers' databases. Moreover, employees can see what they might earn in similar jobs in other companies.

Third, the Internet increases the importance of standards. Indeed, the glue that holds it together is essentially a set of software standards. Their user-friendly simplicity allows people to use the Internet in many different roles – as customers, suppliers, employees, job-seekers – without needing to be retrained. Electronic commerce needs standards in order to make it easy to transfer information between companies with different systems: hence the importance of XML, a programming language. Companies also need rules about what can be bought online: the aggregation of many departments' orders for staplers will save money only if all departments are willing to buy from the same standard shopping list. And aiming for a standard technology, and standard look and feel for customers, is a way to reduce maintenance costs and to measure more easily how customers behave. However, standardisation is a force for centralisation in companies. Once the human-resources department decides it needs a single global website to keep employees up-to-date, head office will want to decide what should appear on it.

This survey begins inside the company to see how the Internet affects the way managers communicate with staff, and staff with each other. It then examines links with suppliers and customers. Lastly, it considers how the shape of the company itself may change, and offers some guidelines for good e-management.

Talking to each other

Corporate culture instilled online

AT ITS MOST basic, the Internet is a wonderful way to communicate. Hit that "send" button and off goes the e-mail, trailing attachments, to everybody in the firm and beyond. No wonder companies find it a perfect way to talk to their staff. No wonder it is so useful – but also so dangerous – when staff want to talk to each other.

Over and over again, the Internet's uses turn out to dovetail beautifully with current trends. As companies become more fragmented and their workers more geographically dispersed, managers need a way to rally the troops. In particular, they need a way to build a corporate culture: that intangible something that binds employees together and teaches them to understand instinctively the defining qualities of the business and the appropriate way to respond to any issue that confronts them. The Internet provides the means to do this.

In a stable, slow-growing and well-established company, a common culture may be easy to maintain. You take each year's new recruits off to boot camp for a fortnight and teach them the company history. But few companies today can afford to be stable or slow-growing. Instability and speed make culture-creation harder.

In Silicon Valley, people count as old stagers if they have been with the same employer for much over a year. But rapid turnover is not the only difficulty. In many companies, the salesforce or the maintenance folk rarely come into the office. A quarter of IBM's workforce, for instance, is now mobile – they spend at least 80% of their time off-site, usually working from home or on the road. Key people may be based in key markets abroad, a day's air travel away from the main office. Mobility goes right to the top: Douglas Daft, chief executive of Coca-Cola, travels 80% of the time. He boasts: "The headquarters office is where I am."

Add in mergers and takeovers, which create a need to proselytise a new bunch of employees and coax them to abandon one corporate creed for another. As companies outsource more and more activities, too, they look for ways to teach their subcontractors to share their values. And the faster things change, the more important it becomes to explain to employees what is happening, and why.

How to do it? "In a rapidly changing and geographically distributed organisation," observes Michael Morris, a social psychologist at Stanford's Graduate School of Business, "you don't have the option of the drink after work." But you do have the Internet. More than any previous technology, it allows companies to ensure that every employee has access to the corporate news, views and vision.

Some companies use it to teach their employees (as well as suppliers and customers) their ethical code. Boeing, for instance, offers an online "ethics challenge" where employees can test their moral instincts on such delicate issues as "acceptance of business courtesies" and "the minister drops a hint". Such applications are a way to spread a common approach throughout an organisation.

But the Internet is also a way for bosses to tell staff where they want the business to go. For example, at Ford, which claims to have the world's largest intranet, 170,000 staff around the world are e-mailed a weekly "Let's chat" note from Jac Nasser, the chief executive. A purpose-built newsroom maintains a website upgraded several times a day, and available to Ford's employees around the world (in English only), as well as to those of its new acquisitions, such as Volvo.

Not only does the Internet allow managers to talk to their staff; it lets them track whether the staff are at least pretending to listen. William Nuti, president of Europe, the Middle East and Africa for Cisco Systems, a high-tech giant, produces a monthly video to send to his staff, explaining where the business is going. What happens if the staff don't choose to watch? Well, the Internet allows you to track who opens an e-mail and when. "I know everyone who clicks on it, and those who throw it away, and I make phone calls to people, saying it's important you watch this." Not surprisingly, Mr Nuti's viewing figures are high.

But all this communication from on high can sometimes cause problems. SAP, a German business-software giant, is another company with an elaborate communications system. It allows material to be broadcast on the car radios of workers on the road, for example. The company found that middle managers objected to the chairman e-mailing all employees. Their authority had rested partly on their role as a source of information, and without it they felt exposed. As so often with Internet-driven changes, the implications of what appeared to be a simple, time-saving innovation turned out to be more complex and politically sensitive.

That sensitivity becomes more acute as communications become increasingly bottom-up as well as top-down. At Siemens, a large

German company, Chittur Ramakrishnan, the chief information officer, has noticed a "very significant number of e-mails to top management. The idea of going through a secretary to get an appointment has changed. People can send e-mails to anyone and expect a response. It is very democratising."

Listen to us

Leaving aside pep talks, companies find all sorts of mundane tasks can be done online with greater efficiency and less expense. As a result, "B2E" – business-to-employee – applications are flourishing. Tim Mead, chief marketing officer for Cambridge Technology Partners, a consultancy, thinks they may be the biggest growth area for Internet applications over the next couple of years. They include many tasks involving staff matters; the creation of an internal job market; and training. These are discussed in detail in the next section. It is one of the strengths of the Internet over previous, proprietary systems that it can be used to provide services to everyone in a company.

Once material of direct interest to workers (say, their holiday entitlement) is available online, they grow used to logging on. Many companies reckon this is a quick way to help their people come to terms with changing business methods. That is why Ford and some other companies, such as American Airlines, are giving their employees computers to use at home. Two-thirds of Ford's employees are hourly workers, who will not be able to use them to do company work from home. But that is not the point. Ford is hoping to get all its people used to thinking online, and to have a direct way to reach them all with a consistent global message.

Consistency becomes important as companies evolve their internal communications. Initially, every department tends to set up its own website, perhaps protected from the rest of the company by a password, and often designed to boost the department's self-esteem. To end such anarchy, or simply to pull all internal information together, a growing number of companies now have a "corporate portal": a centralised home page with links to various services, items of information and titbits to entice the staff to keep looking in. Click, and there is a map of each floor of the office; click again, and there are photographs and personal details about who sits where. Elsewhere on the page there may be links to the online services of the human-resources department, or the day's news clippings, or a page allowing workers to fill in expenses claims, order office supplies or find telephone numbers.

To persuade employees to look at the home page as often as possible, companies think up various inducements. Cisco Systems, keen to attract the attention of its option-owning employees, plonks its share price centre-screen. Other companies post a list of employees with a birthday that week. Scient, an Internet consulting firm in San Francisco, has an area called "Do you want to scream at anyone?", for employees to complain about colleagues who send excessive e-mails. The site shows the daily winner in categories such as "Take a chill pill".

The good thing about such pages is that they are accessible not only to employees in head office, but also to people in distant subsidiaries, on the road or at home (though this can cause culture clashes: Scient's British staff are bemused by its "stream of e-consciousness" site). Increasingly, employees can personalise their page, so that if they are working in the marketing department they do not receive a deluge of news clippings on camshaft design. Companies with lots of old "legacy" computer systems can use the home page as the entrance to a network designed to pull all the old systems together.

In time, these in-house portals may become important sources of revenue for many companies. Some already sell their own products to their employees online. Ford has a scheme to allow the friends or family of an employee to buy a company vehicle at a discount. The employee enters his social-security number, name and address on a website and receives a personal identification number which he can e-mail to his friend. That allows the friend to pick up the vehicle from a dealer. Rival car companies have similar schemes offline that involve lengthy form-filling.

Next, there is the prospect of turning the corporate workforce into a marketplace. It is an advertiser's dream: a stable group of people with regular pay and a known employer. Why not, for instance, offer a link from the page that informs an employee of her holiday entitlements to a travel company with which the company already does corporate business, and which will offer discounts on leisure travel? Why not charge local restaurants for the occasional advertisement?

Indeed, this is already starting to happen. For instance, Exult, a consultancy to which BP subcontracts much of its human-resources work, is discussing just such a proposition with companies offering financial services. But how will businesses feel about encouraging their staff to hunt for a home loan when they should be finishing a presentation? Alan Little, Exult's head of global client relationships, replies robustly that, if employees can work from home at the weekend on their company laptop, then surely they should be allowed to book their holidays from the office on a weekday. They should be judged by results.

The inside story

Better ways to manage your staff

WHEN COMPANIES ARE learning to eliminate paperwork and speed up processes online, they often begin in their own backyard. They notice that many of the jobs that keep their human-resources people busy can be better done electronically. They discover ways to handle employee expense claims online. They create an internal electronic job market. And they put training online to keep dispersed and busy employees in touch with constant innovation.

Most of this is relatively easy to do because it is, on the whole, unthreatening. It may cut out some HR jobs, but it does not alter the main business. On the other hand, it helps to teach companies and their employees about applying the Internet. Techniques honed in the HR department can be readily transferred to the customer-services department, and filling in expenses claims online is a lesson in electronic procurement.

HR departments used to spend much of their time answering questions from employees. The move towards "cafeteria" benefits – a choice of various permutations of pension, health plan, holidays and pay – brings lots of calls from workers asking: "What happens if I ..." Such questions are often more easily answered by a computer than a human being. This has encouraged companies to put their employees' details on a website, protected by a password, and allow their staff to update their personal information or, in refined versions, to experiment with different combinations. The results can be dramatic. Even when staff could use the HR website only to update their records, Ford found that calls to the central help desk fell by 80%.

Once they realise how much of HR can be shifted online, some companies start to think about passing the chores on to someone else. BP, an acquisitive oil company that is digesting Amoco, Arco and Burmah Castrol, agreed in December 1999 to outsource much of its HR work to Exult, a start-up that has recently negotiated a second deal with Unisys, a computer giant. Exult is building a network that will give BP's staff in more than 40 countries information on all sorts of HR issues. It will alert a manager when staff turnover in his unit passes a certain level, for instance, or allow him to see how various permutations of pay for his staff will affect his budget. It will tell a worker how much holiday entitlement he has left, and whether he could roll some over to next year. Or it will allow an expatriate employee to look at terms and conditions for his next foreign posting.

One effect of handing over this project to Exult has been to draw BP's attention to the hundreds of different pay scales, holiday policies and benefit arrangements that have sprung up throughout the business. Deciding whether and where to introduce coherence is a job for BP's own management, not for Exult. But Exult's Mr Little argues that, although 20% of local variations may be justified, 80% are not. Reducing unnecessary diversity brings all sorts of benefits: pay scales become easier to explain and faster to alter. Once again, the impact of the Internet is to encourage simplicity and centralisation.

Expenses claims, too, can switch to self-service. Oracle's Mr Henley has got rid of a quarter of the people in his accounts-payable department who were doing nothing but filling in data from expenses forms. To get employees to submit their expenses claims electronically, the company simply e-mailed everyone to inform them that claims on paper would no longer be paid. There are few better incentives for even the most Luddite employee to learn to use a new technology.

Cisco Systems went through a similar process. Sue Bostrom, in charge of Internet business solutions for the company, says it was costing $50 a time to process a claim. That figure has come down to between $2 and $7. When she returns from a trip, she simply pulls up the record of her company's American Express account, fills in what each payment was for, clicks and submits. She gets paid much more quickly. But there is also, she points out, a further benefit. Cisco has an application called Metro that compares an employee's spending with the corporate average. If employees typically spend $250 on a night in a New Jersey hotel, but this one chooses to spend $350, the system will automatically flag it up and ask him to fill in an explanation. A couple of auditors review all flagged claims; if they disagree with an employee's explanation, they e-mail him and his manager. That not only educates employees about what they are expected to spend, but also makes it easy for managers to check out-of-line claims.

Picking winners
Most companies' top priority is to find the best people for a job, and then to keep their knowledge and skills bang up-to-date. Hence the enormous importance attached to recruitment and training.

Once you have details of employees' work experience on a database, you have a more efficient internal talent market and a faster way to recruit a team to work on a particular project. A manager, using a special password, can examine a potential recruit's work experience, past assignments and willingness to move home, together with the latest job review. Armed with such information, a manager can search the database for a particular set of skills, a job that would once have needed help from the HR department. The result should be a better internal job market.

There are, of course, some snags. Employees may not much like the idea that managers in other departments can sneak a look at their latest job review without their consent. If a company puts up employees' photographs, it may find itself dealing with a discrimination case if a black face is screened out of a suitable job online (the technology will reveal whose records were considered). Most important, though, a database will probably not be able to answer the recruiter's biggest question: is this person any good? At SAP, for instance, Thomas Neumann, director of human resources, admits that his vast database of 22,000 people in 50 countries works "better for skills than for competencies". But the managers that use it still claim it saves them time. It may be a coarse sieve, but for the harassed manager it may well be better than waiting for a telephone call from the HR department.

Recruiting from outside also becomes faster if a curriculum vitae can be sent in electronically: 71% of the Fortune 500 companies now accept applications on their corporate websites, according to a survey by recruitsoft.com. That allows them to be circulated around departments the same way. Faced with a mammoth skills shortage, some companies have found cunning ways to use the Internet as a lure. Siemens, which now gets 60% of its job applications online, was struggling to find clever young engineers as the number graduating from Germany's crack universities was declining rapidly. So, with the help of psychologists, the company designed an online game that would test for the skills it most needed, such as an ability to work in a team. The company ran the game on its website for six weeks, expecting perhaps 2,000 people to play. In fact, says Peter Pribilla, a member of Siemens's corporate executive committee, 10,000 did, many of whom were young engineers. The company interviewed those with the highest scores. Perhaps partly as a result, three independent surveys of engineering graduates in the past year have rated Siemens the best company in Germany to work for.

Once good people are on board, the next challenge is to train them and to keep their knowledge up-to-date. This is not easy when your staff are constantly on the road. And most people end up dozing through much of any training course, because either it tells them what they already know, or it tries to convey something so complicated that an hour's class is not long enough.

For years, companies have dabbled with using computers instead of teachers. The results have generally been dire. Online training programmes are often little better. But some companies now think that they have begun to crack the problems of teaching their staff electronically. At SAP, for example, Rainer Zinow, head of knowledge management, says that his web training programme is the most expensive one he designs, needing between 100 and 200 hours of production time to produce a single hour of material. The important thing, he says, is to realise that "my classroom is a room in a medium-sized hotel in Connecticut", with a dial-up connection and a consultant who will pay attention for at most an hour between 6pm and 7pm. So to work well, an online training programme has to be broken into small bits, be able to grade skill levels carefully, and be designed to cope with irregular sessions.

Any training programme needs to test how well employees do before and after. At Dell Computer, John Coné, the company's head of learning, measures "initial ramp time": how long it takes a newly hired salesman to achieve his full sales quota. Dell used to pack new employees off to boot camp for three weeks to be taught about systems and processors, rules and regulations on selling, and the finer points of Dell's product lines. But that was wasteful: some recruits knew most of it already, others knew nothing. Now the company first tests what people know and then offers part of their training online. It has cut a week off boot camp and two weeks off the normal ramp time.

Managers, too, receive some online training, for example on coaching. An optional one-day old-fashioned class is available at the end of the course, but Mr Coné says that only a small minority signs up for it. "The only thing we can't do asynchronously online", he says, "is to have an individual attempt to display a learned behaviour, and get immediate feedback based on judgment." Dell now delivers roughly 60% of all formal learning online, and hopes to raise that proportion to 90% within two to three years. The main thing to remember, he says, is that online learning is like the microwave oven: it is not a complete replacement for the traditional model, but it does some things better.

Selected, rewarded and trained: the next stage is for employees to learn to work together.

A little knowledge ...

... goes further if you collaborate

COMPANIES THESE DAYS bang on a lot about knowledge management. But what, exactly, does it mean? Some interpret it as training, others as managing an online database. One of the better definitions comes from the Yankee Group, an American consultancy: knowledge management involves efficiently connecting those who know with those who need to know, and converting personal knowledge into organisational knowledge. People are wonderful receptacles of valuable ideas and information, but they tend to move on, taking their knowledge with them. The challenge for companies is to find ways to extract and share the stuff.

That is what many of the most interesting new Internet applications are intended to do. "Collaboration" is a powerful word in Silicon Valley. Companies need their workers to share ideas more than ever before, for a variety of reasons. One is the need for incessant innovation and refinement of new products and processes. That requires an endless stream of new ideas. Another is that, just as workers in an old-economy factory work together physically to build a machine, so workers in an office need to communicate and co-operate to build a service.

But sharing has grown harder, partly because workers on the same project may be separated by long distances and time zones. The stimulating chat around the coffee machine, source of bright ideas and quick fixes, gives way to the international telephone call or e-mail interchange. In addition, the employees who collaborate to produce a given service may work for different companies. As tasks become fragmented among different firms, good collaboration tools become essential.

Daily sharing of information goes on in most businesses, of course, and in geographically dispersed companies some of this has been long-distance for many years. Designers at some companies have been based in different time zones and passed work to each other round the clock. But until recently they have used proprietary networks. The Internet makes such round-the-clock sharing available to everybody in a company, or all the people working together on a particular project. It thus enhances global teambuilding, and encourages the emergence within companies of horizontal communities, bound together by a common function or interest. These communities can now easily float ideas with each other, or gossip, or discuss best practice – around the clock and around the globe.

Managing collaboration requires special skills: less emphasis on individual achievement, more on teamwork. Moreover, just as companies can learn lessons from developing online HR services for their own staff that can be applied to running online support for their customers, so the lessons they learn from collaborating within the organisation can be applied to collaboration with other companies. A company that cannot persuade its own staff to work together smoothly and efficiently is unlikely to do better with its suppliers.

Collaboration also requires appropriate pay structures, designed to reward teamworkers rather than lone rangers. But the motives that persuade people to work together are not exclusively financial. One of the most interesting and inspiring models of collaboration, according to Alan MacCormack of Harvard Business School, is the development of open-source software such as Linux, on which thousands of people around the world who have never met work together – unpaid. The model requires a workable kernel (in this case, the initial 10,000 lines of code written by Linus Torvalds) to which people can easily add; a modular design, so that different people need to understand only the part they choose to work on; and a small team at the top to set broad guidelines and select the best ideas. The most powerful development, says Mr MacCormack, is that by users. Their reward is global recognition – "because software code is a universal language, if I make a good patch, the world knows" – and the satisfaction of seeing their ideas discussed (often on slashdot.org, a website boasting of "news for nerds") and adopted.

Could companies inspire the sort of altruism that has gone into developing open-source software? In some other ways, Mr MacCormack points out, corporate innovation increasingly resembles Linux code-writing. The teams that work on it are often geographically dispersed. And design is increasingly modular rather than sequential: people no longer design the engine and then pass it on to a second group to produce the casing, which discovers problems with the engine design just as the first team has moved on to its next project. Besides, innovation is increasingly delivered not by a single company's research-and-development department, but by a network of companies, each working on a different part of the project.

However, unlike Mr Torvalds, companies need to rely on their employees' ideas for making their money. Biotechnology companies long ago found ways to give staff a share of the rewards for their research. Now even IBM has a scheme to make sure that good ideas bring more than a pat on the back. Developing a patent wins a financial reward; so does authorship of a certain number of articles; and consultants get bonuses for creating and sharing good ideas. The company is also working on a way to encourage people to put their ideas into its knowledge database, by rewarding those who create material that is frequently used, as well as those who review or grade the stored ideas in particularly helpful ways. The aim, says Scott Smith, who helps to run the knowledge side of IBM Global Services, will be to create a "self-rewarding content-grading system", a bit like Amazon.com's way of persuading customers to review the books they buy.

Divided but united

Much of the everyday collaboration that goes on in companies is far more humdrum stuff. But when people are scattered, or working for several different employers, simply co-ordinating their joint efforts can be immensely time-consuming. Keeping track of who has done what is essential, though, if only to avoid legal wrangles when things go wrong.

All sorts of new collaboration tools now allow people to work together on a single project. For example, they may set up a shared website on which any of them can post or update material. That does away with the cumbersome business of sending e-mail attachments back and forth, especially irritating for people working with a laptop in a hotel bedroom. GE Capital, for instance, has something called a Quickplace on which a group of staff members working together can store all the documents, plans, correspondence and other details to do with a project. Ford uses a similar collaboration technology to handle due diligence when it acquires a company. The system was developed when teams in Sweden, Britain and America collaborated over the acquisition of Volvo; after that, it was used in the purchase of Land Rover. Staff can use instant messaging for quickly checking particular points. The advantage of such "e-rooms", says Bipin Patel, head of management systems at Ford, is that they are asynchronous (meaning that people can use them when it suits them); they are always up-to-date; and they do not use much manpower.

Such tools are now being developed further by companies such as Firedrop, a Californian start-up that has devised something it calls a "zaplet". This arrives in your mailbox like an ordinary e-mail, but when you open it, the zaplet becomes a window on to a server, or central computer. The information you see, therefore, is whatever is now held on the server, so you always get the most up-to-date version. The zaplet may also allow you to use an application that sits on the server: a spreadsheet, perhaps, or a way of managing a customer database.

Alan Baratz, Firedrop's chief executive, arrived in July 2000 to find the company concentrating on consumer services. He saw a way of turning it into a tool that managers could use in, say, recruitment. At present, piles of applications pour daily into the recruiting department, which sorts them and sends the best to managers, who indicate the candidates they want to see. The department then telephones the applicants and goes backward and forward trying to find a suitable date for an interview. Firedrop's device, Mr Baratz says, allows the recruiting department to e-mail hiring managers just once. Then they can all look at the same application and write in comments, or compare notes on which time slots are free for an interview. With luck, zaplets should be commercially available during 2001.

Lots of consultants and technology companies now hope to teach firms to collaborate better. At Peoplesoft, which has made it a speciality, Baer Tierkel, who is in charge of worldwide marketing and strategy, argues that effective tools must be based on the Internet: its open standards make it much easier for everybody to work with everybody else. Such intercommunication becomes easier with XML, or Extensible Markup Language (see "Do you speak invoicing?").

In future, Mr Tierkel thinks, software tools will increasingly allow the sort of information normally available only in corporate back offices to be readily accessible to people in the field: "A customer-services manager might get an alert to say that, given the level of orders coming in, she was going to need more staff; or a salesman might be able to find out, as he was walking into a customer's office, whether that customer was happy with the relationship so far." New tools will also increasingly turn a company's customers and suppliers into one large collaborative network.

For the moment, dream on. Most companies are still discovering how to do online procurement, and how to deal electronically with their customers. Both often turn out to be harder than the enthusiasts would have you believe.

Do you speak invoicing?

The joys of XML

FOR COMPANIES TO make full use of the Internet's potential, they need to be able to receive information arriving electronically from a customer or supplier, and pass it through their own systems without having to print it out and manually transcribe data, or change the format. This is what Electronic Data Interchange (EDI) tries, clumsily, to do. Until recently it has not been possible with information that arrives on the Internet.

To understand why, look at the "page source" on your web browser. There you will see how Hypertext Markup Language (HTML) tells your computer how to display a page of material you have found on the Internet. Enclosed between angular pairs of brackets, you will see words such as ‹HEAD› or ‹FONT SIZE›. These are called "tags". What you will not see are any tags that tell you whether you are looking at an invoice, or a set of medical records, or instructions for installing a condenser. The absence of such information helps to explain why, when you search for something on the Internet, you receive vast amounts of irrelevant junk. Search engines cannot tell, from the tags on web pages, whether they have found a page on books *by* Charles Dickens or books *about* Charles Dickens.

Since 1999, companies have begun to use a new language to describe web pages, called Extensible Markup Language or XML. It inserts many of the same standard tags as HTML (such as P for paragraph and IMG for image), but it also allows people who create web pages to add more tags of their own. These, unlike the tags that HTML uses, need not be confined to a few dozen standard words. They can, for instance, say that the page provides information on a book's authorship rather than title; or that it is a purchase order; or that it is in Chinese.

But, if the creator of the web page has invented these tags, how will your browser know what they mean? The answer is that the top of the page will carry all the information your browser needs to understand the rules that the originator of the page has drawn up. It is as though each page was a board game which arrived with a set of rules telling you exactly how to play it. To be comprehensible to your browser, the rules must be set out in a standard way. XML is that standard.

Because XML describes the content of a web page in terms of the type of data it contains, rather than the way the data should look, it helps groups of like-minded people to share information. They simply need to agree on a set of tags that meet their particular needs. So a consortium of fishing fleets might agree on a standard way to describe information about fish catches – the number landed, the species, the average size. They could use their own XML tags to store this information. A search engine could then look for data types rather than just words: for all the fleets that landed cod of a certain size on a certain date, for instance, rather than just any website containing the word "cod".

Like HTML, XML is an open standard: anybody can use it without paying a licence fee. It was developed by a working group from W3C, the World Wide Web Consortium, a standard-setting body. Gartner Group, a research firm, reckoned that by 2001 it would be used for 70% of electronic transactions between businesses.

However, individual industries and other groups still need to agree on the set of tags that their particular trade will use. Otherwise, some fishing fleets might store data on catches with a tag called ‹SIZE› while others might use ‹WEIGHT› instead. These subsidiary standards, crucial for commercial interaction, are being agreed on in various ways. Some will emerge from standard-setting bodies such as RosettaNet, a not-for-profit consortium that mainly works with companies in the electronics industry. Some will be commercial versions, hoping to drive out rivals and achieve supremacy. Microsoft has an initiative called BizTalk which aims to create common ground. Some of these new standards may be proprietary, although most will probably be open. Agreeing on them will often be a tortuous and acrimonious process. However, once these standards emerge, XML will become the core of electronic commerce.

New life in the buying department

Tons of savings on toner cartridges and toilet paper

As OLD-ECONOMY companies go, few are in more robustly physical businesses than United Technologies, owners of a cluster of firms such as Pratt & Whitney (aero-engines), Otis (lifts) and Sikorsky (helicopters). Yet few enthusiasts for the transforming power of the Internet rival Kent Brittan, who heads supply management. The reason: UTC is one of the world's largest online purchasers.

The idea that anybody in supply management should be enthusiastic about anything would once have seemed a contradiction in terms. Buying departments have long been one of corporate life's dead ends, full of people chasing up invoices and trying to work out how many widgets the company bought last year. No longer. These days, the folk who manage purchasing are the shock troops of the business-to-business business, and those far-sighted enough to have gained an MBA in supply management from the state universities of Arizona and Michigan, the only two places that award them, are on an Otis ride to the top. Mr Brittan even has his own human-resources staff, to augment the "legacy" employees of UTC's purchasing department with people who know their XML from their EDI.

Mr Brittan's purchasing people can deliver something that not many managers struggling to introduce the Internet into established firms can hope to achieve: real and speedy savings. To do so, though, they need to persuade this traditionally decentralised group to accept a high degree of central discipline. That means finding ways to make the Internet understandable and interesting to the multitude of employees who think of it mainly as something to play with at home. So the purchasing department has built its own elaborate website, with news articles, employee profiles, a glossary of financial and engineering terms, and information on corporate training courses. Advertisements in the company's lifts and corridors tell people to click on this corporate portal – which also offers a gateway to electronic purchasing.

Corporate buying online seems at first glance a logical development, on a far larger scale, of Internet purchases by consumers. In fact, consumer buying is mainly online catalogue shopping, with a few advantages such as round-the-clock ordering and more flexible pricing. But much industrial purchasing is far more complex. For instance, many products are made to detailed specification, not bought off the peg; they are purchased by teams, not individuals, so that the decision to go ahead occurs at a different time and place from the actual transaction; and they are generally bought under long-term contracts, specifying all sorts of quality, price and delivery characteristics. Sam Kinney, one of the founders of FreeMarkets, a Pittsburgh company that runs electronic marketplaces, explains in *An Overview of B2B and Purchasing Technology: Response to Call for Submissions*, a paper of exemplary clarity written in 2000, that "these complexities eliminate the possibility that simple B2C business models could be successfully applied to business purchases".

Mr Brittan's task is thus a complicated one. But it is made a little simpler by the fact that purchasing falls into two rough-and-ready categories: direct materials that go into end-products (such as parts or chemical feedstocks), and indirect materials, which may be anything from carpets to lubricants to hotel accommodation for travelling staff. The arrival of the Internet changes both kinds of buying, but many of the quickest hits are in the second category. In many large companies the first kind of purchasing has been online for years, although the systems have been proprietary, inflexible and expensive. The second has floated free. Buying the paper for the ladies' loo can sometimes be done on the say-so of the local building manager, and sometimes needs a sign-off from the buying department.

One way, it tends to be extravagant and hard to track; the other, expensive to manage and infuriatingly slow. When the local manager places the order, a big company may use hundreds of suppliers; "rogue" purchases proliferate; and company purchasing policy becomes impossible to enforce. Nobody can track how many pencils are being bought each year, or where, or at what price, or whether a rise in the pencils budget is the result of a rise in pencil prices or in demand for pencils. When the buying department does the ordering, the company spends money on squads of bored clerical staff that might be better spent on a few of those Arizona State MBAS.

All too often, the business of purchasing costs more than the items bought. When companies pause to look at these systems, they are aghast: SAP, for instance, realised that purchases had to go through four levels of approval; UTC found that it was handling 200,000 invoices for 12,500 different items of office supplies. "Maintenance, repair and operation typically account for 20% of a company's purchases but 80% of its orders," says Patrick Forth, managing director of iFormation, a spin-off of the Boston Consulting Group. "The cost of a purchase order is typically $100. E-procuring costs $10."

One of the fastest ways to save money, therefore, is to use the Internet to try to bring indirect purchases under control. That means, first, putting somebody in charge of the task. Next, it involves negotiating centrally with suppliers; drawing up a single catalogue; and insisting that staff either buy from it or explain why they want something different. Even if a company does no more than that, it makes savings: "If you don't get at least 10% out," says Mr Brittan, "something's wrong." If you cut the number of suppliers, you not only get bulk discounts; you save your buyers' travelling time, and you probably save money and space tied up in unwanted stock.

Some companies find that the largest savings come from curbing those rogue purchases. At IBM, purchases made outside the procurement system accounted for 30% of the total in 1995. The figure is now down to 0.6% of total purchases of more than $45 billion a year. Because expenses claims are online, it is easier to check who steps out of line. If staff are to accept such discipline, they need a system that is very easy to use and very quick. IBM cut the purchase process from 30 days to one, renegotiating the contract with Staples, a large office-supplies company, to speed up delivery.

In time, many other benefits emerge. For example, there may be savings in buying legal or accountancy services. This will be harder than reorganising the purchase of paper for the photocopier, because professional services are difficult to specify. Yet even this may turn out to be possible. SAP, for instance, is trying to define different levels of consultant, because consultancy services are its largest single purchase. If it can do so, the company will be able to start managing its consultancy purchases in the same way as it now manages purchases of PCs.

Moreover, quite apart from the gains that come from aggregation and control, there may be savings in the transaction costs of placing orders, especially if the Internet can link the buyer's computer system directly to the vendor's. Until the middle of 1999, SAP's offices bought locally. Now, employees log on to a standard catalogue, click on an item and fill in an online purchase order. This is sent electronically to a specific vendor, with whom a price has already been agreed. The system confirms the delivery date and issues the credit note, closing the whole deal under a set of predetermined terms and conditions. The enormous saving in staff that such automation makes possible has allowed SAP to cut internal purchasing costs by around 80%.

Such standard catalogues are one of the more important innovations that the Internet has made possible. Its low-cost, flexible technology allows them to be used by everyone in a company who has access to a browser, and to carry illustrations or even videos to show how a particular part fits or a tool is used. Even small companies can afford them. They will transform the management of indirect purchases and become one of the most reliable sources of corporate cost savings.

Direct hits

Some savings also turn out to be possible in companies' purchases of direct materials. But these have always tended to be handled differently. No local manager nips out for a rogue purchase of a few thousand motherboards. Instead, expensive software programs called Materials Requirements Planning track the materials and components that a manufacturer needs in the production process and calculates what replenishment is needed on the basis of production orders from the sales department.

However, such programs are no use when a company designs a product from scratch, hunts for suitable components and suppliers, specifies parts and negotiates prices. Nor can they easily be connected to the programs of other companies, to allow a supplier to understand what is happening in a buyer's plant. The Internet and its associated technologies make both these things possible.

The answer here is not a standard catalogue. Most companies buy custom-built parts and components from suppliers with whom they have worked on design specifications and entered into long-term contracts. Instead, Internet-based software makes collaboration easier, both within and among firms. That, as the next section demonstrates, turns out to be useful at almost every stage along the supply chain.

Trying to connect you

A supply-side revolution

EVERY DAY CISCO SYSTEMS, acme of the new-economy corporate model, posts its requirements for components on an extranet, a dedicated Internet-based network that connects the company to 32 manufacturing plants. Cisco does not own these plants, but they have gone through a lengthy process of certification to ensure that they meet the company's quality-control and other standards. Within hours, these suppliers respond with a price, a delivery time, and a record of their recent performance on reliability and product quality.

This process, says Cisco's Mr Nuti, has replaced a room full of 50 agents, who would pull together much the same information with the help of telephones and faxes. The operation generally took three to four days. Now, says Mr Nuti triumphantly, "those 50 people are redeployed into managing the quality of components".

Three aspects of Cisco's supply system are particularly significant. One is the use of a form of electronic market to set prices. Online marketplaces of various sorts proliferated in 2000. One is the exchange of information between buyers and sellers. The Internet's ability inexpensively to increase this flow is altering the whole nature of the supply chain. The third is the extent to which Cisco outsources activities that other companies do in-house. That, again, is made easier by the Internet.

Most companies have moved nothing like as far as Cisco (see chart 2). But they are beginning to realise the essential steps they need to take. These involve both widening the potential pool of suppliers for any given contract, and deepening the relationship with the supplier that eventually wins it.

The obvious way to widen the pool of suppliers is to participate in an online auction of some kind. Many companies are beginning to put contracts for supplies out to tender online: indeed, GE has done so for several years, though on a proprietary network. By the end of 2000, the company aimed to "e-auction" more than 10% of what it buys, using the Internet as a trading floor. GE thinks that eventually 70% of its supplies can be bought this way.

One result has been to widen the geographical range of suppliers, says Mr Reiner. For example, the company has developed new capacity

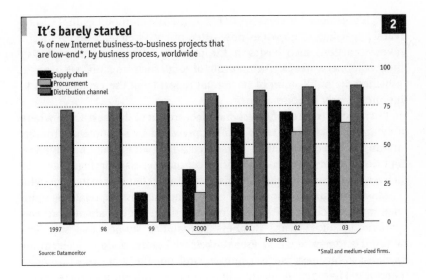

It's barely started

% of new Internet business-to-business projects that
are low-end*, by business process, worldwide

Supply chain
Procurement
Distribution channel

1997   98   99   2000   01   02   03

Forecast

Source: Datamonitor   *Small and medium-sized firms.

to ship supplies from Shanghai, in response to a big rise in bids from Chinese manufacturers. Over time, electronic tendering seems bound to push up the proportion of their supplies that rich-world companies buy abroad, and especially in the developing world. That has important implications. For a start, buying abroad will help to push down prices. "Instead of importing inflation, you import deflation," says GE Capital's Mr Alexander. Moreover, a labour shortage at national level is no longer a problem: "Labour is more fungible than ever before." But casting the net more widely will also increase the need for verification services, of the sort provided by SGS, a 120-year-old Swiss firm that has 60% of the world market for managing trade risk. It has recently launched an online division to protect the interests of buyers and sellers who may never meet.

On the whole, GE has preferred to build its own electronic exchanges rather than use those run by others. However, lots of experiments in online auctions are taking place. As it happens, auction theory has become one of the most fashionable branches of economics, and plenty of academics dream up (and sometimes set up) ingenious auctioneering enterprises of their own. For example, Barry Nalebuff, of Yale University, has launched a company called splitthedifference.com. Buyers and sellers of a given commodity set out their reserve prices; the software then works out the mix of buyers and sellers that delivers the greatest value, and splits the surplus evenly between both sides.

Such ingenious schemes work for standard commodities that can be readily specified. Companies now often put out to tender (in essence, a reverse auction) such basics as the printing of their annual report or their stationery needs. As a result of one such auction, GE Capital switched the publication of its annual report from the United States to Indonesia early in 2000.

But most purchases of direct materials are not of that sort. Even where they are, the bidding process is often preceded by a couple of arduous months of "prequalification", where company and would-be suppliers nail down every variable except price: quality, size, timing, ability to deliver, and so on. This rigorous discipline brings benefits in its own right.

Most companies, moreover, have long-established relations with their main suppliers. Are they right to toss them aside in an auction? Das Narayandas of Harvard Business School has studied five suppliers whose customers went to FreeMarkets, the grand-daddy of electronic business-to-business auction houses, and put the work out to tender. Four of the five customers, he will report in a forthcoming article, "were back knocking on their [old] suppliers' doors three months later". The new suppliers had underestimated, and failed to match, the huge value of customisation in the contract.

At FreeMarkets, the founders agree that such things happen. But, says Glen Meakem, the best auctions do something rather different. They provide the foundation for long-term relationships. In his paper, Mr Kinney adds that "Buyers typically use the auction to determine with whom to establish the market relationship, based on excellent price discovery. But, once the auction is over, production parts are approved and tooling is installed, the working relationship can run for years." That "price discovery" is not to be sniffed at: at FreeMarkets's auctions over the past five years, says Mr Meakem, buyers have ended up paying on average 15–16% less than the previous purchase price.

What am I bid?

Given the frenzy to set up electronic marketplaces, a bit of scepticism is in order. The experience of financial markets strongly suggests that many of the 700 or so e-markets will soon be wiped out or merge, as happened with the multiplicity of regional stock exchanges in the early 20th century. However, whereas some electronic markets will be too puny to live, others have such powerful parents that they worry competition regulators. Covisint, set up (and spun off) by four of the world's biggest car companies, was initially stalled on competition grounds.

In fact, many of these marketplaces were established to provide a common trading standard, so that buyers and vendors can easily exchange information electronically. "It is a convenience for suppliers not to have to build 20 interfaces," says Ron Wohl, head of applications development for Oracle. The bold vision of some trading platforms is to assemble an entire industry, not into a supply chain, but into a network or – to use the most fashionable word in e-business – an "eco-system". For that, common standards are the essential first step.

Enter the eco-system

From supply chain to network

EVERY QUARTER, CISCO SYSTEMS hosts about 1,000 meetings with the top brass of companies around the world. "It used to be the technicians," says the company's Mr Daichendt. "Now it's usually the CEO." At the height of the Firestone crisis, Jac Nasser brought along the entire board of Ford. The same thing happens at Dell Computer. A procession of chief executives make the pilgrimage to Austin, Texas, to learn how to "Dell" their company.

They want to understand how the Internet can transform the management of their supply chain. The main thing they learn is the importance of sharing information. Suppliers benefit greatly when they can see their customers' production schedules and sales data, because they can then plan ahead for the volume and timing of orders. They can react at once, rather than waiting for news to trickle down. Something of the sort was possible before the Internet came along, but only if both supplier and buyer had installed expensive proprietary technology. The Internet (along with its associated applications) allows such communications to take place among many buyers and suppliers, big and small. It also makes information available simultaneously all the way along the supply chain. Once this happens (and it is only just starting), it becomes more appropriate to think in terms of a supply network than a chain.

The Internet has allowed a further refinement. Dell's suppliers know not only how fast Dell is using their components; they know what finished products customers are ordering. When a customer places an order by clicking on the company's website, the software immediately feeds the order into the production schedule, and can thus tell the customer, almost instantly, when the order will be ready for shipment. Once the order is in Dell's system, suppliers can see it coming and start making the appropriate parts. So the Internet turns the company into a sort of portal through which orders arrive for redistribution among suppliers. Dick Hunter, who is in charge of Dell's supply-chain management, explains: "We are not experts in the technology we buy; we are experts in the technology of integration."

In time, says Mr Hunter, "Information will replace inventory." As an

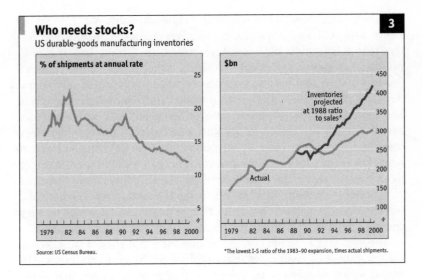

Who needs stocks?
US durable-goods manufacturing inventories

3

% of shipments at annual rate

25
20
15
10
5

1979 82 84 86 88 90 92 94 96 98 2000

$bn

450
400
350
300
250
200
150
100

Inventories projected at 1988 ratio to sales*

Actual

1979 82 84 86 88 90 92 94 96 98 2000

Source: US Census Bureau.

*The lowest I-S ratio of the 1983–90 expansion, times actual shipments.

example, he cites the suppliers who make the metal and plastic boxes for Dell's computers, mostly local firms with factories up to 90 minutes' drive away. They have access to up-to-the-minute information on Dell's stocks and its use of their products, and often keep a truck full of boxes waiting on Dell's site. The moment the first is unloaded, they send another truck. In their own plants, they keep less than a day's worth of finished stock. "If our information were 100% right," says Mr Hunter, "the only inventory that would exist would be in transit."

This point is worth pausing to think about, because it has large implications for the business cycle. Lee Price, chief economist in the American Commerce Department's Office of Policy Development, sounds rather like Mr Hunter when he describes inventories as: "A substitute for information: you buy them because you are not sure of the reliability of your supplier or the demand from your customer." But since the late 1980s inventories have been falling sharply, relative to sales, all over American manufacturing (see chart 3). A report for the Commerce Department in which Mr Price had a hand, "Digital Economy 2000", calculates that this has saved American companies some $10 billion a year – a cumulative $115 billion since 1988. Moreover, leaner inventories should reduce the ferocity of any future downturn. In the past, when demand grew a bit more slowly, inventories would often fall, amplifying a mild deceleration into a recession.

But back to Dell. Three things make the company's build-to-order approach easier. First, it is a relatively young company, which began in direct sales, so it does not suffer the conflicts among competing distributors that plague most old-economy companies. Second, Dell is a one-product company. Old-economy firms, with much more complex product lines, find it hard to integrate even their in-house supply operations, let alone those with myriad outside firms. Third, Dell (like Cisco) has remarkably few suppliers: 200 or so in all, with 30 companies accounting for about 78% of its total purchasing. Conventional large computer companies have about 1,000 suppliers. Dell tries to have at least two sources for every component, if only for safety's sake. Many of those sources are on the other side of the planet: it buys more than half its supplies from Asia. What matters is the closeness of the relationship, not the physical distance. But as Mr Hunter points out, "It would be very complex to have an intimate relationship with hundreds of suppliers."

The next step is to pass on the information that goes to suppliers to the second tier of companies that supply them. Texas Instruments sells digital signal processors to Solectron, which incorporates them into motherboards that Dell buys. Dell is talking to both companies to see how much of the data it passes to Solectron should also go to Texas Instruments, and is having similar talks with other suppliers. The company dreams of the day when all computer manufacturers who buy hard drives, and all suppliers who produce them, make that information available anonymously on an electronic exchange. That would allow the whole industry a clear view of the balance of supply and demand. It might even reduce the gluts that periodically plague the industry, as lack of information leads many companies simultaneously to take the wrong decisions.

For whom the Dell tolls

What benefits spring from collaborating? Dell makes much of its greater capacity to get technological innovations to customers quickly, and the fact that customers can specify exactly the machine they want. But the real gains are those inventory savings. A company building a product the material cost of which drops 1% every fortnight cannot afford to keep more stock than absolutely necessary. Compaq, says Mr Hunter smugly, may have 30–35 days' worth of inventory in its sales channels. "We have none." In total, the company holds about 140 hours of inventory (measurement in days is now old hat), and hopes to cut even that minuscule number in half over the next two years.

Better still, Dell collects the money from its corporate customers 30 days after shipment (or, for retail sales, on ordering), but pays its suppliers after 45 days. As a result, the company is in the delightful position of having what it calls "negative cashflow", which actually means money in the bank. That benefit will not last indefinitely: sooner or later, competition will make sure it is passed on in lower prices. But getting your customers' credit-card companies to provide your working capital is a trick worth knowing.

No wonder so many other companies hope to do the same. Among the most enthusiastic are the car manufacturers. At present, their customers face a bleak choice. In Europe, most cars are now built to order. In the United States, most cars are built for stock. Plenty of permutations are possible, but the customer sees only what the dealer ordered, two or three months earlier. These stocks of finished products clutter dealers' lots, tying up billions of dollars of cash; and yet customers still complain that they cannot find the car they want. A dealer who guesses wrong needs to persuade customers to buy to shift the stock and cut his interest charges, periodically wreaking havoc with margins.

However, applying Michael Dell's bright idea to Henry Ford's legacy is not easy. For one thing, American car companies do not go in for direct sales but have huge dealer networks. Brian Kelley, who came from GE to run e-commerce at Ford, thinks that dealers are an essential part of any new supply chain. "Most customers planning to spend $25,000 to $30,000 on a new product want to see and test it first," he argues. "Besides, 80% of people who buy a new vehicle have an old one to trade in." And cars need servicing from time to time. But Mr Kelley also sees Ford's dealers as delivery channels for a new venture called FordDirect, launched in August 2000. This allows customers to configure, select, price, finance and order a new car or truck through a website and then pick it up from a dealer.

Ford's strategy of deepening relations with intermediaries provides a model for the many old-economy companies that depend too much on physical distribution channels to want to abandon them overnight. Instead, they use the Internet to give intermediaries additional information, to bind them into the distribution channel in the way Dell and Cisco have bound suppliers into their supply network. Thus Honeywell, a computer company that has relied heavily on intermediaries to sell its products, has created myplant.com, a website that solves problems for managers of large industrial plants. As a strategy, this may not necessarily be second-best. Physical channels have some advantages over virtual ones. For example, Wells Fargo,

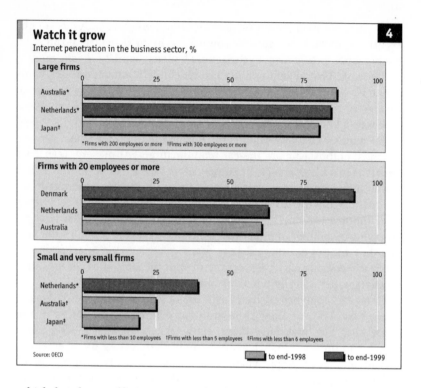

Watch it grow `4`

Internet penetration in the business sector, %

Large firms

Australia*
Netherlands*
Japan†

*Firms with 200 employees or more †Firms with 300 employees or more

Firms with 20 employees or more

Denmark
Netherlands
Australia

Small and very small firms

Netherlands*
Australia†
Japan‡

*Firms with less than 10 employees †Firms with less than 5 employees ‡Firms with less than 6 employees

Source: OECD

to end-1998   to end-1999

which has been offering Internet banking for consumers longer than just about anyone else, still has bricks-and-mortar branches, and finds that they do better at closing sales than anything offered online.

Most of Ford's biggest dealers have signed up for FordDirect; smaller ones, and those in less populated regions, have been much less enthusiastic. But winning dealers' support is only part of the challenge. A bigger problem is that Ford's cars are simply not built like Dell's computers. To take just one example, the colour of a Ford car is determined early in the production process, before the metal is even stamped. That is not the only way to build a car – DaimlerChrysler's Smart car, for its part, has clip-on side panels, allowing a dealer to change its colour in an hour. Eventually, Mr Kelley accepts, Ford will have to refit its plants so that they can build to demand, rather than "to sit on a lot". But, without a way to smooth out peaks and troughs in demand (a virtue of building for stock), capacity utilisation will fall and costs will rise.

The trick will be to make the manufacturing process more modular and less sequential: for instance, the basic platform will be built for stock and then turned into the vehicle the customer orders. Indeed, a change in the manufacturing process is at the heart of true build-to-online-order projects. Frank Piller, an economist at the Technical University of Munich who has made a special study of the way companies customise a basic product line (a process dubbed "mass customisation"), argues that the biggest change is usually in the design and construction of the product. "You need modular production that fits together like Lego blocks," he says.

Ford also hopes to use Covisint, and the common standards that the trading platform will create, to share its forecasts and its inventory information with its suppliers, both the front line and those further along the chain. As Mr Kelley freely admits, "It won't be simple to link the unconnected legacy systems and the internal workings of each of these very large companies." The putative savings, of $1,000 to $3,000 a car, will appear only if all the disparate systems can be joined up. "It will take years to play out," forecasts Mr Kelley.

Between the two extremes of Dell and Ford, many companies will search for middle ground. Hau Lee, director of the Global Supply Chain Management Forum at Stanford Graduate School of Business, sees them moving through four stages. First, there is an exchange of information, such as demand forecasts and sales data that allow better planning. Companies quickly realise that they need to define common standards for things like point-of-sale and inventory information, so that each can interpret the other's material. Next, companies move beyond data to exchange knowledge: for example, when Wal-Mart's Florida stores ran out of mosquito repellent during a heatwave, the company discovered that Warner Lambert not only made the repellent but tracked weather forecasts to spot future peaks in demand. By sharing the information online, both companies could do better.

At the third stage, says Mr Lee, companies exchange the right to take decisions. Often, it is enough for one link in a supply chain to make a move: for example, if Wal-Mart decides to stock more nappies for babies, there should be no need for Procter & Gamble, which makes the things, and for 3M, which supplies the sticky plastic tapes, to have three separate decision-making processes for a single product. So these companies are experimenting with a system that allows one person to reach the decision for all three. The last step is an exchange of work and roles. "The manufacturer becomes a retailer," says Mr Lee, "and retail moves to a support role." For example, companies such as VooDoo Cycles and Cannondale, makers of high-margin sports bicycles, are increasingly taking customers' orders direct and only then building the bicycles. But a costly bike requires much last-minute tweaking before it is fit for the road. Returns by dissatisfied customers are expensive for mail-order manufacturers. And customers picking up a new bicycle may well want to buy a new helmet or some lycra shorts. That is the new role – akin to Ford's repositioning of its dealers – that bicycle retailers may take on. One effect of the supply-chain revolution is thus to change the way companies manage their relations with customers.

The personal touch

Making customers feel special

How do you persuade people not to buy a car-insurance policy purely on price? Progressive, an insurance company based in Ohio, has an answer. If a policyholder has an accident, a claims officer goes straight to the scene, gives him a cellphone and a cup of coffee, pulls out a laptop and, in 95% of cases, hands him a claim cheque on the spot. Some customers say: "I wasn't a member until I was hit by one." The service, not the price, sells the product.

"It creates a 'wow' experience," says Joe Pine, an authority on such things: he published a book (with James Gilmore) in 1999 on *The Experience Economy*. The Internet, he argues, transforms the way companies have to manage relations with customers, partly because it is "the greatest force for commoditisation ever invented". But it also affects companies' relations with customers by cutting the cost of routine transactions, and by giving them new ways to reach and monitor those customers.

For the technology companies that have pioneered many business applications, customer support is one of the first opportunities for change the Internet provides. But the opportunity is to offer a service that is more of a commodity rather than less, by refining ways to deliver help online rather than over the telephone.

Companies will continue to offer telephone help: indeed State Street, a Massachusetts bank which uses lots of voice response to answer calls, is building a new call centre as well as developing the Internet. "Some customers simply want to talk to someone," admits John Fiore, the chief information officer. But the scarcer support staff become, and the longer telephone inquirers have to hold on, the easier it is to encourage customers to look up the answer to their problem online. And there are huge gains to be made from applying the usual 80/20 rule, says Ward Hanson of Stanford Graduate School of Business: if the staff answer only the complicated 20% of questions, their work is far more interesting and productive. ("The grey area", he adds, "is trying to hide the help-desk telephone number from the remaining 80%. I usually go to 'investor relations'.")

More savings are to be had if the customer can be trained to place orders online. "Accepting a simple order on the phone takes about four

Why companies need to get wired	5
% of 60 global companies	
Only 48% of firms know about a problem before a customer does	
Only 43% offer better service to profitable customers	
Only 42% would sell something during a service call	
Only 37% know if they share a customer with another division	
Only 23% of telephone agents can see customers' web activity	
Only 20% know if a customer has visited the website	
Source: Forrester Research	

minutes and costs around $5," says GE's Mr Reiner. "In our higher-tech businesses, it may cost $80. We get 20m telephone calls a year in our appliance business. If the order comes in online, it costs 20 cents." For the customer, too, Mr Reiner insists, it costs less to place an order on the Internet than on the telephone. But just in case the customer does not initially see things that way, GE offers to send out "e-mentors" with its sales staff to visit customers. It also gives sales people a bigger commission on orders that arrive online, so that they share some of the savings. The sales folk themselves know all about working online: in the past four years, the number of sales offices has fallen by half as more of them work on the road, from home or from their customers' premises.

Nor do the savings in online selling come only from lower transaction costs. Dell's Mr Marengi argues that the best thing about moving to online ordering has been to dispose of endless arguments over whether the customer ordered this keyboard or that keyboard. "If the customer puts in the order," he says cheerfully, "that conflict is eliminated."

By exploiting the Internet, companies not only gain opportunities for cost-cutting; they can also create new business opportunities by learning more about their customers (see chart 5). At present, such information tends to be strewn among many different databases, each attached to a particular product line or sales channel. By integrating the data, companies can present a single, coherent face to customers. That task is easier when customers shop online.

It may be valuable for a customer to have a single view of his transactions. "A company may know what it's buying in Boston, but not what its subsidiary is buying in Spain," says Mr Marengi. "Often a company will ask us what it bought globally, because its own internal systems can't tell it." But the seller benefits too. "The web allows companies to draw a graph of a customer's lifetime value," says Mr Hanson. "You can learn how customers end up in your fold, which are the best and the worst ones, and why some abandon their electronic shopping trolley before they make the final transaction."

One effect of such data has been to teach companies much more about the relative costs of acquiring and keeping customers. All of them promptly reach the same conclusion: "There has been a big shift from acquisition to retention," says Harvey Thompson, who runs customer relationship management at IBM Global Services. The reason is partly that a click of a mouse is the fastest way ever for customers to change providers. But companies now also have the tools to exploit what they know about their existing customers.

Among the first businesses to take advantage of those tools have been financial institutions. Wells Fargo, that experienced online bank, finds that attrition is one-third less for online than for comparable offline retail customers. For customers who make bill payments online, attrition is 54% lower than for offline customers. Online customers, says Avid Modjtabai of the bank's Internet services group, tend to keep higher balances, and seem to be much more likely to buy extra products, than the unwired ones. Simply knowing more about your customers makes it easier to keep them.

One obvious way companies could use that knowledge is to offer different customers different prices and levels of service. So far, most have hesitated to do so – or at least have hidden their attempts more carefully than did Amazon.com, which was savaged in 2000 for quoting different prices for the same book. But many companies are starting to understand that they cannot offer the same quality of service to everyone. They know that the true promise of customer data is to help them to discriminate, in service quality and perhaps in price, and to target their services so that they give priority to the most profitable folk on their books. They also know that this will not be easy to do. "I would not want one of my customers bumping into another and saying we had given them a better deal," says Cisco's Mr Daichendt.

But whose customer is it anyway? As companies tie their databases together, and try to cross-sell products to a customer who sees a single common front, they run up against a new version of an ancient problem: how to motivate one salesman in a company to hand over a customer to another. Without incentives to share customers, the most elegantly reconciled data in the world will make no difference. In most companies, the tyranny of the distribution channel will make customer-sharing a hard problem. "In large companies," observes George Colony, chief executive of Forrester Research, a high-tech consultancy, "tremendous political power has built up around these channels. It takes the CEO to force the breakdown of the walls."

Every company now claims to be, in that horrid but popular word, "customer-centric". In fact, most companies always said they were. But the Internet and its associated technologies allow companies to discover whether customers were aware of it, and if not, to do something about it.

From a commodity to an experience

One force for change will be the discovery of the commoditising power of the Internet. Once buyers can readily shop around online, or aggregate corporate demand, or put out tenders on electronic trading platforms, then companies that sell on nothing more than price will be in trouble, unless they are supremely efficient. What is the alternative?

One strategy, described by Philip Evans, co-author of a book with the ominous title of Blown to Bits, is to separate the information-rich part of the business from the commodity part, and sell them separately. He describes a manufacturer of industrial abrasives and drills who decided to split his business in this way. One part now specialises in long runs of standardised products. It has eliminated most of the salesforce, and abandoned research and development. The other part, which is far smaller in turnover but almost as profitable, employs engineers as consultants, solving problems with the "drilling solution". The first part uses the Internet to sell partly through electronic markets; the second, to connect engineers with their customers.

For consumer products, there is Mr Pine's experience-economy type of solution: turn a product or service into an "experience", and you defeat commoditisation. As examples, he cites a company in Minneapolis that repairs computers. It calls itself the Geek Squad, and sends round nerds dressed in white shirts and black ties in new VW Beetles or vintage cars. Or there is Steinway: when a customer acquires one of that company's top-of-the-range concert grands, the company offers to lay on a free concert in the buyer's house, providing a concert pianist, sending out the invitations and passing round the hors d'oeuvres. Mr Pine's own daughter favours American Girl Place, which makes character dolls. It has a sort of indoor theme park in Chicago where small girls and their parents can pay to watch a show about their doll, or pay again for lunch with a special chair for their doll to sit in. What they all have in common is that they add extra value to an easy-to-emulate product by throwing in a corny experience. This produces at least three gains: a higher margin, greater customer loyalty and local jobs.

With a bit of ingenuity, the technology that turns a manufactured consumer good into a commodity can also turn it back into an "experience". Levi, whose famous blue jeans now compete with cheaper copies, offers customers in its Union Square store in San Francisco the chance to be measured by a body scanner. Half an hour later a Levi factory has pulled their vital statistics off the Internet and begun to cut the jeans of their choice. Unfortunately, it then takes ten days to get the finished jeans to the customer. At present, the jeans cost 30–40% more to produce than a standard pair, mainly because as yet the company does not accept online repeat orders. But returns, which usually run to 40% of mail-order sales, are down to single digits. Moreover, the store has found that it learns from its customers when it sells to them this way: for instance, it noticed that those who designed their own jeans wanted them slung low on the hips, months before the average customer stopped buying "high-rise" jeans.

Levi's experience carries several lessons for companies that want to use the Internet to manage customer relations. One is that companies which once thought of every sale as a separate transaction will increasingly make money not from a first sale but from repeat business. That implies creating a continuing relationship with the customer, such as service providers usually enjoy. A second lesson is the importance of being able to involve the customer in development, design and market research. Richer communications make that possible. A third is the need for speed: the Internet, with its round-the-clock, round-the-week availability, raises customer expectations. Lastly, the relationship between factory and retailer changes. The factory may become the retailer's back office. Or, if the manufacturer has a strong brand and an accumulation of customer data (including, in Levi's case, their hip and inside-leg measurements), then the manufacturer becomes a portal for the retailer. Either way, the close contact with the customer fostered by the Internet is the most valuable commercial advantage a business can have.

The shape of the new e-company

Who runs what, and for whom

IN 1998, NORTEL NETWORKS, a Canadian company that specialises in building high-performance Internet networks, took a revolutionary decision. It would move from vertical integration to "virtual" integration. Since then, the company has sold 15 manufacturing facilities around the world that make things like printed circuit-boards. The plants have been bought by large manufacturers such as Solectron, SCI and Sanmina, which were already selling to Nortel and have now signed new long-term supply agreements with the company.

The benefits, says Chahram Bolouri, president of global operations, come partly in the form of lower costs, because these contract manufacturers have a far larger turnover than Nortel alone would have had. They can also afford to keep track of the fast-changing manufacturing technology of the particular components they produce, and invest heavily in their development. In addition, Nortel gains flexibility: if it has a large order from a particular part of the world, it can more easily arrange production nearby. Most important of all, the company can specialise in what it does really well: it has retained the highest-value part of the manufacturing process at seven "systems houses" worldwide, staffed with people skilled in industrial and test engineering.

The reorganisation has also changed the way Nortel deals with its suppliers. Three years ago [1997], says Mr Bolouri, it concentrated on buying; now it deals with technology and planning. It employs a different bunch of people: fewer paper-pushers, more industrial engineers. They spend much of their time talking, not only to the first tier of suppliers, but also to the second and third, about the main constraints in meeting demand from a particular market or consumer, and how they can be eased. Particular teams concentrate on the needs of particular customers, such as WorldCom or Cable & Wireless.

Thus rearranged, Nortel can move much faster than in the past. In the mid-1990s, when it owned most of its suppliers, an order might take up to three months to fulfil. Today, orders for some products take days, and soon that will be hours. Using a newly created Internet exchange, called e2open, the company can circulate an order instantly to a galaxy of 60 potential suppliers. Mr Bolouri devotes most of his time to recruiting and keeping high-quality talent to manage the supply chain, and to making sure that everyone along the chain is kept constantly up to date on what the company is trying to do.

As the Internet becomes built into corporate life, the economic foundation of the company changes. In an essay on "The Nature of the Firm", published in 1937, Ronald Coase, an economist who later won the Nobel prize, argued that the cost of transactions determined the boundaries of firms, making it more efficient for workers to band together in a company than to operate as separate agents. The impact of the Internet has been to reduce those costs. Because almost everything can be inexpensively outsourced, it is possible to create a company from nothing in no time: to go from idea to product in nine months. Many Internet start-ups are, in the neat phrase of Stanford Graduate Business School's Mr Saloner, "plug-and-play" companies.

Many of the barriers to entry that once protected big companies are therefore disappearing – or at least some parts of established companies are vulnerable to being "blown to bits". One effect, as Nortel's case demonstrates, is a move away from vertical integration, as the value chain is broken up into more specialised firms. In that respect, says MIT's Mr Brynjolfsson, we are seeing the substitution of hierarchies by markets.

Competition may not challenge all the things a company does – just the most profitable parts. Brendan McLaughlin, head of e-business at Cambridge Technology Partners, a high-tech consultancy, has a story about how he told one of his clients, a manufacturer of tapered roller bearings, about a Scandinavian firm that made nothing but replacement tapered roller bearings, and monitored other companies' production lines over the Internet to check when the parts were needed. "Our client's jaw dropped. 'But we make all our money from replacements,' he gasped."

Plenty more such competition lies ahead. At Harvard Business School, Bill Sahlman, professor of business administration, observes: "Our students go systematically, SIC code by SIC code, through industries, looking for ones to revolutionise." Moreover, start-ups are starting to understand their intrinsic weaknesses, and to evolve business models that cure them. Julian Lighton, in charge of corporate networks at the Silicon Valley office of McKinsey, a consultancy, notices that the incubators in which some start-ups begin life are becoming "semi-permanent *keiretsu*", or loose confederations, to share the costs of acquiring and managing customers and talent. One incubator, ICG, has a common recruiting department with 20 staff to help its offspring. "Shared talent management" is the heart of these confederations, because the scarce talent of Silicon Valley prefers to work for a network of small companies that pool job opportunities.

Dotcom or dot.corp?

Big, established companies often find it hard at first to respond to such competition. Not every company is as bold as GE, which in 1999 ran an exercise called "destroyyourbusiness.com" to force managers to consider where they were most vulnerable to unexpected competition delivered by the Internet.

Most old-economy companies initially choose not to re-engineer their whole business, but rather to spin off a new division to experiment. IBM's Mr Martinez sees three possible models. One is that of Reflect.com, a start-up by Procter & Gamble, a consumer-products giant based in Cincinnati, Ohio, which sells customised cosmetics over the Internet from the safe distance of San Francisco. That avoids nasty conflicts with existing distribution channels, and gets the expense (and possible losses) safely off the balance sheet.

A second model is that pursued by Staples, a successful office-stationery business, which has created Staples.com as a separate business, but kept the links close. Staff at Staples have options in the dotcom's tracking shares, and the dotcom folk have options on Staples shares. The two businesses share a building in Boston, to make sure that each learns from the other. Staples's own share price has been hit because the company has had to write off this investment as an ordinary operating cost. One of the great oddities of this revolution is that whereas investments in physical assets can be capitalised and depreciated, investments in intangible assets count as expenses.

The third model, says Mr Martinez, is that of IBM itself, where the company has decided that it cannot afford to spin anything off. The firm itself becomes a dotcom – or rather, as Forrester's Mr Colony quips, a "dotcorp". This is hardest of all to pull off, so not many companies try it.

One trouble with spin-offs is that they waste a company's scarcest resource: good people. "That's why I tell companies to take all the juice and put it in one place," says Mr Colony. Another problem is that spin-offs are odd animals. The business model of the genuine dotcom is usually to grab the money and run. For the purchaser, it is a way of outsourcing innovation: the start-up takes the initial risk, and the big firm buys the experiment that has succeeded. But if a big company spins off its most innovative bits, it may lose the chance to recreate itself.

The alternative, for old-economy companies, will be far-reaching change. As Mr Saloner explains, many companies resist innovation partly because it seems expensive, and partly because they cannot bear to destroy their existing, successful business model. "We are designed to do what we do really well," they say. What they need to remember is that this is a transitional stage. It is good news for business schools, he points out: "The demand for executive education is going through the roof as chief executives say, 'The top 300 people in my business just don't get it.' But these people will be needed again, and they know the business."

So how should companies manage this period of dislocation? No easy answers, but the next section offers ten basic rules.

How to be an e-manager

Ten handy hints

ACROSS THE DESK of anybody writing about management these days pours a torrent of books about running an e-business. Most start off by saying that everything is different – and then talk as though everything was much the same. It is true that the Internet changes the skills required from managers, but not fundamentally so. Anyone who is a good manager can also become a good e-manager.

However, some qualities have become even more important than they used to be. Here, for any manager too busy wrestling with the Internet economy to plough through the literature, are the top ten things you need.

1 **Speed.** The list could, perhaps, stop right here. Being quick is more important than being large – indeed, large companies find it hard to be speedy. "There are very few things that the Internet slows down," reflects MIT Media Lab's Michael Schrage. "Companies that take three or four months to reach a decision find that others have redesigned their websites in that time." Production cycles grow shorter; consumers expect service around the clock; companies do things in parallel that they would once have done sequentially. One way to be speedy is to avoid big-bang decisions. Internet-based technology can help. At Oracle, Gary Roberts, head of global information technologies, points out that Internet applications tend to be smaller than yesterday's proprietary systems, and the software is faster to develop. But speed is also a matter of a company's decision-making processes. Bureaucracy is a killer.

2 **Good people.** Human beings are the most important of all corporate inputs. Companies need fewer but better people: "celebrity teams", as Novell's Eric Schmidt puts it. Employees with new talents, skills and attitudes must be made to feel at home. Completely new jobs have sprung up in the past three years: content manager, information architect, chief e-business officer, chief knowledge officer. Companies need new ways to hire and – trickier – retain these people. They also need new ways to measure their performance.

3 **Openness.** The open nature of the Internet drives its success. The economic rewards that come from belonging to a large network will

ensure that the new standards that emerge will remain open. In addition, as the Paris-based OECD pointed out in *The Economic and Social Impact of Electronic Commerce*, a prescient study published in 1999, "Openness has emerged as a strategy." Many e-businesses allow their partners, suppliers or consumers an extraordinary degree of access to their databases and inner workings. To allow another business inside the corporate machine in this way requires trust, and a willingness to expose your weaknesses and mistakes to the world.

4 **Collaboration skills.** The Internet creates many new opportunities for teams and companies to work together. Only as companies learn new ways for their own people to collaborate do they begin fully to realise the opportunities to work with customers, suppliers and partners. Teams may be separated by time zone or by geographic distance, or they may work for different employers: the spread of outsourcing means that companies manage many more alliances. This calls for a different approach from that required to manage competition.

5 **Discipline.** Can this go with creativity and openness? It has to: "The Internet is all about discipline, protocols and standard processes," insists United Technology's Kent Brittan. When a software program replaces human action, the garbage-in, garbage-out principle applies. Unless companies carefully specify the parameters of a procurement order, for example, it makes no sense to invite tenders in an electronic marketplace. Companies need to insist on a standard look and feel for their websites to avoid confusing customers; and they need to insist on common practices within the company on such issues as purchasing to reap real productivity gains from the Internet.

6 **Good communications.** Given the pace and complexity of change, communicating strategy to staff matters more than ever. Few grasp the Internet's breadth of impact. Communications can no longer be confined within the company, or even within the country. What a company thinks of as external information can turn into the internal sort, and vice versa.

7 **Content-management skills.** All those websites that companies design to reach their staff, their customers or their corporate partners almost always start off by carrying far too much information. Companies are not used to being content providers, and the people who know most about the subject on the site frequently do not, or cannot, manage the site. IBM's Pete Martinez recalls asking the manager of one of his company's intranet sites who its audience was, and what they needed to know. "We took 80% of the information off the site, use rose 3,000%,

and the cost of running it fell dramatically." Many corporate managers are simply not used to expressing themselves clearly and concisely.

8 **Customer focus.** New opportunities have opened for companies to deepen their relations with customers. The emphasis has shifted from recruitment to retention, from the commodity to the service and from the mass market to the personalised. Companies are concentrating less on product and process management and more on the customer, treating each as an individual and trying to provide him with precisely the product he wants. This shift, made possible by enriched communications, is altering the whole shape of many companies. On the organisation charts that managers love to draw, the long shapes of product-related "silos" are now criss-crossed with a matrix of lines of functional responsibility. An executive in charge of retail banking or light trucks, for example, might also be in charge of monitoring fulfilment across the business.

9 **Knowledge management.** The communications revolution has raised the importance of pooling the skills and knowledge of a workforce. The development of sophisticated databases and intranets makes it possible for companies to build a core of knowledge that they can draw upon across the globe. But this is not easy. Managing workers of this kind requires a new sensitivity. Getting intelligent people to share what is in their heads takes more than mere money or clever software – although both can help.

10 **Leadership by example.** Plenty of bosses, especially in Europe and Asia, do not know how to use the Internet, and wear their ignorance as a badge of honour. But chief executives who have never done their own e-mail, or bought something online, or spent an evening or two looking at their competitors' websites, are endangering their businesses. "Top-level management must spend real political capital to create an e-business," insists Forrester's George Colony. This is unlikely to happen if they have no first-hand experience of what the transformation is all about.

Armed with these ten essentials, old-economy managers should see the challenge ahead for what it is: the most revolutionary period they have ever experienced in corporate life. It will be frightening and exhausting, but it will also be enormously exciting. It may even be fun.

The material on pages 151–98 first appeared in a survey written by Frances Cairncross in *The Economist* in November 2000.

V

ONLINE FINANCE

The virtual threat

The Internet has already forced wrenching change on the financial-services industry – and the revolution has barely begun

THE MOST REMARKABLE thing about the effect of the Internet on the financial-services sector is not how pervasive it has been; it is how limited a transformation it has so far wrought. Financial institutions, after all, deal in a product – money – that for many of their customers has long been "virtual". Bank-account holders are used to the notion that their cash is represented by a series of numbers on a monthly statement generated by a computer, or by the glowing green figures of a cash machine. And they have become accustomed to making payments using pieces of plastic backed with a clever magnetic strip. The Internet might have been designed for the distribution, monitoring and management of this ubiquitous electronic commodity.

More worryingly for the firms that make their living out of arranging financial transactions, the Internet might also have been designed to do away with them. Banks and other financial firms are intermediaries, standing between lenders and borrowers, savers and spenders. For decades, banks in rich countries have been fretting about how to cope with "disintermediation": lenders dealing direct with borrowers (as many do already in the capital markets), without using a bank's balance sheet to add a layer of cost. The Internet is, potentially, the greatest force for disintermediation the banks have ever had to tackle. Other intermediaries, such as retailers, face the same problem. But money, unlike, say, an item of clothing, is a commodity that can actually be used, transferred and delivered electronically.

Samuel Theodore, of Moody's, a credit-rating agency, believes the banks are currently undergoing their "fourth disintermediation". The first involved savings, and the growth of mutual funds, specialised pension funds and life-insurance policies at the expense of bank deposits; the second saw the capital markets take on some of the banks' traditional role as providers of credit; in the third, advances in technology helped to streamline back-office operations. Now, in the fourth stage, the distribution of banking products is being disintermediated. This process has been going on for some years, with the spread of automated teller machines (ATMs) and, over the past decade or so, telephone banking and PC-based proprietary systems; but the Internet hugely enlarges its scope.

Spotty youth

Yet, except for one activity, share-trading, and one part of the world, Scandinavia, Internet-based financial retailing is, if not in its infancy, then scarcely at puberty. And wholesale banking, although it relies heavily on complex electronic trading systems and information technology, is still conducted mostly on closed proprietary networks. To be sure, there are some signs that the disintermediation the industry fears may be starting. Internet banks, with their low costs – and their dotcom habit of paying more attention to the acquisition of customers than the turning of profits – have drawn deposits away from offline banks in some countries. And in the capital markets, bond issues and share offerings have been syndicated and distributed over the Internet. Some highly rated borrowers have for years been borrowing through their own issues of commercial paper. The Internet can only enhance the appeal of do-it-yourself fund-raising.

But these are just the early signs of an upheaval that is gathering momentum by the day. There are a number of reasons why many online financial services have been slow to catch on, and why they can now be expected to develop faster. Concerns about the security of Internet transactions, a particularly important issue for financial dealings, are gradually being eased. Internet use, even in the rich world, has been patchy, but is spreading fast. And whereas conducting financial transactions online up to now has often been clunky and annoying, the technology is improving all the time. Those technological advances are also liberating the Internet from the confines of the PC.

Most important, financial institutions themselves, which in the past have often resisted change, may now become its most ardent promoters. Having invested heavily in their own systems, banks were understandably reluctant to jettison them for web-based replacements. And adapting their own processes for the Internet has often proved cumbersome and difficult. Moreover, until recently banks faced little pressure from their customers to change what were seen as useful but boring services, much the same as electricity and gas. But soon, in many countries, customers will expect an online service as a matter of course.

The banks' staff, too, have been reluctant to abandon the old ways of doing things. Besides, those old ways have often been extremely profitable, so change threatens not just working habits, but the bottom line too. Now, however, almost every financial firm, from the swankiest Wall Street investment bank to the provider of microcredit to the very poor, has found that it has no choice but to invest in an "Internet strategy". And having invested in it, it will need to persuade its customers to use it. So in areas where the advantages of doing business online may not be obvious to the consumer – notably in retail banking – the banks may find themselves trying to coax, bribe and bully reluctant customers online.

The banks' conservatism, on which they used to pride themselves, has become an embarrassment. It has also been spotted by the new breed of Internet entrepreneur taking aim at the banks' business. The models are firms such as E*Trade and Charles Schwab, discount stockbrokers that found in the Internet a means of challenging even the biggest and most prestigious traditional firms. Now commercial and investment banks, fund managers and financial advisers are all vying with each other to present themselves as Internet-savvy, and boasting about their investment in online services.

All this has created a strange, contradictory world. Clever young things with a bright idea and a few million dollars of venture capital behind them talk cheerily of the demise of traditional banks. Bill Gates, no less, said several years ago that banking is necessary, but banks are not. Now, the story goes, they are irredeemably hampered by their "legacy systems" – their existing management structures, staffing levels and computers – and by their "channel conflicts" – between what they do now, and online methods of sales and distribution. Their bosses simply do not "get it". Or, even if they do, their institutions are so deeply rooted in the old economy and pre-Internet styles of business that there is no point in turning them around.

The dinosaurs in the supposedly stuffy offices of these big banks and securities firms appear unaware that a meteorite may be on its way to obliterate them. On the contrary, resolutely upbeat online-service managers, often rather self-conscious in their tieless, suitless new-economy uniforms, claim they are having the times of their lives. Never has technology revealed so many new avenues for developing the business. It is, says Denis O'Leary, who runs Chase Manhattan's Chase.com, "a golden age".

Not least because, in the industrialised West, many firms have been making bigger profits than ever. Years of economic expansion and bull markets have yielded good income from traditional lending, from trading and from investment. The only obvious cloud in the sky is that banks' share prices seem not to reflect this (see chart 1). Indeed, in some

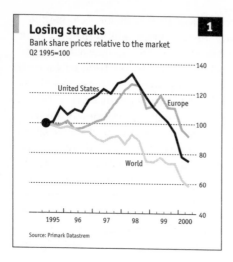

Losing streaks [1]

Bank share prices relative to the market
Q2 1995=100

Source: Primark Datastrem

countries, such as Britain, they imply that the market expects banks' profits to collapse in the next few years. Even the stockmarket seems to believe the dotcom wannabes, and rewards them with much richer valuations than boring old-economy banks.

Still kicking

And yet this survey will argue that many of the older institutions have a good story to tell. The "legacy systems" at which the upstarts scoff have one big virtue: they have tended, by and large, to work. Big banks process trillions of dollars a day. It is almost inconceivable that they might close down for a few hours because some clever Internet saboteur has found a way of snarling up their technology (as has happened to some of the biggest websites). Existing banks have customers in numbers that newcomers can only dream of, and even unpopular incumbents benefit from their customers' inertia.

The Internet also brings established firms huge opportunities as well as threats. To take two important examples, it offers ways of cutting costs and of marketing products much more efficiently. For years, in America, Europe, Japan and elsewhere, the industry has been consolidating: bank after bank has been taken over by or teamed up with an institution in a complementary line of business. Usually, these deals are justified to shareholders by the extra returns that can be generated once overlapping costs are stripped out. The Internet, potentially, offers a way of taking a knife to whole layers of costs. Once a customer is convinced to carry out most of his transactions online, his account becomes much cheaper to administer.

The other much-cited benefit of consolidation is "cross-selling" – of insurance policies to bank-account holders, for example. Yet so far this has rarely been all that successful in practice. The Internet can be a precision-guided marketing tool. For example, if you apply online for a credit card from NextCard, an American Internet operation, you will be offered a choice of three charging structures. To qualify for the most favourable, you have to transfer a certain outstanding balance from your other credit cards. That sum will – fancy that! – be the actual total of your other balances, which NextCard has just ascertained online from the credit bureaus. Or, in wholesale finance, suppose you are a potential investor in a company's initial public offering of shares, and have just finished watching the boss boosting his company's prospects on Merrill Lynch's online investment-banking service. The phone rings. And yes, it is a Merrill Lynch salesman who knows you have been watching, and thinks that now may be the moment to clinch a sale.

But, for banks, each of these pluses comes with a minus. Because costs are so much lower for Internet-based transactions, the barriers to entry are lower as well, which implies that margins will come under pressure. And although the Internet makes well-directed sales pitches easier, that is hardly compensation for the precariousness of online customer relationships. Once your client is on the Internet, he is only a mouse-click away from your competitor, and more and more financial sites, search engines and portals will be pushing competing products at him. That, too, will squeeze margins.

Viewed from this perspective, for many financial institutions the Internet is a double bind. Embrace it, and you may still find yourself losing business, or at least seeing profit margins dwindle. But ignoring it could be terminal. This survey will argue that the pressures for change have become irresistible. It concentrates on places where the process is most advanced – America and Europe – but the same lessons apply everywhere. Big financial institutions are global firms. And on the Internet, change spreads like wildfire. The stockmarket with the highest proportion of Internet trading is not, as you might think, in New York, but in Seoul.

In public, no bank boss these days would admit to anything less than whole-hearted enthusiasm for the online adventure. In private, however, some still see it as just another distribution channel, perhaps less important than others, such as the telephone. A few still cling to the dream that it is a fad they have to indulge because their shareholders seem to like it. Even such non-believers, however, are being forced by the market to formulate an online strategy. If they are too slow, or get it wrong, the consequences for their firms could be deadly. And if they still need convincing, they need only look at what has happened, in just a few years, to stockbroking.

Going for brokers

Online stockbroking is one of the Internet's big success stories

ALONG WITH E-MAIL and pornography, investment is one of the "natural" uses of the Internet. So thinks Daniel Leemon, chief strategy officer at Charles Schwab, a San Francisco-based stockbroking firm. Schwab should know. It boasts of maintaining the World Wide Web's largest encrypted site, and of doing more business over the Internet than any other firm, anywhere. Of its nearly 7m customer accounts, more than half are online and "active", in the sense that the holders have visited in the past few months. They contain about a third of the $700 billion-plus in assets sitting in Schwab accounts. Early in 1999, Schwab's stockmarket valuation overtook that of Merrill Lynch, which at the time controlled three times as much in customers' assets.

Schwab is by no means a "pure" Internet broker. It had a large offline business in pre-Internet days, and still has more than 350 branches. It also offers its clients both touch-tone and speech-recognition telephone services. But it has been in the forefront of a charge online that has already transformed stockbroking in America, and is rapidly doing so in Europe, Asia and elsewhere. The growth has been phenomenal. E*Trade, for example, an online broker, began Internet trading only in 1996. Now it claims to be the world's "most-visited online investing site". In 1999 its number of customer accounts nearly tripled. By the end of March it had reached 2.6m.

The online-broking market is now fiercely competitive. It is still dominated by discount brokers, led by Schwab and E*Trade, and including rivals offering even cheaper services, such as Ameritrade, TD Waterhouse and Datek. For years, the big Wall Street firms either snootily ignored the sniper fire from the Internet, or were openly hostile. In 1998, one of Merrill Lynch's bosses famously called online-trading firms "a serious threat to America's financial lives", and reassured Merrill's thousands of brokers that his house would never go down that path.

We didn't mean it

Now Merrill and all the other heavy artillery – the big old banks, securities houses and investment firms – are rumbling to the front as fast as they can. Merrill is offering its clients online dealing at a price of $29.95 a trade, compared with the $100–400 that clients of full-service brokers have been accustomed to forking out. Merrill's online clients have access to some of its research, and can watch videos of analysts' daily briefings. Fidelity, the world's biggest fund-manager, is also making a big push. It has renamed its online broker "Powerstreet" to make it sound more exciting.

This competition is changing the online brokers. The market leaders, such as Schwab and E*Trade, are now under pressure from both ends of the market. In 1999, despite enjoying a 48% rise in customer assets and a 44% increase in revenue, Schwab's share of the online market by assets fell from 28% to 23%. Further downmarket, firms are vying for the business of the most active investors, the famed day-traders. In response, in 2000 Schwab acquired CyBerCorp, a day-trading broker. It also cut its standard commission in half, to $14.95 a go, for investors who trade more than 60 times a quarter. Further upmarket, the threat comes from the traditional houses with their established brand names, fabulous resources and global reach.

What began as a battle largely about the cost of individual transactions has become one about the quality of the technology – especially its ability to deliver rapid and efficient execution of trades – and of the ancillary services the brokers offer. Schwab, for example, has dropped its old squeamishness about giving sharebuying advice. It used to distinguish itself from the longer-established houses by the independence of its stance, accusing the older firms of conflicts of interest between their analysts' share tips and their investment bankers' hunger for deals. Schwab will now give advice, so long as it is "unconflicted". It has also, like E*Trade, taken a stake in an investment bank.

What else can you do?

There is, of course, a mind-boggling array of financial data, newswires, advice, gossip and so on available on the net at no charge at all (and some of it extremely dangerous, see "Rampant abuse" on page 213). So the brokers are having to expand their services in other ways too. These may include, for example, alert systems, which will inform the customer by e-mail, or message to a mobile phone, pager or PDA, when a particular share price has reached a trigger level. The brokers are also competing by offering banking and money-market accounts. Schwab, for example, already provides an electronic bill-payment service. In 1999 E*Trade bought Telebank, a telephone and Internet bank, and is now offering a current account on which it pays interest, bill payment and (within limits) free ATM use.

In any forecast of likely winners in the Internet-finance stakes, firms such as Schwab and E*Trade are likely to figure prominently. They have the customers, the name-recognition, and the ability to expand energetically into new areas of business. However, all is not rosy even in this booming corner of cyberspace. E-brokers have spent heavily on advertising. In the Internet world, such splurging on marketing is seen as an essential "customer-acquisition cost", which will one day be translated into profits.

For example, E*Trade, whose television advertisements graced the commercial breaks in American football's "Superbowl" in 1999, reported that in the first quarter of 2000, its acquisition costs per net new account were $256 (down from $294 in the previous quarter). This was, it notes, one of the lowest levels in the industry, but it is still an expensive way of adding business. In those three months, E*Trade just broke even on net revenues of $407m. To stay in profit, each new account will have to be highly profitable, and stay with it for some time. Yet so fierce is the competition that pricing levels for online broking may, for a while at least, be slashed to madly loss-making levels, enticing active traders to hop from broker to broker.

It is a myth that online-brokerage accounts are held mostly by hyperactive day-traders. Schwab's analysis of its own customer base suggests that they are a tiny minority. Mr Leemon thinks day-traders will be "a footnote in the history of online investment". It is true, however, that online accounts tend to be more active than others. Research by Fidelity's Powerstreet suggests that the group of investors it calls "pioneers" (those who have been trading online for more than three years) make, on average, 6.5 trades a month. By contrast, 57% of Fidelity's "traditional" account-holders had no plans to deal at all in the year ahead.

The next volume

This leads to another worry. Online broking is a volume business that has matured during a long and, at times, roaring bull market. The surge in technology and Internet shares in the second half of 1999 and early 2000 pushed up online trading even further. As Credit Suisse First Boston (CSFB) noted in a research report, this entailed levels of account activity that are "unsustainable, and will inevitably lead not only to slowing transaction volumes but also to sharp drops from existing levels". If online brokers are losing money even at times of market frenzy, what might happen in calmer periods, let alone a prolonged bear market?

Of course, many of the factors that have encouraged online trading will continue to apply even in a falling market: most obviously, that it offers significant cost savings over traditional trading methods. There are other, more psychological, attractions too. Schwab quotes research in America in which half the people surveyed said that talking to a broker about their investment options made them feel ignorant. The web, on the other hand, is a wonderful place to ask stupid questions, or to make blunders in apparent privacy.

Broking for goers 2		
Online equity trading, Q1 2000		
	Average daily trades, '000	Market share, %
Charles Schwab	245.7	19.9
E*Trade	201.5	16.3
TD Waterhouse	156.8	12.7
Ameritrade	131.2	10.6
Datek Online	121.3	9.8
Powerstreet	120.0	9.7
DLJdirect	45.3	3.7
Brown & Co.	39.0	3.2
ScoTTrade	32.8	2.7
Suretrade	23.3	1.9
Others	117.9	9.5
Total	**1234.8**	**100.0**

Source: Hambrecht & Quist

Schwab's Mr Leemon explains his firm's success by its appeal to the baby-boomer generation, which he says is unwilling to compromise and needs to feel "empowered" and in control. As he puts it, a baby-boomer confronted with an easy-to-use Apple computer is not lost in admiration for the clever people who designed and built it. He thinks: "I am a computing genius!" Online broking can turn them into investment geniuses as well.

And not just American geniuses. From a later starting point, online investing is also showing spectacular growth in other countries. In parts of Europe, for example, it is growing faster than in America, even though far fewer adult Europeans own shares (35m, or 12% of the total, compared with 104m Americans, or 50% of the adult population, according to estimates by J.P. Morgan, an investment bank). In January 2000, Datamonitor, a research outfit, estimated that an average of 466 new online accounts were being opened in Sweden every day, 685 in Britain, and 1,178 in Germany. It forecasts that the number of online-brokerage accounts in Europe will reach 7.5m in 2002. J.P. Morgan puts the total even higher, at 10.5 m. It estimates that in 1999 the number of specialist discount online brokers in Europe increased from about 20 to over 50, and the number of their customers more than doubled.

More than half of these were in Germany (see chart 3), where in the final quarter of 1999 the four biggest discount brokers already accounted for 13% of all stockmarket transactions – a proportion not far short of

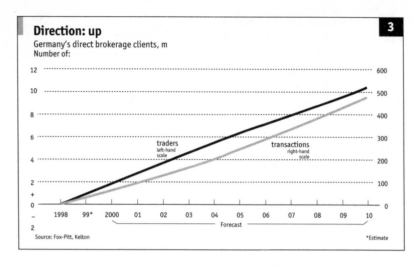

Direction: up

Germany's direct brokerage clients, m
Number of:

Source: Fox-Pitt, Kelton   *Estimate

that in America (about 16% by number of transactions). Measured by the number of executed orders, three of the four biggest Internet brokers in Europe are German, led by Comdirect, owned by Commerzbank, and ConSors, 73% owned by Schmidt, a small bank. It seems puzzling that Germany should have taken such a lead over, say, Britain, which has more Internet users (12.5m compared with 10.4m), and, since the privatisation splurge of the 1980s, a much higher level of share ownership (25.1% of the adult population, compared with 7.1% in Germany).

Online trading is growing at a tremendous rate in Britain, too. In 2000 about 10% of all trades on the London Stock Exchange (LSE) were being conducted online, up from almost none a year earlier. But some brokers think that growth in Britain could be faster still, and is being held up by some of the peculiarities of the British market. The securities industry as a whole is united in blaming high stamp duty (0.5%) on individual share transactions as a serious brake on the growth of the retail market. There are other complications; for example, trades in about 150 shares known as "residuals" are not settled electronically, but by the physical movement of paper, which can take weeks or even months. According to James Marler, boss of E*Trade's British operation, the residuals include some of the small, high-growth technology companies favoured by online investors, so this too is slowing the growth of the market.

One reason for the even faster growth of Internet broking in Germany may be the inchoate nature of the country's equity culture itself. Share ownership and Internet penetration seem to be taking off in tandem. In America the growth of online broking has been led by the migration of existing investors from other channels. Discount brokers such as Schwab had several years of experience in drumming up business, and employees' 401(K) pension plans encouraged individuals to manage their investments more actively. In Germany, says Gerrit Seidel, of the Munich office of Arthur D. Little, a firm of consultants, online brokers have "created a new market" rather than merely convincing old customers to conduct their business in new ways. Their success at attracting online customers may have surprised even the brokers themselves. As everywhere, the new technology has had its teething problems, including systems failures at busy times.

As Internet usage and individual share ownership continue to rise across Europe, so will the competition for online-brokerage accounts. Already the multinational brokers have moved in. These include the Americans, such as Charles Schwab, E*Trade and DLJdirect, which see Europe as a large potential growth area, both in its own right and as a part of the putative round-the-clock global market most participants expect to emerge in a few years' time. Within Europe, e-cortal, the largest French Internet broker, has launched a pan-European service; Britain's Barclays Stockbrokers has started operations in France; and ConSors has bought small brokers in France and Spain. Unofirst, the result of a recent merger between a Spanish and an Irish-based Internet bank, also plans to offer low-cost Internet broking.

Just the job?

Perhaps the biggest potential for growth in online broking lies neither in Europe nor in America, but in Japan. At present it lags not only the rich western countries, but also regional neighbours such as Taiwan and South Korea. Both those markets are dominated by retail investors and enjoy a strong technological infrastructure. South Korea has also been boosted by an extraordinary bounce-back from economic near-collapse over the past two years. Taken together, these factors have given it the highest proportion of online trading in the world, about 30% of stockmarket turnover.

There seems no obvious reason why Tokyo should not start to catch up with Seoul and Taipei. Indeed, there are a number of grounds for expecting it to (see "Japan's online adventure" on page 214). On October 1st 1999, the Japanese stockmarket underwent its own "big bang", which ended all fixed brokerage commissions and opened the field to new discount brokers. Internet penetration in Japan, at more than 10% of households, is high and rising fast (though still far below American levels). Japanese consumers are quick to adopt new technologies, and Japan has a marked lead in two of them – mobile telephony linked to the Internet, and Internet-games consoles – that will be vehicles for online investment. Traditional stockbrokers have a poor reputation, making it easier for new competitors to poach their business. And few Japanese workers feel free to trade shares at the office, adding attraction to any system that allows investors to do so at home.

Most tantalising of all, there is the huge mountain of Japanese personal savings, estimated at ¥1300 trillion ($12 trillion), much of it stacked up in the postal savings system, and ¥106 trillion of which is expected to mature over 2001–02. As these deposits made at relatively high interest rates a decade ago mature, the post office will be able to offer only nugatory yields on reinvested or new deposits. Some of this money is expected to find its way into the stockmarket, so the financial industry is scrambling for a chunk of the assets.

But there are big obstacles to the spread of online trading. Many individual investors in Japan have an acute aversion to risk, which suggests that a big part of that pile of savings may stay in low-yielding but safe deposits. Moreover, regulation has failed to keep pace with technological change. At present, every single online transaction is duplicated offline. The rules recognise only paper and seal. So every share trade generates a paper confirmation, and prospectuses for, say, bond issues or investment trusts have to be distributed by post or fax as well as on the Internet. Changes to the law and its interpretation are under way. Masamichi Kono, of the Financial Supervisory Agency, the main regulator, claims the FSA is moving towards "accommodating and encouraging" online trading. For frustrated online brokers, however, prevented by current practice from making large cost savings, change is too little and coming too late. They worry that online stockbroking will become one more story of Japan's unfulfilled financial potential.

Rampant abuse

FOR ONE SORT of participant in the financial market, the Internet is indeed just another distribution channel, and a huge improvement on the older models. Fraudsters love the anonymity, the reach and the interactive give-and-take of Internet newsletters or chat rooms.

America's market watchdog, the Securities and Exchange Commission, set up an "Office of Internet Enforcement" nearly two years ago. Matthew Moro, the office's deputy chief, says it receives about 300 referrals a day, but has not so far come across anything that would need new regulation, just the "same old frauds in a new guise": attempts to raise money for bogus companies, for example, or schemes known variously as "scalping", "pumping and dumping" or "ramping".

One example involved three men who allegedly bought more than 300,000 shares in a bankrupt company, NEI Webworld, for between 0.5 cents and 13 cents each. The SEC claims that one weekend they went to the biomedical library of the University of California in Los Angeles, where they had access to the Internet. Using 47 different names, they posted more than 400 messages on Internet bulletin boards maintained by Yahoo!Finance, to spread a rumour that NEI was about to be taken over by LGC Wireless, a privately owned Silicon Valley company, which would push up NEI's share price. Under yet more assumed names, they also discussed the rumours on other public message boards, including one run by Raging Bull, a Boston-based financial-news service.

By the time the Nasdaq over-the-counter market opened on the Monday morning, NEI's shares had risen to $8, and 30 minutes later they had advanced to nearly twice that. The SEC says the three men sold all their shares, making a profit of $360,000. When it became clear the rumoured takeover was a myth, the share price collapsed. Some investors, having failed to put a price limit on their purchase order, lost heavily.

Controversially, the SEC has been looking at systems for monitoring websites for suspicious activity. You might think that anybody buying shares on the strength of Internet gossip is less raging bull than barking mad. The sad truth is that the Internet's "democratisation" of investment has provided plenty of inexperienced prey for conmen.

Japan's online adventure

OKI MATSUMOTO INTRODUCES himself with the disarming claim that he is insane. As evidence, he cites his decision to quit his partnership in Goldman Sachs, an American investment bank, before it floated on the stockmarket in 1999, making its partners extremely rich. He compounded this act of folly by devoting himself to what many of his colleagues thought a crazy venture: setting up an online brokerage, Monex, which he now runs, and owns in a joint venture with Sony. In Japan's hidebound business culture and change-resistant financial system, all this does indeed seem a little mad.

Or maybe not. Online broking has been growing at a hectic pace in Japan. By mid-2000 Monex claimed 57,000 accounts, producing 16,000 orders a day. Its staff of 40 are casually dressed, packed in a cramped office as if trying to replicate Silicon Valley in downtown Tokyo. Monex contracts out most administrative work, so in total about 140 people are employed full-time on its business.

The marketplace is already crowded. Besides Monex, there are other "pure" online brokers such as E*Trade Japan, a joint venture between E*Trade and Softbank, a Japanese technology investor. And there are discount brokers moving online. The leader in this category is Matsui Securities, which has been offering online trading for since 1998.

Then there are the established securities houses. Of these, Nikko, in which Citigroup intends to increase its shareholding to 20%, has set up a specialist online brokerage, Nikko Beans. Hideyuki Omokawa, head of sales at Beans, admits the firm's name means even less in Japanese than it does in English. It is meant to sound trendy and appealing, qualities not associated with traditional brokers.

As it happens, the firm that claims to be the leader in online broking is just such a throwback: Nomura, which boasts a 26% market share, or 150,000 accounts, growing by 1,100 a day. As so often on the Internet, however, the numbers are open to dispute. Rivals accuse Nomura of signing up clients for Internet accounts that they never intend to use. They also point to a serious channel conflict for Nomura: the competing interests of its online brokerage and its branches with their armies of brokers.

Yoshihiko Kan, a manager in Nomura's retail strategy department, denies both charges. He says the number of "zero-balance" online accounts is no more than 10% of the total, and that there is no "cannibalisation" of existing business, because branches are in control of the process of putting customers online. The Internet merely offers a way of appealing to younger investors (the average online customer is 13-14 years younger than an offline one). But Mr Kan also concedes that for Nomura's customers, trading on the Internet is only about 20% cheaper than using the branch network. Its competitors are offering a far cheaper service.

Nomura and other traditional houses seem to be avoiding the worst of the mistakes made by their American counterparts as the Internet began to transform their business in the late 1990s: they are not shunning online trading altogether. But by seeking to promote it merely as an alternative, slightly cheaper channel, they still seem to be trying to have their rice cake and eat it too. The American experience suggests that, as online trading gains popularity, they will have to slash their commissions to remain competitive.

Branching out

Many financiers expect online banking to grow as fast as online broking. It is not obvious why

BY THE SPRING of 2000, Internet fever had begun to spread from stockmarkets to bank boardrooms. In Europe and America, as banks announced their annual results – which for the sector as a whole were excellent – they also unveiled new plans to conquer cyberspace. Many announced they were to pour fabulous sums into their online businesses. This was where their future lay, they said (at least in public). But are they right?

As yet, most Internet banks are tiny, and for most big banks online banking is still a trivial part of their retail business. True, in both Europe and America, Internet banks have been springing up all over the place. Some are – separately branded – subsidiaries of existing, rather stodgy-sounding institutions. Others, especially in America, are pure start-ups. One online finance directory now [May 2000] lists 74 Internet banks in America alone. But most are small to the point of invisibility. Rory Brown, the boss of one much-trumpeted start-up, VirtualBank, says he would be satisfied if it attracted 10,000 customers by the end of 2000. The biggest of the "pure" online banks in America, Wingspan, a subsidiary of Bank One, has about 100,000. By comparison, the big American banks are already giants online as well as offline. Citigroup has around 500,000 online customers in America; Bank of America has more than 2m; Wells Fargo nearly as many, and the most-visited bank website.

But even these numbers are not enormous compared with those involved in online share trading. CSFB estimates that, whereas 16% of American equity trades are now conducted online, only 3% of American households bank online. According to a survey cited by VirtualBank, by Cyber Dialogue, a research firm, 3.2m people in America did sign up for online banking in the 12 months to July 1999 – but 3.1m closed their accounts.

Other surveys come up with less spectacular but nevertheless alarming results. One in *American Banker* magazine in 1999 found a "churn" rate among online-account holders of 35%. Nevertheless, VirtualBank also cites projections that, by the end of 2003, there will be 100m online bank accounts in America. Similarly, the International Data Corporation forecasts that about 40m American households will bank online by then. Jupiter Research predicts a mere 25m households. In Europe, J.P. Morgan expects 40% of European consumers to use some online financial services by 2003.

To gauge how realistic these expectations are, it is worth asking a few questions:

- ◪ **What are the savings on offer?** In online share-trading, the savings made by using an online discount broker are substantial and immediate. They could amount to $100 a trade, a useful sum even for an investor who deals just half-a-dozen times a year. In contrast, the cost savings for the consumer of maintaining a bank account online are not huge, and not always transparent. Banks tend to pay below-market interest rates on current (cheque) accounts, and even on savings deposits. But precisely for that reason, investors, especially in America, tend not to keep much money in such accounts. Some Internet banks, notably in Britain, are offering above-market rates, but most do not differentiate between online and offline deposits. In America, where stockbrokers also compete for money-market accounts, the scope for offering better terms online is limited. So few bank accounts will move online just to earn a better interest rate. "With broking", says Kathy Levinson, president of E*Trade, "the incentive was economic; with banking it will be convenience."
- ◪ **Is it worth the bother?** Online banking offers only a restricted range of services. "Smart cards", which can be used to store and spend cash, are not yet in wide use, so most people still need their bank for one thing that their PC cannot provide: folding money. Branches in which to queue up may be dispensable, but an ATM is not. Indeed, providing cash machines (along with telephone banking) has enabled banks to offer something that stockbrokers could not deliver without the Internet: constant availability, the "24 x 7" service that online businesses see as essential. For some customers, avoiding a visit to a hole-in-the-wall or a branch is inducement enough to bank online. But Internet banking does not offer many additional banking functions. You can check your balances, and you can transfer money from one account to another. Even that service might become redundant if the banks offered an automatic "sweep" from low-interest current accounts into higher-yielding deposit or

money-market accounts. But since that would cut their income from the "float" of cheap money in low-interest accounts, few offer such a service. The one obvious application that is indeed becoming more widespread is paying bills (see "Scandinavian models" on page 221).

◪ **How much inconvenience will it cause?** "Switching bank accounts", notes Moody's, in a report on the Internet and American banks, "is difficult, time-consuming and disruptive – the financial equivalent of root-canal work." So, in the absence of compelling incentives, people are reluctant to move – good news for incumbent banks, which should be able to help their account holders make the transition online as painlessly as possible. In Britain, analysts at Deutsche Bank have attempted to calculate the value of this inertia to banks in terms of how much more it enables them to charge each customer than if they had to match the best deal in the market. They put the figure at £350 ($540) per average customer.

In another important difference from stockbroking, there is no huge new market for basic banking services ready to be tapped. Whereas the Internet has made share-trading cheaper and more accessible for millions who might not have invested in equities before, most people who want a bank account already have one. (Sadly, those who do not are rarely the sort of customers banks want.)

It is not surprising, therefore, that until recently many banks were reluctant to promote their online services among their offline customers. Why should they? Offline customers have been the source of the banks' handsome profits. And unlike brokerage clients, who have long had a clear choice between paying top-dollar for a full-service account and using a discount service purely for execution, bank customers are often used to paying concealed charges for all sorts of services they may not use. Trying to make them bank online before they are ready risks antagonising them; it may even open their eyes to some of the competing services available on the web. It has seemed, in short, a recipe for cannibalisation.

Cheap and cheerful?

Nor is it certain that the Internet can really offer significant cost savings. Online financial folklore has it that Internet-based transactions cost a bank only one cent – a fraction of the cost of other channels. But this figure seems either far too high or far too low. The incremental cost of an online transaction is negligible: "How much does it cost to move a few electrons in our database?", asks Elon Musk, boss of X.com, a Silicon Valley firm. But if all costs are fully allocated – incremental IT expenditure, regulatory compliance, credit control and accounting, let alone "customer acquisition" – one cent is hardly likely to cover the bill. Finland's MeritaNordbanken

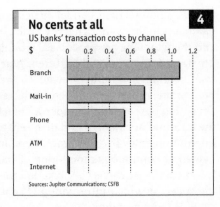

No cents at all 4
US banks' transaction costs by channel

Sources: Jupiter Communications; CSFB

(MNB), where costs of adapting "legacy systems" were probably unusually low, estimates the average cost per Internet transaction at about €0.11 ($0.10). Lehman Brothers cites one estimate that it costs just 14% less to service an online account than an offline one. Peter Duffy, who runs Barclays' online-banking operation, agrees the savings are "not enormous".

Moreover, in another difference from stockbroking, retail banks make their money mainly from interest income and the sale of particular products, not from charging customers for transactions. The other fly in the ointment is that transaction costs are saved only if customers use the new channel instead of, rather than in addition to, existing ones. Online banking encourages people to visit their bank more often (an average of once every six days, says Barclays' Mr Duffy). But unless the bank actually cuts back on its other resources – branches, call centres and people – the Internet simply adds to its costs.

Banks around the world are indeed cutting back on staff and branches, but usually to tidy up after mergers or acquisitions, rather than as a direct response to online banking. Until recently, few banks have been prepared to admit freely that they were closing branches and sacking people because they were moving on to the Internet. Where they do, they risk a storm of disapproval from customers and politicians, as Barclays found when it shut 171 branches in rural Britain in April 2000.

Banks also have expensive trouble in adapting their existing IT systems for the Internet. This is why Rory Brown, of VirtualBank, believes he can compete on price with outfits such as Citigroup, which he says has spent $500m on integrating its website and its "legacy systems".

Many traditional banks find it difficult to link their mainframe computer databases to the web. Often the problem is twofold: disjointed systems, and the way data are processed. A bank maintains different processing programmes to deal with different types of transaction, and may separate them by geography as well, so there could be dozens of separate programmes running independently of each other. Worse, data about deals done are gathered under a "batch-processing" system, in which the information is stored and then sent to the mainframe in batches. The Internet operates in "real time", here and now, offering the enormous benefit of continuous monitoring of balances in a variety of accounts, for example. However, because Internet-based systems cannot control the speed at which information is sent to the server, they can be susceptible to overloading.

Scandinavian models

Finland and Sweden have shown that online banking can be a popular success

TO SEE HOW obstacles in the way of online banking might be over-come, it is worth examining the one part of the world where it has taken off in a big way. Most outsiders, after a moment's thought, have no trouble accepting that Finland, Norway and Sweden should be global leaders in mobile-phone and Internet usage. After all, they have sparse populations, scattered over large desolate areas which for much of the winter are dark and bitterly cold. Better to log on or talk remotely than to go out. Those who know their Strindberg may even have bleaker surmisings about Scandinavian difficulties with interpersonal relationships.

Natives, of course, tend to offer more pragmatic explanations: good education systems that lead to a high degree of computer literacy; enlightened governments that were quick to spot the potential of the online world – and to encourage it, as Sweden's does by allowing com-panies to provide their staff with a home computer as a tax-free benefit in kind; widespread knowledge of the international language of geeks (English, of a sort); and the presence of world-beating technology com-panies, such as Finland's Nokia and Sweden's Ericsson.

Whatever the reason, the institutions housed in the traditionally imposing, solid and rather dour bank buildings clustered around Stock-holm's Kungstradgarden (King's Garden) are now, improbably, vying with each other for the title of the world's most advanced Internet bank. Claiming the prize by the plausible measure of "log-ins per month" is MNB, soon to be renamed Nordic Bank Holdings. It is the result of one of Europe's very few cross-border banking mergers, between Finland's Merita and Nordbanken of Sweden. Finland, on many measures, is ahead of even its Scandinavian neighbours. It leads the world in mobile-phone-usage per head of population and comes second in Internet pen-etration (see chart 6); it also has the highest number per head of payments by credit, debit and smart card, and the lowest ratio of cash-in-circulation to GDP of any rich country.

So almost inevitably, Finland also has, proportionally, the world's biggest online-banking population. By early 2000, MNB had 1.2m

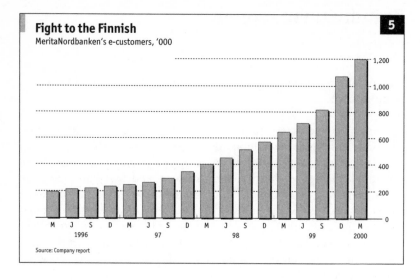

Fight to the Finnish 5

MeritaNordbanken's e-customers, '000

Source: Company report

Internet customers – two-thirds in Finland, the rest in Sweden – who between them visited their virtual bank more than 3m times a month. The bank started to move from "wired" to "wireless" – that is, to mobile-phone banking – in Finland in 1999, and is now planning to do likewise in Sweden. And it is already plotting its next advance: the integration of mobile-phone banking with smart cards.

The other big Swedish banks are all scrambling to hold their own in this market. Svenska Handelsbanken, one of the largest, was slow off the mark, but is racing to catch up. Others are credible rivals to MNB for leadership. Skandinaviska Enskilda Banken (SEB), for example, was the country's first Internet bank, having gone online in December 1996. It now says that a quarter of its total customer base, and half its "genuine" customers (those whose primary banking relationship is with SEB), use its Internet bank – nearly 400,000 people. SEB calculates that these "early adopters" are twice as active and generate, on average, 2.5 times as much revenue for the bank as the offline laggards who, for their part, seem pleasingly docile. In February SEB announced it would be closing 50 branches across the country, a quarter of its total. There was little fuss.

Tomorrow, Europe

SEB also has 1,000 corporate clients for an impressive operation it calls Trading Station – an Internet-based trading mechanism for foreign-exchange dealing, stock-index futures, and Swedish treasury bills and government bonds. Helped by its acquisition last year of BfG, a German bank, SEB now wants to expand its Internet-banking presence from the Nordic and Baltic markets to the whole of Europe. Some 40% of its foreign-exchange deals by number are now transacted through the net (though the proportion by value is much smaller because the huge interbank market uses non-Internet systems).

Both SEB and MNB, however, were beaten in 1999 in a ranking of the world's "leading Internet banks" compiled by IBM and Interbrand, a consultancy. Second only to Citibank worldwide came ForeningsSparbanken, a relatively obscure hybrid of Sweden's main agricultural bank and its mutually owned savings institutions. Known abroad as Swedbank, it has 550,000 online customers and gets more visitors to its website than any other Swedish bank.

That its inhabitants are so tech-savvy is not in itself enough to explain why online banking has done so well in Scandinavia. To tap the potential demand for Internet services, and to try to make money out of it, the banks have also had to meet three conditions: to provide customers with products they want; to make them comfortable about online security; and to persuade them to pay a commercial price for online banking.

All the banks have linked online stockbrokers. SEB's, for example, now does more than 30% of its business online. But the basic online-banking activity is paying bills, which lends itself naturally to integration with e-shopping. MNB, for example, has an online "mall" of more than 900 shops which accept its "Solo" payment system. You visit the website, choose a purchase and select the "Solo" button on the retailer's screen as a method of payment, which is then made directly and immediately from your bank account. Swedbank offers a similar system called "Direct".

Safe and sound

The banks have so far avoided the big security scares that have afflicted online-finance providers elsewhere. Besides using advanced encryption technology, they have also adapted a basic but effective system – known as "challenge response logic" – from telephone banking. In its simplest form, at MNB, this involves a list of code numbers sent to every online client and used in sequence, in combination with their password or personal identification number (PIN). This gives each transaction a unique code, and has so far proved safe. SEB and Swedbank use a more

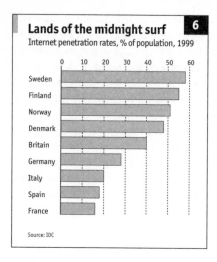

Lands of the midnight surf 6

Internet penetration rates, % of population, 1999

Sweden
Finland
Norway
Denmark
Britain
Germany
Italy
Spain
France

Source: IDC

sophisticated version of the same technique, issuing customers with a credit-card-sized device used to generate the unique number each time the customer logs on.

From foreigners' viewpoint, the most exciting aspect of these three banks' online successes is that they have been able to offer these services without slashing prices. This is partly a tribute to their quickness off the mark, and to their ability to keep their customers happy. In Scandinavia online banking is seen as a premium service, not as a low-price, cost-cutting alternative.

But the limited competition must also be playing a part. Almost all of the banks' online business comes from their existing offline customer base. Jacob Wallenberg, SEB's chairman, is unembarrassed to acknowledge that Swedish banking is an oligopoly: the big four banks control 89% of total banking assets. Margins are beginning to come under pressure in some areas. There are signs, for example, that online competition is bringing down the price of mutual funds. Home loans may also be vulnerable. In February 2000, EuropeLoan, a Belgian bank that wants to be Europe's first pure Internet mortgage provider, began business in Sweden. But on the whole the banks have been able to move online without having to slash their margins.

That Sweden and Finland are, in themselves, small markets is a help in a number of ways. It has delayed the entry of foreigners; it has helped in the introduction of proprietary systems such as "Solo" and "Direct" (when there are so few banks, adding one or two such payment options to a website is not that cumbersome); and it has allowed the banks to expand their online businesses at rapid but manageable rates. Swedish and Finnish bankers like to describe their markets as "laboratories" where clever new technologies and practices can be tried out, and have enjoyed a measure of cloistered calm to conduct their experiments. In other markets, bankers do not enjoy that luxury.

Paying respects

The banks fear that the Internet might elbow them out of the payments system. They need not worry – yet

AMONG ITS VARIOUS claims to world leadership, Sweden's Swedbank prides itself on having been the first bank to introduce "electronic bill presentment and payment" (EBPP). In America, this facility is now spoken of as the "killer application" that will inspire the multitudes to bank on the Internet. Swedbank now handles 2m bill payments a month. However, Hakan Nyberg, who runs its Internet service, concedes that "presentment" – the delivery of bills for payment online – has still not reached critical mass. There are only about 20 "live" billers, and only a few thousand bills presented each month.

Swedes, like Finns, long ago stopped using cheques. In America, where direct debits are less common, 70 billion cheques are written each year. Mr Nyberg regards America's continuing attachment to these bits of paper as of the "stone age". Since Swedes were already used to issuing direct-debit or wire-transfer instructions by post or telephone, the move to the Internet was a relatively small step. It brought obvious advantages of convenience, which are even more clear-cut in cheque-based cultures. An estimated 3m Americans currently pay some of their bills electronically, and if the forecasts are right, this figure will grow exponentially. But just as important for the banks as the benefits to individual customers are the big advantages to their bill-issuing corporate clients. The savings in paper, handling and postage are estimated at well over a dollar a bill, and can be much more. Since 3m firms in America issue about 21 billion bills each year, the sums of money involved are huge.

Efforts to move bill-presentment online have largely coalesced into two rival systems (though Bank of America, America's biggest online bank, operates a third). One is offered by CheckFree, which in February 2000 bought its nearest competitor, TransPoint (a joint venture of Microsoft, Citibank and First Data, the Atlanta-based parent of Western Union, a money-remittance firm). Both companies, having established a relationship with the biller (from whom they earn their income by charging a fee for each bill paid), transmit electronic bills to the payer's chosen vehicle, which might be a bank or an Internet personal-finance service,

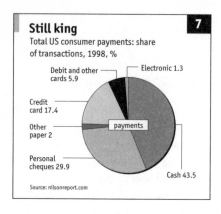

Still king | 7

Total US consumer payments: share
of transactions, 1998, %

Debit and other cards 5.9 — Electronic 1.3

Credit card 17.4

Other paper 2 — payments

Personal cheques 29.9

Cash 43.5

Source: nilsonreport.com

such as Microsoft's Money-Central or Intuit's Quicken.

The other effort, Spectrum, is a joint venture of Chase Manhattan, First Union and Wells Fargo, three big banks, and seems to be inspired by the fear that CheckFree and others like it might squeeze the banks out of the EBPP market. It differs from CheckFree in that it is an open system, allowing any biller and any financial institution to use it. Chase.com's boss, Denis O'Leary, is convinced that this "open architecture" will make it the winning model. It could, in theory, coexist with CheckFree, but once it adds a bill-payment function, promised for the second quarter of 2000, the two systems will be even more obviously in competition. Since both sides have such strong backers, and since EBPP is seen as such an important area, the battle could be bruising.

Also at risk of injury if the banks and billers manage to come up with a widely used standard are the firms that approach the issue from the other end – that of the harassed bill-recipient, willing to pay someone to handle the lot. They can, for example, have their bills routed to PayMy-Bills.com, a firm that will scan them into a computer and e-mail them to the customer for approval or simply pay them up to an agreed limit.

For the moment, only a small proportion even of bills that individuals run up on the Internet is paid by direct online transfer. In America, 95% of online payments are made by credit card, compared with only 18% in the real world (see chart 7). The Internet, in other words, is a bonanza for three organisations that play a role in the vast majority of online transactions: American Express, MasterCard and Visa. But as e-commerce develops, so will the demand for other means of payment. The chart suggests that, given the choice, many online shoppers would prefer not to use credit cards. That is also true of many retailers, who have to fork out set-up and transaction fees plus 2–3% of every payment.

Even the American government has criticised the reliance on credit cards for online transactions. The Treasury is experimenting with e-cheques and smart cards which, for example, are being used to pay soldiers in Bosnia. Treasury officials complain that for e-commerce, credit cards are too costly and inflexible, and call for a uniform mechanism for electronic payments.

Where cards won't do nicely

In particular, credit cards are unsatisfactory for three sorts of users:

- **Small spenders.** Credit cards are uneconomic for "micro-payments" of a few dollars or even cents. Yet as Internet companies start to worry about how they will ever make money, charging very small amounts is one obvious starting point: for example, a newspaper might demand a small payment for each article viewed; or some sites might charge a bit for access to games.
- **The young.** Jupiter Research predicts that America's "kids and teens" population will balloon to 42m in 2003. Offline, youngsters are big spenders in their own right, as well as important influences on their parents' spending. Yet most credit-card issuers exclude the young (as well as many adults).
- **Individuals.** Credit cards are also unsuitable for most person-to-person transactions. The popularity of online auctions, such as eBay, has shown this up. Cheques take too long for many buyers.

A number of other payment systems are now being tried out. Some, such as Flooz, are gift certificates, usable at online retailers. Beenz offers a kind of loyalty programme resembling an air-miles scheme, and is now planning to provide debit cards allowing users to convert their beenz into cash. Billpoint, a service now owned by eBay and Wells Fargo, enables individuals to accept credit-card payments. But perhaps the most successful of the new systems – at least in terms of numbers of accounts held – is that operated by X.com, a Silicon Valley online bank, and PayPal, which merged in March 2000. Their service allows account-holders to e-mail money to each other, and then to download it to a bank account. It already has more than a million accounts. It is popular with some eBay users (X.com claims that a tenth of eBay transactions are settled through its service). Cynics, however, say its popularity derives largely from its practice of paying customers $5 for opening an account, and $5 for referring new business. Many college students are said to have spent their golden welcomes on beer. But Elon Musk, X.com's chairman, points out that, as "customer-acquisition costs" go, X.com's are small beer. Besides, many account-holders do not bother to download their cash. This "float" left in cyberspace is a nice interest-earner for X.com. But even Mr Musk concedes that, in terms of overall payment flows, X.com is "a grain of sand on a very long beach". For the next few years, credit cards will hold sway. Jupiter sees their dominance shrinking, but only to 81% of transactions by 2003.

New ways of dealing with micro-payments are emerging, such as by deduction from a prepaid account with a computer company, Internet-service provider or mobile-phone company. And as online bill-payment becomes more common, direct transfers out of bank accounts may also be used more widely for other purchases on the Internet.

Virtual rivals

The Internet is presenting retail banks with plenty of new competition for their business

IN MOST RICH countries, unlike Sweden and Finland, all manner of institutions are snapping at the banks' heels. In addition to their traditional rivals in the high street, banks now also have to contend with new Internet banks, with non-bank websites seeking to enhance their allure, and with other institutions for which the Internet seems to offer a low-cost way of entering the financial marketplace.

The firm with the strongest claim to be the world's first Internet bank is Security First Network Bank (SFNB), based in Atlanta, which was allowed to offer federally-insured deposits from October 1995. In 1998 it was taken over by the Royal Bank of Canada. Its home page – an image of a real, if ultra-modern bank lobby, complete with tellers – seems to want to have it both ways: a virtual institution with real smiles. And indeed in March, 2000 Royal Bank agreed to buy Prism, an American mortgage bank, which gives SFNB access to 150 branch offices. SFNB, like Virtual Bank and Telebank, is regulated by the Office of Thrift Supervision (OTS), rather than the Office of the Comptroller of the Currency (OCC), the main banking regulator. It offers its own deposit accounts and credit cards, but also has partnerships with other financial-service providers.

CompuBank, which began operations in 1998, was the first "pure" Internet bank to receive a national bank charter, though, like SFNB, it also offers a telephone service. It boasts an impressive list of shareholders, including Goldman Sachs, Japan's Softbank, and Marsh & McLennan, the world's largest insurance brokers. Its original focus was on consumer banking, but it is now also offering cash-management services to small businesses. Net.B@nk (not be confused with Netbank AG, a German online bank), originally the Atlanta Internet Bank, opened in 1996, and claims to be the world's largest Internet-only bank. It is still America's only quoted online bank, offering higher interest rates and lower service charges than offline banks.

On a wingspan and a prayer
Despite Net.B@nk's claims, it has probably been overtaken as America's

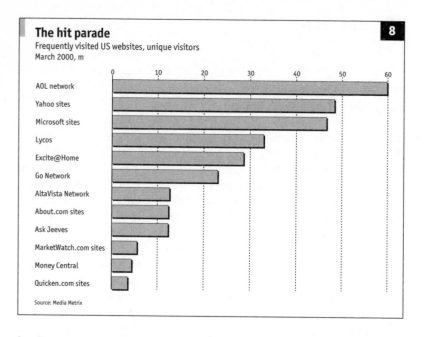

The hit parade   **8**
Frequently visited US websites, unique visitors
March 2000, m

Source: Media Metrix

leading Internet-only bank by both E-LOAN and Wingspan. E-LOAN started in 1997 to concentrate on home loans, and has since expanded its offerings to include other sorts of lending, making it America's leading exclusively online lender. Wingspan is a subsidiary of Bank One, an established Chicago-based bank that is America's fourth-largest in terms of assets, and also owns First USA, a troubled credit-card division with extensive online operations. Oddly, Bank One itself is a substantial Internet bank, offering a comprehensive online banking service. Lehman Brothers, a Wall Street investment bank, calls Wingspan "one of the first true product-segmentation efforts by a bank in the online universe".

Wingspan is, in other words, one possible answer to banks' worries about "channel conflicts", and about cannibalisation of their existing business. Wingspan's customers benefit from lower fees and higher interest rates. They have free access to Bank One's ATMs, but not to the branch network. Bank One's customers have the full "multi-channel" service. To get it, they pay more for some services, such as online bill-payment. They also earn less interest on their deposits. Clearly, the strategy relies on keeping the two sorts of customer distinct. So Bank One's and Wingspan's websites are both discreet about their connection

– there is no cross-linkage. Confusingly, the Bank One site also offers an Internet-only option that is remarkably similar to Wingspan's.

Everbank, a St Louis-based joint venture formed by the Wilmington Savings Fund Society, says it wants "to create a bank that consumers will love". Its bosses, according to its website, are very nice people, with nary a "pin-striper" among them. The chief executive skis and mountain bikes, coaches soccer and plays a "mean game of ice hockey". So darn nice are these people that they offer a "100% satisfaction guarantee" on their e-checking account – if you do not like it, you can close it after three months, and they will pay you $50 for your trouble.

Similarly, VirtualBank aims to earn the sort of loyalty enjoyed by the credit unions operating in some workplaces. In particular, its targets are the employees of technology companies with high disposable incomes and an appreciation of online technical wizardry. Its boss, Rory Brown, thinks that the mistake made by outfits such as Wingspan has been to concentrate on attracting deposits, and to source their lending business from other operators – for example, from E-LOAN for mortgages. This meant they were "competing in a market nobody cared about", because deposit rates were much the same anywhere. Better, argues Mr Brown, to take advantage of a banking licence to build a complete financial relationship with the customer. So VirtualBank will concentrate on lending, offering credit cards and loans for buying cars, homes and shares. It hopes to make these attractive with cheap rates and easy approval.

VirtualBank has attracted a lot of interest because Microsoft has used it in its advertising campaigns. Mr Brown thinks the software giant may be using it as a marketing tool in its efforts to sell systems to other banks. But rivals also wonder when and how Microsoft is going to make a big move in banking, and ask whether VirtualBank might be its vehicle to challenge myCFO, a venture launched by the firm's old nemesis, Jim Clark, of Netscape fame (of which more later).

In Europe, most "Internet banks" are in fact subsidiaries of bigger offline institutions. Two of the most interesting are based in Spain. Bankinter, a smallish bank with 380,000 customers and a market capitalisation of around €5 billion, intends to transfer its existing business entirely to the Internet. In 1999 a third of its transactions were electronic. Another 16% were carried out over the telephone. With Internet usage in Spain still among the lowest in Europe, the bank is probably right to see huge potential for its services.

Nurturing greater international ambitions is BBVA, which in February 2000 announced a strategic alliance with Telefonica, the formerly state-owned telecoms monopoly. The two were to increase their existing cross-shareholdings and to co-operate in e-business, mobile-telephone services and payment mechanisms. That entailed Telefonica raising its investment (through Terra, its Internet-service provider) in BBVA's Internet bank, Uno-e. The following month, Uno-e announced a merger with first-e, a Dublin-based bank that began operations in Britain in 1999. First-e is a subsidiary of enba, an Internet-company incubator, whose imposing list of shareholders include Morgan Stanley Dean Witter, a bulge-bracket investment bank; Intel, of computer-chip fame; Capital Z, an American venture fund; and CGU, a British insurer. At the time of the merger, first-e claimed 110,000 registered users, of whom 50,000 were active.

These numbers must have been disappointing. Like Egg, first-e offers attractive interest rates on deposits, in its case accrued daily and paid monthly. But unlike Egg, first-e has no British household name behind it. Rather, it acquired its banking licence through Banque d'Escompte, a small one-branch Parisian firm. The European Union's banking "passport" scheme allows it to do business in other EU countries, but as the British press has pointed out, being regulated by the French authorities, first-e is excluded from the ombudsman scheme run by Britain's Financial Services Authority. Unfairly, that makes it look risky in a business where, as first-e's Richard Thackray puts it, "trust and respectability are big issues."

"No bank can be a portal," according to Bo Harald, head of Internet banking at MeritaNordbanken. He rejects the suggestion that MNB, with its much-visited website, might challenge the Yahoo!s of this world and provide not just financial services on the web, but an entire framework for its customers' Internet life. If Finland's leading Internet bank is so dismissive of the prospect, it is fair to assume that banks elsewhere will be even less sanguine about assuming a comprehensive non-financial role. One reason for this is the banks' refusal, so far, to embrace "open finance", offering products from other suppliers as well their own.

Passing the portal

But can portals become banks? Obviously they cannot take deposits or lend money. But many in the industry see portals such as NetCenter, Yahoo! and AOL, and online personal-finance sites such as Microsoft's MoneyCentral and Intuit's Quicken, as the biggest long-term threat to the banks. These sites have visitors in their millions, and many of them have established a reputation for reliability.

The ideal would be for an individual to be able to deal with all his finances on one easily accessible site. Bank accounts, insurance policies, share portfolios and tax returns would all be available there, completely up to date, and the user could do whatever he wanted without having to leave the site or go through any additional security firewalls. A customer would have a "universal password" which would give him access to all his online finances. Pursuing this ideal are firms such as Yodlee and Vertical One, known as "aggregators" or, more rudely, screen-scrapers. Yodlee's site offers to set up just such a personal balance sheet. But, like Vertical One, it also licenses its technology to other Internet companies – selling "picks and shovels for the gold-rush", as Vertical One's boss, Gregg Freishtat, puts it. By this month it claimed 110,000 accounts and growth of 10–15% a week. Yodlee has a deal with Altavista, one of the biggest portals and search engines.

The aggregators have attracted some controversy. To function properly, they require the customer to provide access to all his accounts. Some banks – and others concerned about the effect of the Internet on personal privacy – were uneasy at impersonal strangers browsing through data held the wrong side of their security firewalls. The authorities are taking an interest, too. In February 2000, the OCC, America's main banking regulator, sought comments on a number of electronic-banking issues, such as whether specific new rules were necessary both for bank websites that offered access to non-bank Internet businesses, and for non-bank sites, such as aggregators, that linked to banks.

Perhaps most worrying for the banks, the aggregators also have the capacity to become active deal-seekers, trawling the net for the best available offers. Mr Freishtat denies there is any conflict with the banks, pointing out that his firm is a subsidiary of S1 Corporation, a software firm that works with more than 50 of the world's biggest banks on creating their own "financial portals", and that SFNB, an Internet bank, is a VerticalOne partner. Yodlee's Sukhinder Singh concedes, however, that there is a "chicken and egg" problem in the failure to date of "tier-one players" to offer aggregation. Big financial institutions are understandably reluctant to provide easy access to competitors' products.

More positively, the banks may calculate that people still prefer to deal with a trusted financial institution when it comes to their own money; and that only a minority of web traffic these days is mediated through the big portals. But that is partly because people are becoming more familiar with the Internet, which could be ominous for the banks: how loyal, or more cynically, how inert, will their customers prove when better offers are available at the click of a mouse? Or when they arrive, unsolicited, by e-mail from your friendly financial intermediary on the Internet? Already in America some aggregators (for example, InsWeb for insurance policies, or LendingTree for home loans) will hunt down the best deal available on the Internet. The habit is bound to spread.

Because it lowers the barriers to entry – in particular the need for physical infrastructure – the Internet is speeding up another trend in global banking: the invasion of non-banks. All over the world, people seem to think that banking is money for old rope, and that if even bankers can manage it profitably, good businesses should be able to coin it. In America, Wal-Mart, a big retail chain, last year bought a small savings bank in Oklahoma that might help it get into Internet banking. Nordstrom, another retailer, has applied for a banking licence, as have more than 30 American insurance companies, car makers and other businesses, including a large coffin maker.

Anyone can play

In Britain, the process is already quite advanced. Marks and Spencer Financial Services was set up in 1985 to run the clothes-and-food chain's proprietary charge card. Virgin, Richard Branson's diversified retail and travel firm, entered the market in 1995. Three big supermarket chains – Sainsbury, Tesco and Safeway – all have their own branded banks. In France, too, some of the leading retailers, such as Carrefour, have banking operations of their own. Volkswagen, a German car maker, has owned a finance company since 1949, which it turned into a bank in the 1990s. It now has more than 250,000 customers, and conducts almost all its business by Internet and telephone.

Japan's banks, which stand out for their dubious loans and their egregious mismanagement, are obvious targets for non-bank aspirants. Eight out of nine "city banks" offer basic online services. But only one, Bank of Tokyo-Mitsubishi, offers 24-hour access. This leaves them vulnerable to more go-ahead financial firms, as well as to manufacturers and retailers. The biggest chain of convenience stores, Seven-Eleven, has already applied for a banking licence. Its idea is to provide Internet shopping at home, or at terminals in its shops. The purchases will then be collected and usually paid for in the shops themselves. Convenience stores already process huge numbers of cash payments for utility bills: in 1999 Seven-Eleven handled ¥600 billion ($6 billion) worth of them. It will now also offer the service for Internet purchases, and not just those made at its own shops. The main attraction of a banking licence for

Seven-Eleven, however, is that it will enable the company to install its own ATMs in its shops, to replace those of other banks.

Other Japanese giants contemplating a move into banking include Softbank, an Internet investor, Toyota, a car maker, and Sony, an electronics firm, which may either buy an existing bank or get its own licence to set up an online bank. This would not be its first financial foray – Sony is already a partner in Monex, an online broker, and in 1999 it set up a non-life insurance subsidiary. With its considerable marketing skills and vast customer base, Sony could make a big impact.

A consortium led by Softbank is taking over the nationalised Nippon Credit Bank. Yoshitaka Kitao, head of Softbank's finance arm, says the new NCB will use Internet technology to concentrate on serving medium-sized businesses. Softbank is a core investor in many of the more interesting financial sites on the Internet: E*Trade, Insweb, eCoverage, not to mention Yahoo!. Before, says, Mr Kitao, his idea was to invest in Internet companies in Japan and elsewhere, to exploit the expertise his firm has acquired in America. Now that Internet companies around the world are "like luxury hotels, they cost the same everywhere", he plans to invest in non-Internet companies and take them online. What Mr Kitao calls his "new arbitrage strategy" should be enough to alarm established banks.

The case for the defence

What "old" banks are doing to fend off the Internet challenge

NONE OF THE new Internet entrants to the banking market as yet poses any serious threat to established banks. Britain's Egg has come closest to it, but only by offering a loss-making service that it surely cannot sustain indefinitely. Britain's supermarket banks also took a big chunk of the savings market when they first opened, but have lost market share since, partly to Egg. Many of the other online banks boast fabulous growth rates, but only because they start from such a low base.

Yet all banks are having to redefine themselves as Internet banks, and that message has spread across the Atlantic to Europe, and to some extent across the Pacific to Asia as well. In France, BNP Paribas said in February 2000 it had earmarked €700m ($690m) for investment in its online operations over the next three years. In the same month, Germany's Deutsche Bank said it would gradually increase spending on Internet-related services to $1 billion a year, and Dresdner Bank by €500m, although when the two banks in March announced a (soon-aborted) plan to merge, they indicated they would scale this back. ING, a Dutch bank, hopes to splash out €2 billion on online banking over the next three years. Credit Suisse is aiming to become "Europe's premier financial-services brand" with a mere SFr1 billion ($600m) outlay on Internet banking and e-commerce. Datamonitor, a research outfit, put total spending on Internet banking systems in 1999 at $362m, and estimates that the amount will quadruple by 2004.

There are at least four reasons why banks are rushing to recast themselves as online businesses. First, nobody doubts that Internet banking will spread; the only question is how fast. Second, the banks need to forestall any threat from new operations; in particular, the worry that specialised, low-cost online operations will succeed in "cherry-picking" some of the banks' most attractive business. Third, and more important, they need to compete with their current offline competitors.

Lastly, the banks' shareholders are demanding online strategies. As noted, banks' share prices in many markets are low in relation to the rest of the market. This is in a part a reflection of the business cycle. In America and Europe, if interest rates continue their recent rise, and bad-debt levels also climb, both of these will hit bank profits. But low share prices also

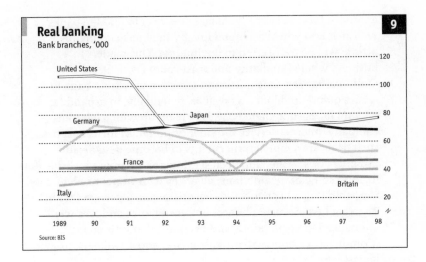

Real banking
Bank branches, '000

9

Source: BIS

reflect the markets' concern about the banks' ability to weather the storms of new competition, including, most important, in the online world.

Broadly speaking, the banks handle their Internet business in one of four ways:

- **"Pure" Internet banking**. Few go as far as America's Wingspan, which keeps the offline parent and its Internet-banking subsidiary completely separate. But a number of Internet banks have names that do not even hint at their parentage. Egg is the most famous. Whatever its long-term prospects, it has certainly started a trend in Britain for staid old-economy banks to give their new Internet subsidiaries quirky names: Marbles (HFC), Cahoot (Abbey National), Smile (Co-operative Bank) and IF (Halifax). Non-British examples of separate branding include Uno-e (BBVA) and e-cortal (BNP Paribas).
- **Online hybrids**. More often, banks simply extend their existing brand to the web, either in its most obvious form (wellsfargo.com, bankamerica.com), or in a jazzed-up version. Chase Manhattan, for example, has made an effort to give a distinct identity to its online service, chase.com. In Britain, Barclays has sought to capitalise on the familiarity of its brand in marketing its online service ("its new-fangled name is 'Barclays'"). American Express has taken a similar line ("online banking from a company that's been around longer than a week"). The hybrid

approach has obvious advantages: it is easy to find the banks' websites, and whatever brand loyalty the bank commands offline will be transferred to the Internet. This matters in a business where familiarity and trust count for a lot.

◪ **Online alliances.** One quick and effective way for a bank to secure online credibility, as well as, potentially, to expand far beyond its existing client base, is to team up with a telecoms company, Internet-service provider or portal. At their most basic, such deals are simply arrangements to become designated providers of specified financial services for a popular website. But in 2000 some more ambitious alliances were announced: in Italy, Sanpaolo IMI, a big bank, was negotiating an agreement with Tiscali, an ISP. Similar alliances between banks and ISPs were announced in Spain (BBVA and Telefonica) and Germany (Comdirect, a Commerzbank subsidiary, and T-Online, Europe's largest ISP).

◪ **"White labelling".** Also called "private labelling", and growing fast, this involves banks' becoming silent partners. They might, for example, provide unbranded back-office services to enable another firm to run a bank. In Britain, Sainsbury's Bank is actually a 55:45 joint venture with the Bank of Scotland. Royal Bank of Scotland [RBS NatWest] is behind another supermarket bank, Tesco's, and also handles processing for first-e, an Internet bank.

Separate lives

There are advantages to establishing stand-alone Internet banks beyond projecting a zippier image. It allows the growth of a separate, more entrepreneurial management culture. It may also make it easier to attract and retain talented staff, who might otherwise defect to chance their arm in an Internet start-up. Some banks, for example, plan to issue "tracking shares" in their dotcom subsidiaries. But that may not go down well with offline bankers, aware that their T-shirted colleagues earn far more even though their venture may be losing money hand over fist.

So long as the stockmarket valued dotcom companies more highly than it does traditional banks, that provided another reason to separate out an online bank. Not only might this insulate the parent bank's shares from the market's negative view of their sector; it also offered some protection from the frenetic volatility in dotcom stocks. Eventually, it might also allow it to profit from high prices by floating the subsidiary on the stockmarket. Even if the dotcom share-bubble bursts, the market may continue to penalise banks whose online efforts are bundled up with their offline businesses, viewing the spending on the Internet business as an expensive investment in IT rather than as an essential customer-acquisition cost.

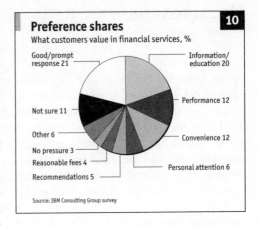

Preference shares 10
What customers value in financial services, %

Good/prompt response 21
Information/ education 20
Performance 12
Not sure 11
Other 6
Convenience 12
No pressure 3
Reasonable fees 4
Recommendations 5
Personal attention 6

Source: IBM Consulting Group survey

Despite the difficulties of integrating Internet-technology systems and Internet culture into old-economy financial institutions, incumbents start with huge advantages over new entrants. But as competition heats up, the banks will not be able to rely on loyalty or inertia to retain business indefinitely; they will also have to offer competitive online services. And in order to offer a wide range of top-quality products, they may have to think the unthinkable and take on the aggregators at their own game (see "Open sesame" on page 240).

In particular, they will need to compete in those areas of retail finance that are only just beginning to be sold and distributed online. Huw van Steenis at J.P. Morgan in London identifies three: mutual funds, insurance (of which more later) and mortgages. E*LOAN in America has already shown the potential for online mortgage distribution. The danger for the banks is not so much the direct threat of new lenders using the Internet to originate mortgage lending, but the indirect one of opportunistic mortgage refinancing, which (unlike the original transaction, which requires some offline checks on the house) can be conducted almost wholly electronically. Also, as sellers of all sorts of big-ticket items have found, the Internet is a wonderful place for comparison-shopping. More and more homebuyers who apply for a loan to their bank or broker are likely to have checked out the rates available on the Internet. And the Internet will make it easier for all kinds of sellers and brokers, including estate agents (property brokers) to offer financing as part of their sales package. In turn, that will put banks under pressure to provide other services connected with buying a home – perhaps help with finding a property, or removals, or legal advice. Otherwise they may lose out to other intermediaries.

Open sesame

THE BANK OF the future has already been built – in Stamford, Connecticut. It is called Open Bank, and it offers video tellers who dispense cash through translucent pneumatic tubes. A huge market-price display clatters away to remind visitors that they are part of the great global capitalist system. A coffee-bar counter is staffed by technical wizards who will sort out your mischievous laptop. A series of high-tech booths offer material on day-to-day money management, mortgages, insurance, investment, and financial planning.

This is a flagship branch, one of the few the bank still maintains. Down the street, it rents space for its super-ATM machine in a former branch that it has sold. Further down the road is a drive-by ATM, where you can stop on the way home. Back in your living room, a screen the size of a small cinema displays your favourite Impressionist painting, until you tell it to show you your diary, or personal balance sheet, or that web-page or video you simply must see now.

Branch, street, ATM and home are all, in fact, indoors, inside the same nondescript single-storey building housing the "Merlin Centre", a marketing and educational tool for John Ryan, a retail-banking consultancy. It is designed to help visiting bank executives to think constructively about the changes affecting their industry. The project assumes that an ever-increasing proportion of basic banking is going to be transacted on the Internet. For the banks that means, in the words of the John Ryan presentation, that "mastering cross-channel delivery is the issue".

But it also raises a difficulty that should by now be all too familiar to the banks. When a new channel for conducting business – the ATM, say, or telephone banking – comes along, the bank invests in it, and then hopes its customers will "migrate" to it. In practice, they often use it in addition to the other channels. What makes the Internet different from other distribution channels is the way it shows up with utter clarity that the products banks sell – however dressed up – are commodities which can be bought just as easily elsewhere.

Open Bank's message is that, to survive, the bank has three choices: to become a specialist in whatever it does best; to try to be, as now, a vertically integrated retailer; or to embrace "open finance" wholeheartedly, and sell other firms' products as well. Bankers who visit the future may not like it much.

How pleasant it is

The rich are more likely to use the Internet than the poor. Money managers have noticed this

IN THE CRISP sunshine of an early morning in spring, Shoreline Park is a vision in natural and man-made beauty, a refuge from the choking traffic that drones along Highway 101 through Silicon Valley. On the golf course, buggies potter through the vestigial mist. On footpaths, roller-bladers career serenely along, passing young mothers jogging behind their infants in their luxury prams. On the boating lake, terns, ducks and gulls swoop and paddle. The air is filled with a delicious fragrance.

It is the smell of serious money. According to Roberta Baxter, marketing director of myCFO.com, in March 64 people in the San Francisco Bay Area were becoming millionaires every day. At times, stockmarket turbulence may partially reverse that process, but more than 25% of the personal wealth of the United States is concentrated in the region, home to hundreds of thousands of the newly rich. What they have in common is that they did not grow up wealthy, and that they are for the most part comfortable using the Internet. Not surprisingly, looking after their money is one of the most contested battlefields in Internet finance. Or, as the practitioners themselves might put it, developing a sticky application for high-net-worth individuals (as even offline bankers call people richer than themselves) is a very attractive space to be in.

The rich are being wooed both by firms that provide a mass-market service on the Internet, and by traditional private banks now obliged to offer an online service too. In January 2000, Charles Schwab announced that it was paying $2.7 billion for U.S. Trust, a firm specialising in private banking for the affluent. In March, J.P. Morgan Private Banking, which boasts an 8% market share among America's "very wealthy", unveiled a fancy online service designed for accounts containing $1m or more. For such blue-blooded traditional firms, the Internet may offer a way of cutting costs, and therefore of catering for rather less wealthy investors too. In April Merrill Lynch and HSBC, a British bank, announced an online joint venture aimed at clients who have $100,000–500,000 to invest.

As they try to expand their client base downmarket, they will find themselves in competition with discount brokers and fund managers moving upmarket, along with many of their clients. Besides offering analysts' research, some brokers are now offering relatively sophisticated asset-management tools for ordinary investors from firms such as Financial Engines, co-founded by Bill Sharpe, a Nobel-prize-winning financial economist.

The wealth effect

Across the continent from Silicon Valley, Tracey Curvey, of the online brokerage at Fidelity Investments, is also impressed by the amount of money around. "There's so much wealth out there," she exclaims, pointing to the panorama of Boston framed by her office window. And a lot of that money is finding its way into mutual funds, of which Fidelity is the world's biggest supplier. It also, since 1999, has the world's biggest fund "supermarket", selling other firms' funds – a position it took, by a whisker, from Charles Schwab.

In America about a tenth of mutual-fund sales are now made online. Europe may soon catch up. Martin Jameson, of Booz Allen & Hamilton, a consultancy, estimates that in France, Germany, Italy, Spain and the Netherlands about 11% of mutual-fund sales are sold direct, by phone or Internet, and thinks that proportion will rise to 35% by 2005, of which 15% (worth about $130 billion) will be by Internet. In Britain, a survey by the Fund Managers' Association found that looking five years ahead, more than 80% of its members expected to distribute some of their products over the Internet.

Online fund supermarkets are also catching on in Europe, which has lagged America partly because it took longer to set up electronic links within the industry. J.P. Morgan expects that most new funds will soon be sold through such supermarkets. In France, BNP Paribas's online broker, e-cortal, offers 1,100 funds; in Germany, Deutsche Bank's Bank 24 sells 1,500. Britain's Egg offers 150 to its account holders. Fidelity's British mutual-fund operation was planning to launch an online supermarket in 2000.

For the very rich, the Internet offers access to even more sophisticated wealth-management techniques than before. But the same is true for the mass-market investor as well, including the humble mutual-fund buyer. This is one place where finance is already "open", and firms like Fidelity, which were suspicious at first, have found it not too bad.

Breaking cover

The insurance industry is still some way from offering a premium online service

ANYONE WHO HAS phoned around for insurance cover for their car might think that here is an industry just waiting to be shaken up by the Internet. The friendly insurance-company salesman will ask a lot of questions, and you can imagine him tapping the answers into his computer terminal. He will ask you to wait a few moments, and then give you a quote. You can even pay over the phone.

It seems quite an efficient process. Unless, that is, you decide you need to do some comparative shopping. There are plenty of insurers who will be happy to provide a competing quote, but the chances are that after two or three phone calls you will become fed up with answering the same questions over and over again. It sounds like just the job for the Internet, which can be used to automate processes that involve providing similar information to different organisations. Given the large number of motorists, and the fact that car insurance is a mandatory purchase, here, surely, is a lucrative market waiting to be tapped.

But insurance has been quite slow to move on to the net, for a number of reasons, and no insurance company has yet established a big online name for itself. In America, for example, a survey by Harris Interactive found that online insurance companies were less familiar to Internet users than firms in almost any other sector (see chart 11). Forrester Research forecasts that by 2003, $4 billion-worth of insurance will be sold online in America – a huge increase on the paltry amount now being written on the Internet, but still less than 0.5% of the total market.

Slow migration

Elsewhere – in Britain, for example – prospects look rather better, although, as so often in the online world, companies are rather coy about giving figures. Certainly, most British insurers already provide online services, offering discounts on offline premiums. Halifax, a building society (thrift) turned bank, is to set up an online insurer called esure. It is the brainchild of Peter Wood, founder of Direct Line, a telephone-insurance subsidiary of the Royal Bank of Scotland. (He is also a director of The Economist Group.) Having started business in 1985 as a

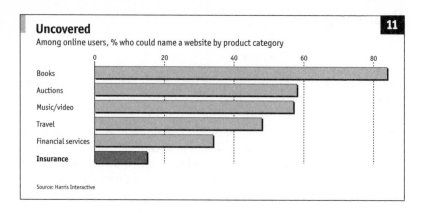

Uncovered `11`

Among online users, % who could name a website by product category

	0	20	40	60	80
Books					
Auctions					
Music/video					
Travel					
Financial services					
Insurance					

Source: Harris Interactive

car insurer, Direct Line transformed the industry. By 1993, it had become Britain's market leader, and has kept adding products (such as home and travel insurance, and mortgages). That may have prepared the market for online distribution.

In explaining why they have been relatively slow to adopt the Internet, insurers point to the complexities of their products. It is true that many consumers prefer to talk through their insurance needs rather than buy them off a virtual shelf, and also that underwriting rules and criteria vary. But there are two bigger factors that are holding insurers back, particularly in America.

The first is that established insurance companies moving online suffer from channel conflict in spades. Most have networks of agents to whom the Internet is a direct challenge. The agents' commissions typically account for 25% of the cost of the insurance, which is precisely where direct insurers hope to make their savings. The second set of obstacles lies in the divergent regulatory regimes between countries, and, in America, between states. These make it more complex to buy insurance than many other financial products.

Every big insurer, like every big bank, now has an online strategy. After all, they are often divisions of the same institution, following the wave of mergers in recent years that have created "bancassurance" giants, such as Citigroup in America and ING in the Netherlands. And, like the banks, the insurers have been spurred into action by the activities of troublesome online start-ups. In a third parallel with Internet banking, these start-ups, broadly speaking, fall into two categories: aggregators (in old-economy language, brokers), and virtual insurers.

Examples of the first are InsWeb, Quicken and Insurance.com, a

Fidelity affiliate. All are online markets that boast of being able to seek out the cheapest insurance policy from a wide range of suppliers – though the stockmarket punished it severely in April 2000 when one of the most important, State Farm, withdrew its participation. InsWeb prides itself on its independence. This has the advantage of offering price savings, but it does make application cumbersome, because it has to cater for the requirements of all potential suppliers. ECoverage, a California-based Internet insurer, uses this disadvantage in its own sales pitch, which stresses the simplicity of its service. It points out, for example, that to insure one driver with a clean licence for one car, it would need to ask only 16 questions on two web pages. InsWeb would ask 72 questions spread over five pages.

But eCoverage's business model also has considerable drawbacks. InsWeb argues that by simplifying the process so much, the insurer may end up with a lot of bad risks, which would eventually have to be reflected in premiums. And eCoverage has faced an uphill struggle in securing regulatory approval to operate in other states. In March 2000 it announced an agreement with the American arm of Royal & Sun-Alliance, a London-based insurer. RSA would provide eCoverage with a "shell" for its operations in 48 states, and eCoverage would provide access to Internet-based information and processing.

This approach is an attempt to marry the interests of the start-up with those of the established giant. But in insurance, as in other financial services, the use of the Internet simply to support existing distribution channels is unlikely to be a long-term solution. Disintermediators and intermediaries can co-exist only for so long.

Earthquake on the Street

Technology is revolutionising wholesale financial markets

SO FAR, THIS survey has concentrated on how the Internet is affecting the way financial firms do business with their retail customers – what in Internet jargon is known as the business-to-consumer, or B2C, market. But that is only one part of a broader transformation of the way those firms do business with each other and with their corporate clients. The whole industry is in a state of flux. Technology has made possible a global digital financial market where, in theory, any security in any currency could be traded anywhere at any time, and where that trade could be settled instantly. But the world is not ready for such a market. Not only do national regulations get in the way; so do the structures of the financial markets themselves, and the competing interests of the various players.

Hence the bewildering reels currently being danced by the world's investment and commercial banks, stockmarkets and other exchanges. It is obvious to all that dislocating change is coming, but not yet what shape it will take, nor which institutions will emerge at the top of the pile. So every day, it seems, some new merger or alliance is being announced, or there is news of a forthcoming platform for trading American Treasury bonds, or euro-denominated shares, or whatever, that "will revolutionise the market". Often the same investment banks, hedging their bets, have invested in several competing platforms. Allies one day in one business, they are competitors the next in another.

As a result, the distinctions between the different sorts of institution are blurring. For years, mergers and takeovers have been creating institutions that bring together different specialisations: commercial and investment bank, stockbroker and trader, insurer and fund manager. Now, in America, the growth of electronic communications networks (ECNs) has seen broker-dealers turn into exchanges, and two have applied to be regulated as such.

For investment banks, and retail banks in their wholesale businesses, the Internet adds another twist to this already tangled skein. It offers enormous improvements in the distribution of information and products, but also makes it easier for the users of financial services to do without the banks' intermediation, and for new entrants to the market to challenge the big players' oligopoly.

Having been exhorted for so long to emulate the gung-ho adventurousness of American capitalism and the seemingly bottomless depth of its capital markets, investors in boring old places such as Europe and Japan have recently encountered another, less enviable trait of the American way of doing things: the ludicrously oversubscribed initial public offering (IPO) of shares in small technology companies. Typically, these companies' share offers are sponsored by a blue-chip investment bank. Only a small proportion of the shares is offered, at a very high price. Demand is such that the price soars even further, enabling the underwriters (said blue-chip bank and its chums) to "flip" the shares for a tidy profit in the grey market, before official trading begins. Retail investors often have no access to shares at all, or if they do, they find they are unable to sell them until off-loading by the underwriters has already depressed the price. In March 2000, for example, this mechanism left some British investors in a much-publicised Internet start-up, lastminute.com, sitting on losses.

Inordinately Profitable Offerings

Scott Ryles, chief executive of Epoch Partners, a new online investment bank in San Francisco, says that in an American IPO, an average of 80% of the shares are held by institutions. A year later, however, that proportion has come down to only 27%, and most of the shares are in the hands of retail investors, many of them clients of discount online brokers. Often, especially in an Internet start-up, the online brokers' clients are allocated little of the IPO itself. The sense of grievance this causes led three of the brokers (Schwab, Ameritrade and TD Waterhouse) to link up with three venture-capital firms to found Epoch.

A handful of such online investment banks, concentrating on technology companies, are now up and running. Their market share is still tiny, but Ron Readmond, a founder of one of the biggest, Wit Capital, says that they are causing the beginnings of a panic among the "cabal" of banks that dominate the IPO business, which realise that the newcomers are threatening their business. Mr Ryles, who worked at Merrill Lynch until 1999, agrees that the "cost side" of the IPO business on Wall Street has to change. But the big firms are once again hampered by channel conflict. Online distribution bypasses their branch networks.

Yet those old firms are themselves being forced to compete. At Merrill Lynch itself, for example, Michael Ryan, the firm's global head of equity capital markets, is proud to show off the "new-issue application" of "Direct Markets", an Internet-based syndication and distribution tool.

He concedes that Merrill had some trouble establishing itself in the niche it thought it deserved in technology IPOs. Now it has the capacity to put almost all of the IPO process online – from the prospectus to videos of the company's management – instead of the traditional "road-show" tours to sell its services to investors.

Curiously (or perhaps not), one aspect of the "cost side" that neither Mr Ryan nor the new banks expect to come under immediate pressure is the standard fee charged for an IPO. But when companies have been able to raise capital at such a high price to the investor (ie, so cheaply to themselves), what is a 7% flat fee between friends? If capital gets more expensive, however, that juicy component of its cost will surely come under more scrutiny.

Such fees are important because so much bank income in recent years has derived from volatile sources: trading securities and venture-capital investments. So investment banks are devoting enormous efforts to finding ways of protecting their profits from online encroachment. J.P. Morgan, for instance, has formed LabMorgan to give a high-profile unifying theme to its e-finance ventures.

The fees investment banks earn for arranging bond issues are also vulnerable. In fact, though they would never admit it, so obligatory is e-enthusiasm these days, in private some firms in the bond business probably curse the Internet. It is set to change the whole bond market, from underwriting to secondary trading. Already, top-ranking bond-arrangers need an online syndication system, if only because some of the world's most important borrowers will demand it. "All our bonds", claims Afsaneh Mashayeki Beschloss, of the World Bank's treasury department, "will be e-bonds."

The beauty of e-bonds

Ms Mashayeki Beschloss, who concedes that some borrowings with complicated structures may prove e-exceptions, was delighted by the experience of the Bank's two online bond issues in 2000. In both, arranged by Goldman Sachs and Lehman Brothers, part of the deal – 30% of the first, 50% of the second – was offered direct to investors over the Internet. Usually, such issues are syndicated among the small (about 400-strong) group of global investment houses, which place them with their clients or hold them in-house.

From the borrower's point of view, the online transaction offers two big advantages: transparency, in that the borrower can see who the end-investors are, and what commitments they are making; and access to new investors. The World Bank, although well-known in the international markets, was less familiar to the American retail investors it wanted to reach. The online issues gave it access to that market (although "retail" is a relative term: the smallest purchase was $1,000, but one investor turned out to want $250m-worth of one bond).

In the secondary market, too, upheaval is already under way. The market, which covers 4m separate issues of widely differing kinds, many of them rarely traded, has been less susceptible to electronic trading than have the stockmarkets. Bond-trading has long been dominated by institutional investors. They have placed their orders through dealers, whose pricing is opaque. But now America's Bond Market Association has counted 39 different firms and alliances that offer – or plan to – secondary electronic trading. These include proprietary systems, such as Goldman Sachs' Web.ET; trading networks sponsored by single firms, such as State Street's Bond Connect, which hopes to create liquidity by allowing participants to trade bundles of different bonds; electronic interdealer-brokers, such as BrokerTec, aimed purely at firms that already make markets in bonds; joint-venture arrangements, such as TradeWeb, an ECN set up in 1998 by Goldman Sachs, Lehman Brothers and others for trading in American Treasury bonds; BondClick, a European equivalent for government bonds launched in 2000; Tradebonds.com, an Internet-based system offering access to prices for thousands of Treasury, municipal and corporate bonds; Securities.hub, which will allow investors access to research and price quotations from six big firms (Goldmans, Lehmans, Merrill Lynch, Morgan Stanley Dean Witter, Salomon Smith Barney and J.P. Morgan) without having to log on separately; and LIMITrader.com, an online marketplace for trading corporate bonds.

TradeWeb, which operates in the very heavily-traded treasury market, has been the most successful of these so far. Larry Buchalter, of Goldman Sachs, thinks that the level of liquidity in particular markets will continue to determine their development in the future. For much-traded issues, such as Treasury bonds, a central liquidity pool will form to create, in effect, an electronic exchange. For less liquid bonds, however, more is needed than the systems that have emerged so far. Most are matching or "crossing" networks, where sellers hope to meet buyers, but have no market-maker to guarantee them a price. What is needed, says Mr Buchalter, is "crossing with capital". But in the bond markets, as elsewhere in finance, a revolution is under way. "It is a once-in-a-career thing," says Mr Buchalter. "It's going to redefine our business."

anyfinancialproduct.com?

Since the Internet offers such obvious advantages for investors and lenders, it is tempting to ask if it will become the medium through which all financial products are traded. Certainly that seems to be the way some other industries are going – witness the proliferation in recent months of online business-to-business exchanges, in everything from motor parts to plastics, bandwidth to chemicals.

But finance is different, in the sense that many of its markets are highly automated already, but using non-Internet systems. For large interbank dealings in deposits and foreign exchange, for example, there is no obvious reason to move to the web when existing systems are doing the job perfectly well. Indeed, new ones are being introduced, such as FX Connect, part of Global Link, an "extranet" run by State Street, an American bank that leads the world in providing custody services. It specialises in catering for large investors, many of whom would prefer to rely on established services rather than take a chance on a new website. In Sweden, SEB's web-based foreign-exchange service is used by corporate clients, not by the much larger interbank market.

Nevertheless, various online exchanges and aggregators aimed at the wholesale market are in operation or under construction. Creditex, for example, founded by two former Deutsche Bank traders, is designed as a platform for trading credit derivatives. These form a rapidly expanding segment of the derivatives market, enabling the credit risk associated with a loan or bond to be traded separately from the underlying asset. Creditex aims to provide an Internet-based marketplace where traders can negotiate anonymously (up to the point of agreement on price). The aim is to improve liquidity, standardisation and price transparency in a new and fragmented market.

Trying to reach a much broader market is CFOweb.com, due to be launched in 2000 by Integral, an American software developer. Its target market is people running corporate finances. It plans to offer a range of financial products, such as spot and forward foreign-exchange dealing, currency and interest-rate swaps, by linking corporate buyers of these products with a range of providers. By April 2000, it was claiming 2,300 registered members and a range of "providers" that included Bank of America, ABN Amro, a big Dutch bank, and the derivatives-trading arm of American International Group, a huge insurer. Clearly the viability of such a system depends on the number of firms using it, and changing existing corporate behaviour will take time.

However, over the next few years some such model seems likely to

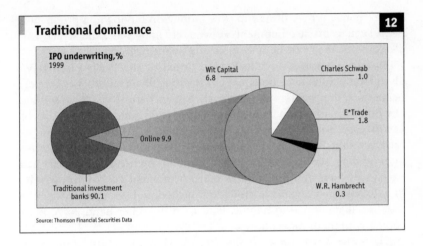

Traditional dominance   `12`

IPO underwriting,%
1999

Wit Capital 6.8

Charles Schwab 1.0

E*Trade 1.8

Online 9.9

Traditional investment banks 90.1

W.R. Hambrecht 0.3

Source: Thomson Financial Securities Data

succeed, including perhaps a big increase in direct trading of financial instruments between non-bank corporations. If, for example, a large online market in currency and interest-rate swaps were to develop, then banks' role as intermediating counterparties may be called into question. So wholesale banking, too, may eventually have to move towards "open finance".

The same goes for the more mundane bread-and-butter businesses of commercial banks. The web threatens their hold on their core corporate customers (especially the smaller ones) seeking access to business loans or trade finance. Many small businesses have, understandably, tended to deal with their local bank branch. Many banks have taken advantage of this to charge them near-monopoly rents. In Britain, for example, it is the need to retain competition in this sector that has been the biggest obstacle in regulators' minds to mergers between banks with overlapping branch networks. The Internet offers such small firms access to far wider sources of credit.

Some banks have recognised this danger, and tried to turn it into an opportunity. In America, Citigroup has a website, bizzed.com, which describes itself as a "portal of business services", aimed specifically at the small business customer. Similarly, in Britain, Barclays is working on a B2B site in conjunction with Oracle, a software firm, and Andersen Consulting. Ian Arthur, of Andersen, argues that banks "need to control the point of relationship with their corporate customers". In other words, in this market banks do need to be portals. Barclays's idea is to expand into the e-procurement field, using their site as a "horizontal"

B2B exchange where firms can save money on buying non-production items such as office equipment, or even recruiting staff, and which will also provide the settlement services and financing if required.

Big Internet banks, such as Bank of America and Wells Fargo, have also identified the large potential in providing financial services for participants in the proliferating "vertical" B2B exchanges (ie, those serving one industry), of which there are already several hundred. Steve Ellis, of Wells Fargo, for example, says that his bank, a leading agricultural as well as online lender, has been combining these two areas of expertise in providing services to an online almond exchange. Although B2B exchanges have the potential to cut procurement costs impressively, they expose participants to credit, transaction and settlement risks that they are ill-equipped to handle. Banks, on the other hand, have been coping with them for centuries.

The same is true for trade finance, a huge market worth trillions of dollars each year, which used to entail the destruction of great swathes of forest – in letters of credit, bills of exchange, certificates of origin, invoices, bills of lading, insurance certificates and so on. Often it still does, but big banks have managed to automate much of the process, for example by scanning documents into their computer systems. Now a number of systems are trying to take this process further, and are vying to become the online standard for trade processing. Bolero.net is a joint venture of SWIFT, a global payments network, and The Through Transport Club, a shipping insurer. It provides a system for trade data to be exchanged via the Internet. Another approach comes from TradeCard, which in March 2000 signed an agreement with Thomas Cook, a travel and financial-services company, to offer an electronic alternative to traditional letters of credit. Thomas Cook will act as an intermediary for payments around the world.

This market is so inefficient that it is not unusual for trade debt to be priced at a much bigger margin than the same borrower is paying for, say, a bond issue. The founders of LTP, a London-based boutique consultancy set up by trade financiers formerly at Deutsche Bank, believe that besides streamlining the cumbersome process of documentary credits, the Internet may also help shake up their financing, through its natural flair for syndication. At present, many trade-finance assets are hardly traded at all. Often they simply sit on a bank or exporter's balance sheet, subject to the scantiest of risk analysis.

One world, ready or not

In the future, it will be possible to trade any financial instrument anywhere, at any time. But how to get from here to there?

FOR LUDDITES EVERYWHERE, April 5th 2000 was a day of exquisite *Schadenfreude*. After an American court had found Microsoft guilty of anti-competitive practices on the previous day, shares in high-technology companies in America gyrated frantically. Investors in the London market, which had also been enjoying a high-tech boom, might have wanted to trade, especially since it was the last day of the fiscal year, and therefore the last chance to take advantage of tax allowances against capital gains. But they could not trade on the London Stock Exchange (LSE). It was shut for nearly eight hours because its computers had gone down.

It was a salutary reminder that it is not just institutional resistance that is holding back the future, but that the wonders of technology are also constrained by the failings of electronic infrastructure. This particular disaster hit a sophisticated custom-built system. The Internet itself is not yet ready to support a global financial market.

Nevertheless, "In a few years, trading securities will be digital, global and accessible 24 hours a day. People will be able instantly to get stock-price quotations and instantly to execute a trade day or night, anywhere on the globe, with stockmarkets linked and almost all electronic." Thus said Frank Zarb, chairman of Nasdaq, the electronic stockmarket that has been the nurturing ground for the high-tech revolution, and which merged with the American Stock Exchange in 1998. Most in the industry would agree that such a market, at least for shares in a few hundred of the world's biggest companies, is desirable and attainable. But Mr Zarb's own road-map for reaching this Eldorado – "an interconnected platform" of Nasdaq and its affiliates around the world – is only one of several on offer.

Indeed, activity in the world of financial exchanges is perhaps even more hectic than among the firms that use them. New "Alternative Trading Systems" (ATS) seem to emerge almost daily, particularly in Europe and America, and new cross-border alliances form or dissolve. Meanwhile, the established stock exchanges are advertising radical re-mouldings of themselves to cope with the combined forces driving

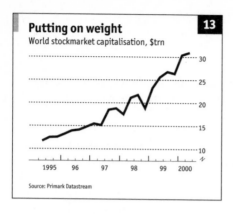

Putting on weight `13`
World stockmarket capitalisation, $trn

- 30
- 25
- 20
- 15
- 10

1995 96 97 98 99 2000

Source: Primark Datastream

technological advance: globalisation, the growth in Internet-based trading and, in many rich countries, the demographic pressure of older, longer-living populations that invest in stockmarkets more than ever before.

They are modelling themselves on the way some derivatives exchanges round the world have coped with this brave new world, using new technology and a network of international agreements to turn themselves from national, floor-based exchanges into global electronic markets. The paradox is that whereas the general direction of change is clearly towards consolidation within national markets as well as across borders, getting there has seemed to require the reverse: a proliferation of competing marketplaces. New small fry keep appearing even as established exchanges team up. Even the world's two biggest exchanges – the New York Stock Exchange and Nasdaq – have at least discussed merging.

It is odd that electronic communications networks (ECNs), which are often described as "pools of liquidity", should give themselves names such as Island and Archipelago, after the solid bits surrounded by water. Mr Zarb insists that, since the prices they offer are all reflected on Nasdaq computer screens, "they are part of our market". But the success of the ECNs in handling a growing share of Nasdaq business (and to a lesser extent that of the New York Stock Exchange, NYSE) has led to worries about fragmentation among different trading systems (see chart 14), and accusations that the ECNs are indeed islands, not part of the ocean.

The growth of the ECNs stems partly from the spread of Internet trading, and partly from changes made in early 1997 to SEC rules. These now oblige Nasdaq market-makers to display and execute their customers' limit-orders (ie, to sell or buy above or below a set limit) when the price is better than the market-maker himself is offering, or adds to the volume offered. So the market-maker's own customers can compete with him, which has had the predictable effect of narrowing the spreads between bid and offer prices. Most ECNs are matching systems, allowing limit orders to be filled without going through a market-maker.

This leads to worries that liquidity is being siphoned out of the main exchanges, and that ultimately this might be unfair to some market participants by denying them the chance to trade at the best price. It is feared that price formation might become unreliable. Some of Wall Street's biggest firms, such as Merrill Lynch and Goldman Sachs, would like to see a more unified market in which prices available in every trading mechanism are open to the whole market, and the trade is done at the best price.

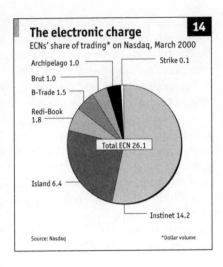

The electronic charge

ECNs' share of trading* on Nasdaq, March 2000

Archipelago 1.0 — Strike 0.1
Brut 1.0
B-Trade 1.5
Redi-Book 1.8
Total ECN 26.1
Island 6.4
Instinet 14.2

Source: Nasdaq   *Dollar volume

Some of the online brokers, which save money by executing a large number of trades internally, argue that it should be enough simply to match the best available price. Kathy Levinson of E*Trade thinks the proposal for a "central limit-order book" is "a scary concept", because "it flies in the face of competition". It is, she says, "a solution in search of a problem": an ECN is only as good as its liquidity, and the market can determine which will survive. As it happens, both E*Trade and Goldman Sachs are shareholders in Archipelago.

Whatever the outcome, this debate highlights a problem that the NYSE and Nasdaq have in common with stock exchanges elsewhere. They are mutually owned organisations, the interests of whose members are at best misaligned and at worst in direct competition with each other. So both the NYSE and Nasdaq, like the Stockholm and Sydney stock exchanges before them, and like London, are moving towards shareholder structures. This, it is hoped, will equip them to move fast enough to cope with the myriad challenges they now face.

In Europe, those challenges include not only the spread of electronic trading and the arrival of new competitors, but, since 1999, the single currency too. In response, national stock exchanges in Europe are forming alliances or merging.

Europe gets it

In their domestic markets, Europe's exchanges have so far proved less vulnerable to upstart ECNs than their American counterparts. However, the launch in London in March 2000 of E-Crossnet, a matching system backed by some of the world's biggest fund-managers, caused some concern. Like Posit, another crossing system, its advantages for users are, first, to cut dealing costs, and second, to provide anonymity, avoiding the "market-impact" costs of large trades (ie, the effect they have on the price).

European exchanges' squabbling has slowed down their efforts to provide efficient cross-border trading. This has opened the door to new competitors, including Easdaq, based in Brussels, Tradepoint, an electronic exchange set up in 1995, and Jiway, a joint venture of OM (which runs the Stockholm exchange) and Morgan Stanley.

In Europe, much of the difficulty lies not so much with the exchanges as with the plethora of clearing and settlement systems attached to them. There, too, consolidation is under way. In 1999 Deutsche Börse's clearing arm merged with Cedel, traditionally a bond-clearing house, to form Clearstream. Euroclear, another leading bondmarket clearer, is to buy Sicovam, the French settlement system. And the London Clearing House and Clearnet, owned by the Paris Bourse, are to form a joint venture to build a single European clearing house.

As European exchanges are consolidating, the stakes are mounting. The prize is not just to be the leading European centre for trading European shares; it is to be the most important European leg in the 24-hour global market that will one day emerge. Jos Schmitt, the former chief executive of the Belgian derivatives exchange and now a partner in the Capital Markets Company, a financial-technology consultancy, thinks that within three to five years investors will have direct access to markets around the world, and will be able to trade and settle where they live.

The digital future

Within a few years, the Internet will become the instrument of choice for managing our finances

CALL IT THE curse of *The Economist*, the vengeance of some gremlin angered by the newspaper's long characterisation of the market in some technology shares as a bubble. In the course of the research for this survey, one of Europe's leading banks, which claims to be a path-breaker in Internet finance, found it impossible to log on to show off its system in its own headquarters. A highly regarded Silicon Valley start-up could not even reach the World Wide Web. One *Economist* journalist reported that an online bank had mislaid £4,000 of his money. Another found herself cursing her computer screen after failing, for the third time, to persuade an Internet broker's system to accept her form for opening an account. The dirty secret of online finance is that, as yet, the technology is too often slow, inconvenient and unreliable.

To those for whom glasses are always half-empty, this is cause for deep scepticism about how much the Internet will change financial services. To the half-full brigade (including this survey), it conveys a different message altogether: if so many people are already using Internet services, just think how many will use them when they actually work properly. Already, Internet sites have become vastly faster and better than they were in the dark age of 1995. The technology continues both to spread and to improve. In particular, once constant, broad-band access becomes affordable for the average home, conducting your financial affairs on the Internet will come to seem as normal as watching television. It will become the medium through which most of the rich world manages its finances most of the time.

For financial-services providers, this will speed up a number of trends already felt:

- **Boundaries will blur.** The distinctions between banks, brokers, insurers, fund managers, pension-providers and so on will become less and less important as the process of buying financial commodities is electronically disintermediated. Partly for this reason,
- **Consolidation will continue.** The takeovers and mergers that

257

have transformed the financial sector will carry on. In Europe, a number of banks have already said that they intend to sidestep the difficulties of cross-border mergers by Internet-based expansion. This is a passing phase, because:

■ **Clicks and bricks need each other.** There will be convergence between online financiers acquiring an offline presence, and the physical banks turning themselves into virtual providers. This will imply even more closures of branches than have been seen in recent years. Customers want to know that some personal contact is available, but call centres and branches will have very different roles, including the provision of:

■ **Open finance.** Competition from non-bank Internet sites will eventually force most financial-services firms to offer supermarket-style services where their clients have access to their rivals' products. This will hurt. But it is better than the alternative, which to some extent will be inevitable:

■ **Disintermediation.** In an earlier chapter, this survey reported that Moody's Samuel Theodore regards the Internet as just part of a process which he calls the "fourth disintermediation". One of the characteristics of this process is that banks now require less of a physical presence than they used to. But Mr Theodore also raises the possibility of a "fifth disintermediation", in which the Internet and related technologies will bypass banks in payments, clearing and settlement.

That day, however, is still a long way off. Many big financial institutions, despite the sniping of the dotcom upstarts, are in fact quite well placed to meet the challenge of the Internet. They have trusted brands, which may have an even greater value in cyberspace than offline; they have customer bases that start-ups can only dream about; they have made huge investments in state-of-the-art technology; and many of them are making profits that will enable them to spend even more. The only risks are their own complacency, arrogance and sluggishness. If they can avoid those sins, the Internet becomes less of a virtual threat than a virtual promise.

The material on pages 201–58 first appeared in a survey written by Simon Long in *The Economist* in May 2000.

POSTSCRIPT

THE BURSTING OF the dotcom bubble inevitably reverberated through the world of online finance, with a number of predictable consequences. First, like other dotcoms, online-finance start-ups faced a much tougher fund-raising environment. Profits, or at least revenues, came back into fashion, and a number of loss-making ventures quietly folded. Second, there was a lull in the cut-throat pricing war. The biggest impact of this has been in the one area of business where the online revolution seemed to have succeeded – share-trading. Losing money as a way of gaining market share seemed reasonable when bulls were roaring. The bear market, however, has exposed "customer-acquisition costs" as a life-threatening drag on even the biggest online brokers.

All of this reinforces the conclusion of the survey: that big financial institutions were, and are, fairly well placed to withstand the "virtual threat". The benefits of having a solid physical presence, of a diversified "clicks-and-bricks" business model, and a chunky capital base to invest in new technologies have become more obvious. Indeed, the end of dotcom euphoria has given these big institutions another advantage: time. What seemed to be an overnight revolution has turned out to be part of a long process of electronic disintermediation of retail and wholesale financial markets. The revolution has not been cancelled; it is just spinning more slowly.

VI

MOBILE TELECOMS

The Internet, untethered

The Internet is going mobile. To succeed, it must learn from the mistakes made in the fixed-line Internet boom

NIGHT-TIME ON the neon-lit streets of Tokyo's Shibuya district, and the scene seems like a vision of the future. Here are the colourfully illuminated skyscrapers and huge video billboards, there are the throngs of exotically clad teenagers. And everywhere there are people talking and typing into astoundingly advanced mobile phones – tiny devices that fold up like colourful make-up compacts, are incredibly light, and have small, vivid colour screens capable of displaying sophisticated graphics. Most important of all, these phones enable their users to access the Internet while on the move. In Japan, the Internet has gone mobile.

In this respect, say prophets of the mobile Internet, the scene in Shibuya is indeed a foretaste of the future. The convergence of the two fastest-growing communications technologies of all time–mobile phones and the Internet – will, they say, make possible all kinds of new services and create a vast new market as consumers around the world start logging on from Internet-capable phones. Market-research firms and consultancies predict that by 2004, the number of mobile-Internet users will rise to around 1 billion, from 200m today.

It was big numbers like this that prompted mobile-network operators around the world collectively to pay more than $100 billion in 2000 for licences to operate "third-generation" (3G) networks (see "Generation game" on page 282). Unlike existing second-generation (2G) networks, 3G systems are designed to handle data quickly and efficiently along-side voice calls, and are thus expected to be one of the key technologies that will underpin the mobile Internet. The enormous sums being spent on 3G amount to the largest bet in business history on the introduction of a new technology. The operators have been trying to justify this spec-tacular gamble with a torrent of hype.

But the downturn in the technology sector, and elsewhere in the world's economies, has cast a long shadow over 3G's future. The huge debts incurred by operators in buying their licences, and their need to cut spending in order to service their debts, were arguably one of the downturn's causes. On top of the cost of acquiring the licences, operators will have to spend another $400 billion or so to build their 3G networks, so they are teaming up to share infrastructure and reduce costs. There are also fearsome technical problems to overcome. This has delayed the start of 3G services: the world's first 3G network was launched in Japan on October 1st 2001, five months late, and 3G networks in other parts of the world will arrive months, even years, later than originally planned.

Generation game

First-generation (1G) mobile phones, which have been around since the 1970s, use analogue technology to transmit voice calls. Sound quality is generally poor, use of radio spectrum is inefficient, and calls can be intercepted quite easily. Of the world's 800m mobile-phone users, around 70m, mostly in the developing world, have 1G phones.

Second-generation (2G) mobile phones use digital encoding. Communication between the handset and the base station takes the form of an encrypted data stream, making eavesdropping almost impossible. As well as voice calls, 2G phones can also send and receive data, so they can provide limited data services such as text messaging and Wireless Access Protocol (WAP) browsing. Most mobile phones in use today are 2G.

Enhanced second-generation (2.5G) phones, which have recently become available, extend 2G technology to offer improved data capabilities, such as higher transmission rates and always-on connections, so these phones can support more advanced data services.

Third-generation (3G) phones will offer high-speed, always-on data connections, as well as support for applications such as videotelephony and advanced data services with full Internet access. 3G networks are also designed to support large numbers of users more efficiently than 2G networks, to allow for future expansion.

At the same time, expectations have been scaled back. Operators are no longer talking of watching video clips on the train, or videoconferencing in a taxi; instead, they are concentrating on more realistic goals, such as using phones to access e-mail, download news and weather reports, and call up location-specific information. Meanwhile, many operators have already launched so-called "2.5G" networks – upgraded 2G networks that offer some of the benefits of 3G (in particular, an "always on" data connection) but cost a lot less. This means that 3G will not be the sudden miraculous transformation originally touted, but a gradual evolution from today's systems. And the truth is that a great deal is already possible with existing mobile networks.

Indeed, amid all the carnage in the technology sector, an important transition is under way. Even allowing for the late arrival of 3G, it will not be long before the number of Internet-connected mobile phones exceeds the number of Internet-connected PCs. Nokia bullishly predicts that this will happen in 2002; Ericsson, another handset maker, says 2003; even pessimistic forecasts put the date at 2005. It will probably be 2.5G handsets, rather than 3G ones, that will overtake PCs to become the world's most prevalent Internet-access devices. But one way or another, phones will soon become the predominant means of access to the Internet.

The obvious recent analogy is with the explosive emergence of the fixed-line Internet in the mid-1990s. In many ways, the mobile Internet is at the same stage of development as the Internet was in 1995. There are hundreds of start-ups, and nobody really knows which technologies or business models will win, or what consumers or corporate users want. There are plenty of lessons to be learnt from the mistakes made on the fixed-line Internet. This survey will argue that heeding those lessons will be crucial if firms are to prosper on the mobile Internet.

A less obvious but more useful analogy is with the switch from the electric telegraph to the telephone in the last quarter of the 19th century. The telegraph, like the Internet, was a revolutionary communications technology that transformed social and business practices, but it could be used only by skilled operators. Its benefits became available to the public at large only when the telegraph evolved into the telephone – initially known as the "speaking telegraph". The Internet is still in a telegraphic stage of development, in the sense that the complexity and expense of PCs prevents many people from using it. The mobile phone thus promises to do for the Internet what the telephone did for the telegraph: to make it a truly mainstream technology.

Because it used the same wires, the telephone was originally seen as merely a speaking telegraph, but it turned out to be something entirely new. The same mistake is already being repeated with the Internet. Many people expect the mobile Internet to be the same as the wired version, only mobile, but they are wrong. Fixed-line web pages will not fit on to phones or handheld computers. Instead, the mobile Internet, although it is based on the same technology as the fixed-line Internet, will be something different, and will be used in new and unexpected

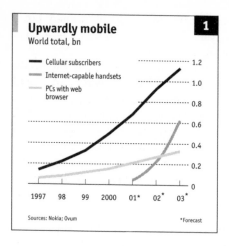

Upwardly mobile 1
World total, bn

— Cellular subscribers
— Internet-capable handsets
— PCs with web browser

1.2
1.0
0.8
0.6
0.4
0.2
0

1997 98 99 2000 01* 02* 03*

Sources: Nokia; Ovum   *Forecast

ways. It will rely on content and technology from the Internet, but instead of the leisurely browsing and searching familiar from the PC, it will concentrate on sending and receiving timely, relevant nuggets of information. Meanwhile, of course, the Internet in its current form will still be around.

The term "mobile Internet" itself is problematic. "Mobile data services" might be more like it. Other names that are bandied about include "the wireless web", "mobile e-services" and "mobile online services". But this survey will refer to the emerging mobile data environment as the "mobile Internet", even though with the benefit of hindsight it may prove to be as much of a misnomer as "speaking telegraph". One attraction of the term is that the word "Internet" technically refers to a network of networks, and that is exactly what the mobile Internet will consist of. It would be a mistake to equate the mobile Internet with 3G and assume that, because 3G is in trouble, other mobile-Internet services are too. Instead, there will be many overlapping and interconnected wireless networks. There will also be a variety of access devices, including laptops, handheld computers and other, as yet undreamed of, gizmos. Different networks and devices will be used in different situations by different kinds of users in different parts of the world. But it seems clear that the Internet-capable phone will predominate as the mobile device of choice.

Combining the Internet with mobile phones will pose technical, business and cultural challenges. For a start, there is a clear conflict of attitudes between Internet and mobile-phone users. Internet users expect things to be free, and are prepared to accept a certain degree of technological imperfection. Mobile users are accustomed to paying, but expect a far higher level of service and reliability in return. Those on the Internet side of the fence complain that wireless firms don't really understand data networking; those on the wireless side complain that Internet technology is flaky.

But the differences between these two worlds also present an opportunity. Content providers see the mobile Internet as a way to start charging for their wares. Wireless-network operators see themselves as potential gatekeepers to the mobile Internet, and may be in a position to grab a share of online commerce revenues, which fixed-line Internet-access providers have failed to do. Hardware and software companies see all sorts of new opportunities in products to knit the Internet and

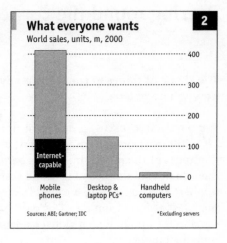

What everyone wants
World sales, units, m, 2000

Internet-capable

Mobile phones

Desktop & laptop PCs*

Handheld computers

Sources: ABI; Gartner; IDC   *Excluding servers

mobile networks together. In short, the mobile Internet is a chance to build a new network, and get it right this time–by learning from the mistakes made by all those failed dotcoms.

The biggest gamble in business history; control of a vast new medium; the opportunity, at last, to monetise the Internet: clearly, a great deal is at stake. Some say it is all just wishful thinking. But in many parts of the world – not only Japan – millions of people are even now using phones and other handheld devices to communicate on the move. All over the globe, the foundations for the shift to more advanced services are already in place.

Peering round the corner

The shape of things to come is already becoming apparent, if you now where to look

WILLIAM GIBSON, a science-fiction writer, once observed: "The future is already here – it is just unevenly distributed." If so, then the future of the mobile Internet is particularly highly concentrated on the 29th floor of the gleaming Sanno Park Tower Building in Tokyo. This is where executives from the world's mobile-phone companies go to see the latest gadgets from NTT DoCoMo, Japan's dominant mobile operator, and the unquestioned global leader in the fusion of mobile phones with the Internet.

The firm's claim to fame is the runaway success of its i-mode service, which pipes cut-down web pages on to mobile phones with small colour screens. DoCoMo, whose name is a play on the Japanese word for "anywhere", has 60% of the Japanese mobile-telephony market. Over two-thirds of its 37m mobile subscribers have also signed up for i-mode, which was launched in February 1999. This allows them to use special i-mode phones to send and receive e-mail, read the news, access weather forecasts and horoscopes, and download ringing tones and cartoon graphics from nearly 50,000 i-mode-compatible websites. The most advanced i-mode phones double up as music players and can download and run small pieces of software, including games. Other popular uses for i-mode include mobile banking, stock trading and reserving airline tickets (see chart 3).

This is impressive from a technological point of view, but what is even more remarkable is that i-mode's 27m users are prepared to spend money on these services. On top of a monthly charge of ¥300 ($2.50) to access the i-mode service, they pay ¥2.4 (2 cents) for every kilobyte (thousand bytes) they download. In addition, around 500 DoCoMo-approved i-mode sites are subscription-only, requiring users to pay monthly fees of up to ¥300 each; DoCoMo collects these fees as part of the monthly phone bill, takes 9% commission, and passes the rest to the sites' publishers. About 50% of i-mode users subscribe to one or more of these sites. And despite claims that i-mode appeals only to teenage girls, half of all i-mode users are 30 or over, and a quarter are over 39 (see chart 3).

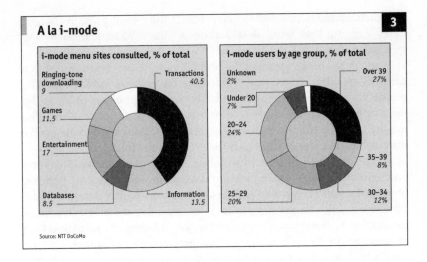

A la i-mode

i-mode menu sites consulted, % of total

Ringing-tone downloading 9
Transactions 40.5
Games 11.5
Entertainment 17
Databases 8.5
Information 13.5

i-mode users by age group, % of total

Unknown 2%
Over 39 27%
Under 20 7%
20–24 24%
35–39 8%
25–29 20%
30–34 12%

Source: NTT DoCoMo

All these data whizzing around produce extra revenue for DoCoMo, and have other benefits too. The company has found that i-mode users are more lucrative than other subscribers. On average, they spend 15% more on voice calls than non i-mode subscribers, and generate 25–30% more revenue overall. They are also far less likely to switch to other operators.

Most important, as competition forces down revenue from voice calls – from an average of $100 per subscriber per month in March 1997 to $65 in March 2001 – data revenue amounting to an average of $17 per user per month helps to make up the difference. Operators and content providers the world over are keen to find out how they can emulate i-mode's example.

Those who have made the pilgrimage to DoCoMo's 29th-floor demonstration room have been rewarded with what is, to most people in the industry, a wondrous sight: 3G phones that actually work. DoCoMo originally intended to launch its 3G service, called FOMA (for "freedom of multimedia access") in May 2001, but teething problems with the new technology forced the company to impose a five-month trial period, and delay a full launch until the beginning of October. FOMA now offers i-mode access at far higher speeds than existing phones, as well as snazzy new features such as videotelephony and the ability to use data and voice services at the same time.

DoCoMo's 3G phones look and feel much like its standard i-mode phones, though the higher data rates mean they are hotter in operation, and their batteries run down much faster; they need to be recharged every day. But look more closely at one of the Panasonic models, and you will notice that it has two green "call" buttons, and a small rotating aperture in the hinge between the phone's two halves. The aperture is a camera, and the second "call" button is for making video calls. The phone is held at arm's length, and the camera can either point at the user, or be rotated 180 degrees to transmit what the user is seeing. The quality of video calls is impressive, at least to anyone used to the grainy moving postage stamps that pass for video on the Internet.

However, there is more to the mobile Internet in Japan than DoCoMo and i-mode. Japan's other two mobile operators, KDDI and J-Phone, have their own i-mode-like services. J-Phone (in which Britain's Vodafone, the world's largest mobile operator, has a big stake) is every bit as innovative as DoCoMo. Although it has yet to launch its 3G network, in summer 2001 J-Phone was the first Japanese operator to launch a phone with a built-in still camera, which allows users to send photos. J-Phone has also pioneered "location-based" services, which allow users to call up information relevant to their location.

Why has the mobile Internet been such a success in Japan, and can that success be duplicated elsewhere? Certainly Japan is a special case in many ways. For a start, PC penetration is low for cultural reasons. It used to be regarded as demeaning for managers to have PCs on their desks; typing was for secretaries. In addition, NTT, the incumbent fixed-line telephone company, charged high rates for dial-up Internet access. And the Japanese have traditionally been enthusiastic about small technological gadgets.

Why Japan has it made

There are other factors at work, too. The three Japanese mobile operators each operate incompatible proprietary systems. DoCoMo users, for example, cannot send short text messages to KDDI users. J-Phone's location-based services are available only to J-Phone subscribers. And if you want to switch from one operator to another, you have to buy a new phone, since Japanese handsets are operator-specific. This absence of open standards may seem old-fashioned to foreigners, but it has been a key factor in getting the mobile Internet off the ground in Japan. When a Japanese operator wants to launch a new service, such as picture messaging or videotelephony, it can specify in detail how that service will work, ask manufacturers to build the appropriate handsets, and ensure that these are available when the service is launched.

Operators in other parts of the world are too numerous to be able to boss the handset makers around in this way. Instead, they face a chicken-and-egg situation. There is no point in their launching an innovative service unless handsets to support it are available; but until such a service has been launched, manufacturers have no reason to produce handsets that support it. This explains why, for example, handsets with colour screens are still a rarity in Europe, though they are commonplace in Japan. Colour screens make phones more expensive, and consumers will not pay more for them unless there are compelling services that use colour. No such services exist, so nobody buys colour handsets, so there is no reason to develop colour services.

Even so, it should be possible to make the Japanese model work in other countries. DoCoMo has great hopes for establishing i-mode as a global standard, and has formed a number of alliances with companies including AT&T and AOL in America; it also owns minority stakes in mobile operators in Europe and Asia. But its attention is currently concentrated on getting 3G off to a good start in its home market, so its plans to launch i-mode elsewhere have not got very far yet.

Besides, the real reason for i-mode's success is not so much the technology as the business model. By giving content-producers a means to charge users, i-mode ensures that there is plenty of content available; indeed there is a growing waiting list of content-providers awaiting DoCoMo's approval for their sites. That wealth of content attracts users who, in turn, attract more content. The result is a virtuous circle. Another important factor is the management of expectations: users understand that i-mode is different from logging on from a PC.

Textbook success

I-mode or no i-mode, the use of phones to send short text messages has grown explosively. The Japanese call them *oyayubizoku* (the thumb tribe), but the sight of people walking down the street frantically clicking at their handsets has also become a familiar sight in many other countries, notably in Europe and some parts of Asia. Some schoolchildren, forbidden to use the phones in class, have taught themselves to input text by touch alone, so that they can send messages while keeping their phones hidden. In the Philippines, the country where text messaging is most popular, the use of the technology by protesters is credited with helping to overthrow the country's former president, Joseph Estrada.

Text messaging is a booming and hugely profitable industry; globally, over 30 billion messages are now sent each month, bringing in over $30 billion a year in revenue for operators, according to figures compiled by Simon Buckingham of Mobile Streams, a consultancy based in Newbury, England. In some parts of the world, text messaging accounts for 10% of operator revenues.

Normal text messages are limited to 160 characters, costing an average of 10 cents each to send, but text messaging can be used to do more than just send quasi-telegrams to other people. Premium-rate text messages can also be used to request new ringing tones and logos. In Finland, it is possible to buy soft drinks and chocolate bars from some vending machines using text messages. In some Scandinavian countries text messages can be used to pay parking or car-wash charges. Increasingly, text messaging is being integrated with the Internet to allow messages to be sent to and from websites or desktop PCs. Hence the claim by Peter Vesterbacka of Hewlett-Packard in Espoo, Helsinki's high-tech suburb, that Scandinavian countries, along with Japan, provide "a partial crystal ball" for the future of the mobile Internet.

By contrast, Wireless Access Protocol (WAP), the nearest thing to i-mode outside Japan, has dismally failed to catch on. It is used by fewer than 10% of mobile subscribers in Europe, and accounts for less than 0.5% of operator revenue. WAP's failure has a number of causes, not least an excess of hype. An infamous advertising campaign launched in Britain by BT Cellnet used fancy computer graphics of a digitised figure whooshing around on a surfboard. The reality of accessing the Internet from a mobile phone, users soon discovered, was far less glamorous.

WAP, which displays cut-down web content on phones, is painfully slow: just establishing a connection takes up to half a minute, and downloading anything requires the patience of a saint. There are numerous WAP versions that are not fully compatible. But worst of all, there is very little content, because there is no way for content-providers to charge for it. Operators collect a per-minute fee for WAP usage, and some of them pay content-providers to produce material to encourage traffic, but there is no virtuous circle of the i-mode sort. WAP is crap, goes the industry refrain. But in essence the problem is the business model, not the technology, particularly now that 2.5G networks are speeding up the service.

When it comes to the adoption of mobile data, the odd country out is not Japan but America. The United States and Canada are the only countries in the world where PCs are more numerous than mobile phones; everywhere else the reverse is true (see chart 4).

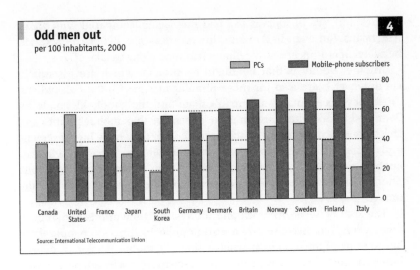

Odd men out
per 100 inhabitants, 2000

PCs | Mobile-phone subscribers

Canada | United States | France | Japan | South Korea | Germany | Denmark | Britain | Norway | Sweden | Finland | Italy

Source: International Telecommunication Union

The American exception

America's enthusiasm for PCs, and lack of enthusiasm for mobile phones, is due to a combination of factors, but mainly to the relative prices of fixed and mobile calls. In Japan, expensive access to the fixed Internet drove users to mobiles; in America it was mobiles that priced themselves out of the market. Local calls are free, and PCs are cheap. Mobile phones, on the other hand, suffer the huge drawback that users are charged to receive calls, so Americans tend to leave their mobile phones switched off to avoid having to pay for unwanted incoming calls. Mobile network coverage is patchy and rates until recently were high. So teenagers who want to keep in touch with their friends are given their own phone line, a PC and a dial-up account with AOL, the world's largest Internet-access provider. Soon they are happily sending and receiving instant messages through their PCs.

Their counterparts elsewhere in the world, in contrast, are brandishing mobile phones and sending text messages. Mobile phones have become even more appealing to teenagers in Europe since the advent of "pre-paid" phones, which can be replenished with vouchers on sale at newsagents and corner shops. Pre-payment, which originated in Italy to exploit a tax loophole, means that even people without bank accounts (such as children) can have mobile phones. It also enables users to control their spending.

Sending text messages is cheaper than making voice calls, which helps to explain their success. According to figures from Gartner, 47% of

273

Swedes and 39% of Italians use text messaging, compared with 2% of Americans. But even in America, lower prices and better co-operation between operators are now helping text messaging to take off.

Another oddity is that handheld computers, which are essentially cut-down PCs, are also far more popular in America than elsewhere. These devices – notably Compaq's iPaq, and the handhelds made by Palm and Handspring – are far more expensive than mobile phones and lack wireless connectivity as standard, though they have bigger screens and more processing power. Americans widely perceive wireless-enabled versions of such devices, rather than Internet-capable mobile phones, as the logical means of accessing the Internet while on the move.

The split between America and the rest of the world is apparent in other ways, too. Because PCs are so popular in America, people there like to think of the mobile Internet as simply a mobile version of the PC-based Internet. This has fired their enthusiasm for a wireless networking standard called 802.11b, or Wi-Fi, which allows suitably equipped laptops within a few hundred feet of a base station to access the Internet. Wi-Fi coverage is spreading fast in many American offices, universities, hospitals, schools and airports, and its advocates claim that it could even do away with the need for 3G networks (see "Cat among the carrier pigeons" on page 276).

Joe Manget, an analyst at Boston Consulting Group, thinks American enthusiasm for Wi-Fi can be explained partly by another cultural difference with the rest of the world. In Europe and Asia, he notes, many people commute to work on public transport, and like to peck at their mobile phones while they are on the move. In America, on the other hand, people commute in cars, which rules out the use of wireless hand-held devices. Instead, they want wireless access for their laptops when waiting around in places such as hotel lobbies and airport lounges.

Martin Dunsby of Deloitte Consulting points to yet another difference. Thanks to general enthusiasm for information technology, American firms are far more likely to have adopted customer-relationship management and enterprise-resource-planning systems (which keep track of customers and inventory, respectively) than companies in other countries, and now want to extend those systems to wireless devices. So in America it is the corporate market that leads the mobile Internet, says Mr Dunsby, whereas elsewhere it is the consumer market.

But whatever the differences, the fact remains that the mobile Internet, in one form or another, generates enthusiasm across the developed world. As with the Internet in 1995, it is clear that both consumers and businesses would like to adopt this new technology over the next few years. Last time round many of the firms that rushed to exploit the demand for everything to do with the Internet got things seriously wrong, as the demise of the dotcoms shows. The builders of the mobile Internet evidently believe they can avoid making the same mistakes. So what makes wireless different?

Cat among the carrier pigeons

Which of several competing standards for wireless data transmission will prevail?

"**Y**OU SEE, wire telegraph is a kind of a very, very long cat," explained Albert Einstein. "You pull his tail in New York and his head is meowing in Los Angeles ... And radio operates exactly the same way: you send signals here, they receive them there. The only difference is that there is no cat."

The difference between the fixed-line Internet and the wireless version is of the same order, only there will be several overlapping kinds of non-cat. On the Internet, information travels around in packets of data, which can be sent from one place to another using a variety of technologies. The one that is currently causing the most excitement in America is based on a standard called 802.11b. This is such a terrible name that the industry has started calling it Wi-Fi (for wireless fidelity, which is not much better).

To set up a Wi-Fi network in your home or office, you simply buy a base station, plug it into a phone socket or a high-speed Internet connection and hang it on a wall. Using unlicensed radio spectrum, the base station communicates with computers within 45 metres (150 feet) or so, provided they are equipped with a suitable plug-in card. It also enables the computers to communicate with each other. In effect, Wi-Fi lets you flood a building with wireless connectivity. This has a number of advantages. It is often cheaper and easier to link several PCs together wirelessly than to run messy cables all over the place.

Using Wi-Fi is astonishingly liberating. Home users quickly discover the joys of picking up their e-mail in the garden. Offices and university campuses have also been quick to adopt the technology, which allows users to keep information at their fingertips during meetings and seminars. Companies are deploying Wi-Fi in warehouses, to link portable computers to inventory systems. Wi-Fi "hotspots" providing free access have been springing up in airport lounges, hotels and coffee shops. Scandinavian Airlines is testing the technology as a convenient way to provide Internet access within airliners, without the need to run cables to every seat. In big cities including San Francisco, London and Seattle, enthusiasts are building "guerrilla networks" in an attempt to provide blanket Internet coverage; firms such as MobileStar and Wayport in America and Jippii in Europe are doing the same, but on a commercial basis.

Is Wi-Fi a killer?

Work is already under way on new versions of the 802.11b standard that will improve the range and transmission speed of Wi-Fi equipment. Some analysts have suggested that Wi-Fi might kill off the high-speed mobile networks, such as the third-generation (3G) networks now being built around the world.

This seems highly unlikely. For a start, Wi-Fi is designed for use with fully fledged computers, and transmits high-speed data over short distances. 3G is slower, but was designed to handle a far larger number of users, and to work with small, handheld devices over longer distances. Wi-Fi is intended to provide coverage in hotspots; 3G is intended to provide blanket coverage. Wi-Fi cannot be used for voice calls, and currently consumes too much power to be used in phones or small handheld computers for more than a couple of hours. It also has serious security shortcomings; its built-in encryption standard has been shown to be flawed. In short, although computer users will happily use Wi-Fi to access the Internet from their laptops when it is available, they will still need a cellular connection if they want to roam beyond Wi-Fi hotspots or make voice calls.

All this means that the technologies are widely seen as complementary rather than competitive. Many equipment vendors, such as Nokia, Nortel and Ericsson, make both cellular and Wi-Fi equipment. A recent report from Analysys, a consultancy based in Cambridge, England, suggested that mobile-network operators should consider building Wi-Fi networks alongside their 3G networks and facilitate seamless switching between the two.

In short, multiple standards will prevail: Wi-Fi, 3G, 2.5G networks, and specialist data-only networks such as Mobitex. Bluetooth, another wireless standard, is intended to allow devices such as handheld computers and printers, or phones and headsets, to communicate over very short distances. And already there is talk of 4G networks. The mobile Internet will exploit a tapestry of wireless networks, all of which will coexist without a tangled cable in sight – or indeed a cat.

Why mobile is different

For a start, people are used to paying for it

HOW DO YOU make money on the Internet? In the late 1990s, this was the question on everybody's lips. The answers bandied about included "building communities", "ensuring stickiness", "B2C", "B2B" and many others. Buzzwords came and went, and eventually nearly everyone went bust. The problem was that advertising revenue was insufficient to keep most sites running, and there was no standard way to charge for things on the Internet. There still isn't. Getting people to type their credit-card details into a web page raises security concerns, and makes purchases of less than a few dollars impractical. A handful of sites selling books, CDs, flights and holidays look as though they will survive. But most news and content sites are losing money, and many are now trying to introduce subscription fees. Many more have folded. Why should things be any different on the mobile Internet?

Mobile is different from the fixed Internet in three important respects. First, a mobile phone is a far more personal device than a PC. It is likely to be used by only one person, who will probably have the phone with him for most of his waking hours. Whereas e-mail messages go to a machine sitting on a desk, text messages go directly to the mobile phone's user. Often the network operator knows exactly who that user is, including his name and address. In order to route calls to and from the mobile, the network operator also needs to know where it (and therefore probably its user) is at all times.

Second, network operators can determine what menus and services appear on their users' phones. Whereas on PCs users have lots of scope to play around with the settings, on mobile phones all they can easily change is the ringing tone and the screen logo. The ability to set the default portal – the starting page that users see when they connect to the mobile Internet – is a big advantage, because it allows operators to act as gatekeepers.

This will cost you

Last, and most important, people know that using mobile phones costs money, and there is a mechanism for the network operator to charge them for that use. What is more, users seem prepared to pay a "mobility premium" to do things while on the move. Sending an e-mail or instant message over the Internet from a PC is essentially free; sending a text message from a phone costs an average of 10 cents, but users are prepared to pay because they regard it as good value, or because it makes their lives easier. And even when text messaging is more expensive, people still use it. In some places, sending a text message home while "roaming" in a foreign country can cost as much as €1 (92 cents). Such charges are currently under investigation by the EU's competition commission. But compared with the cost and hassle of buying a postcard and a stamp, this still seems reasonable enough to many people.

In short, if you have a mobile phone, the network operator knows who you are, where you are, can direct you to the portal of its choice, and can charge you money. This is a very different world from that of the fixed Internet.

Mobile has some drawbacks, of course. Mobile devices have more limited screens and keyboards than PCs, and slower connections. Also, says Niklas Savander of Nokia, mobility makes people much more impatient. Researchers have found that a five-second delay to access something on an Internet-capable phone seems far longer to users than a five-second wait to call up a web page. "With the same response time, people rate mobile as slower," he says. "So we have a slower connection, but users want a faster response."

For me, here, now

But the combination of personalisation, location and a willingness to pay makes all kinds of new business models possible. Tomi Ahonen, head of 3G Business Consulting at Nokia, gives the example of someone waiting at a bus stop who pulls out his Internet-capable phone to find out when the next bus will arrive. The information sent to the phone can be personalised, reflecting the fact that the user's location is known, and perhaps his home address too; so bus routes that run from one to the other can appear at the top of the list, saving the user from having to scroll and click through lots of pages and menus. A very similar service, which allows users to find out when the next bus is due by sending a text message from a bus stop, is already available in Italy.

Would-be providers of mobile Internet services cannot simply set up their servers and wait for the money to roll in, however, because the network operators – who know who and where the users are, and control the billing system – hold all the cards. This has changed the balance of power between users, network operators and content providers. On the fixed Internet, the network access provider acts as a "dumb pipe" between the user's PC and, say, an online bookstore or travel agent. The access provider will not know how the connection has been used, and there is no question of claiming a commission. Mobile network operators, on the other hand, are in a far more powerful position. "Wireless is a smarter pipe," says Chris Matthiasson of BT Cellnet. This means that operators are much less likely to be disintermediated.

Having avoided one mistake made on the fixed Internet, however, wireless operators may be tempted to make another, by setting up "walled gardens" of services and content. In theory, restricting users to a handful of approved services will enable operators to capture a much larger chunk of the expected bonanza in data revenues. In the 1990s, online services such as AOL, Compuserve and Prodigy operated on the walled-garden principle; but as soon as one of them offered unfettered Internet access, the others had no choice but to follow suit. The walled-garden model will turn out to be just as unsustainable on the mobile Internet, because users get annoyed by it.

Furthermore, unlike Internet access providers, wireless operators charge by usage, either for every minute spent online, or for every byte downloaded. This means they make money on transporting data come what may, so it makes sense to offer users the widest choice of content possible to encourage them to run up transport charges. That is how i-mode works; the vast majority of DoCoMo's data revenues come from transport, not the sale of content (though the firm does take 9% on the sale of other providers' content). A typical i-mode user spends ¥2,000 (about $17) per month on data-transport fees, and only ¥400 on content subscriptions.

Operators therefore generally offer a selection of approved services through their own chosen portal, and also give subscribers the option of going elsewhere. This is what AOL does with its dial-up Internet service; it offers services such as instant messaging, chat-rooms and e-mail, as well as access to the web. But surveys show that most users still spend most of their time within what used to be AOL's walled garden. The best way for operators to keep users within their walled gardens, says Katrina Bond of Analysys, is to offer attractive services. The fact that operators know who and where their users are – and may be able to keep this information to themselves – can give their home-grown or approved services a valuable advantage.

The upshot is that the operators need decent content and services to drive traffic; and the content providers need the co-operation of the operators if they are to charge for their wares. A number of business models have emerged to govern the relationships between the two.

Show me the money

The simplest one of these involves sharing revenues from text messages. Lycos, a web portal, provides a service that allows PC users to send text messages from a web page and receive the replies on their PCs. The effect is to stimulate text-message traffic between mobile phones and PCs. The PC users do not pay to send messages, but the mobile users do; and through agreements with mobile operators, Lycos gets a cut. There are other services, such as mobile games, that encourage mobile users to send text messages; the content provider gets a share of the extra revenue generated. Sometimes the operator also charges for the messages at a higher rate.

Another model involves the use of premium-rate text messages as a means of charging for one-off lumps of content, such as ringing tones, logos or horoscopes. Users send a text message to a special number, are charged accordingly, and have the content delivered in the form of a text-message reply.

More elaborate is a model sometimes called "reverse billing", in which services are charged directly to the user's phone bill. In effect, the operator bills the user on behalf of the content provider, and then hands over the money.

In theory, reverse billing could be used as a means of payment for online commerce; a book, CD or cinema ticket could be charged directly to the user, as though it were an expensive phone call. And since mobile operators are used to handling a large number of small transactions, their systems can handle such transactions at around a tenth of the cost of a bank or credit-card transaction. This means that micropayments, which have never taken off on the fixed-line Internet, are feasible on the mobile one.

But users may prefer to pay lumpy subscription fees rather than a small charge for every morsel of information they access. Following the example of i-mode, whose sites work on monthly subscription fees, T-Motion, a mobile portal owned by Deutsche Telekom, has decided to try that model for WAP content, starting from November 1st 2001. Subscribers to its T-Motion Plus service pay €10 ($9) a month for a bundle of free ringing tones and text messages, plus news, weather, financial updates, sport reports and games; this revenue is split 50/50 with the content providers. T-Motion will track the popularity of the content, and

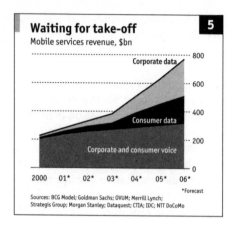

Waiting for take-off **5**

Mobile services revenue, $bn

- 800
- Corporate data
- 600
- 400
- Consumer data
- 200
- Corporate and consumer voice
- 0

2000 01* 02* 03* 04* 05* 06*

*Forecast

Sources: BCG Model; Goldman Sachs; OVUM; Merrill Lynch;
Strategis Group; Morgan Stanley; Dataquest; CTIA; IDC; NTT DoCoMo

will replace the least popular services every three months. With this model, the paid-for services cannot be given away free on other portals, or users will not be prepared to pay for them; the effect is to produce a walled garden of sorts, with premium services that only subscribers can access.

The most radical model is the "mobile virtual network operator", or MVNO, in which a network operator acts as a wholesaler of airtime to another firm, which then markets itself to users just like an independent operator with its own network infrastructure. Virgin Mobile, a British mobile operator, is in fact an MVNO that resells voice and data airtime on the network belonging to another operator, One2One. The MVNO allows content providers to get their hands on transport fees, but operators feel ambivalent about the concept. On one hand, MVNOs can brand themselves to appeal to a wider range of customers, and thus boost overall use of the network; but on the other, MVNOs turn network operators into dumb pipes, giving them a smaller piece of the action. For the time being, most operators have chosen to deal with customers direct, rather than become wholesalers to MVNOs.

In various combinations, all of these models are in use now, but operators are still struggling to implement new billing systems. Most operators, says Nokia's Mr Savander, have between 20 and 40 separate billing systems to handle different kinds of services; one has 54. Software firms are competing to offer consolidated billing systems that will support any or all of these business models.

Which model will prove most successful remains to be seen, but there is certainly money sloshing around on the mobile Internet. Unlike on the fixed-line Internet, people are prepared to pay for content and services they really want. But what exactly might those be? As on the fixed Internet, there are two distinct markets: consumer and business. Although it is still early days, there are already signs of a "killer application" in each.

Looking for the pot of gold

What do consumers want from the mobile Internet?

W AY BACK IN 2000, in the heady days before the dotcom crash, it all looked so easy: to make money from the mobile Internet, you simply created a mobile version of what worked on the fixed-line Internet. The vogue at the time was for business-to-consumer e-commerce, so the obvious target was mobile e-commerce, or m-commerce. Surely anything that could be sold over the conventional Internet to PC users could be sold over the mobile Internet to mobile subscribers. And because mobile users have their phones with them at all times, they might be expected to do more shopping than stationary customers. Best of all, since the mobile operators have a firm grip on their users, and know about things like billing, they should be able to claim a piece of the pie that eluded the fixed-line Internet-access providers.

In March 2000, just before the dotcom bubble burst, Jeff Bezos of Amazon, the leading online retailer, predicted that by 2010, all of his firm's customers would use wireless devices to make purchases. Describing m-commerce as "the most fantastic thing that a time-starved world has ever seen", he predicted that it would change the way people shop, since they would be able to make impulse purchases anywhere, at any time. Within five or ten years, he claimed, "almost all of e-commerce will be on wireless devices." Analysts queued up to make rosy forecasts of m-commerce revenues. With such a bonanza apparently around the corner, is it any wonder that mobile operators paid so much for those 3G licences?

But as the Nasdaq crashed and the dotcoms started going under, it became apparent that making money was hard enough even on the conventional Internet, where the technology is mature; the prospect of buying things on phones, with their tiny screens and keyboards, suddenly seemed far-fetched. Surveys showed that consumers found the reality of m-commerce hugely disappointing. Horror stories abounded: one trial found that it took over 40 minutes to order a book by phone; making a booking on one mobile hotel-reservation system required 37 clicks.

Users paying by the minute to access the mobile Internet were not prepared to put up with that. Research carried out by Boston Consulting found that during 2000, one-third of early users in Europe abandoned m-commerce after only a few tries. And a worldwide survey by A.T. Kearney found that the number of mobile users who said they intended to use their phones for m-commerce fell from 32% in June 2000 to 4% in June 2001.

Many of the consumers' concerns, such as security and privacy, and difficulty with navigation, are reminiscent of worries in the early days of e-commerce, and may eventually be overcome. But there are broader problems with using handheld devices for shopping. Compared with PCs, which have large colour screens, handheld devices are hopeless for browsing. Scrolling through lists is cumbercome, and features and prices are hard to compare. Moreover, there seems little point in making an impulse purchase of a book or CD if it will then have to be delivered by post. Rather than spend ages pecking at a phone keyboard, why not wait until you get home and order in comfort from your PC? "The unique selling point is mobility," says Declan Lonergan, an analyst at Yankee Group. M-commerce should not try to compete with e-commerce, he says, but should stick to sales of small-value mobile items, such as cinema tickets or paying for parking, that people are likely to want while on the move.

Location, location, location

People may not want to buy things by phone, but they are more enthusiastic about paying for information, as the success of downloadable ringing tones and logos demonstrates. Furthermore, focus-group research carried out by Nokia found that users expect purchases on mobile devices to depend on where they are, rather than the desire to buy a particular item. The buzzword now is "location-based services". One helpful factor has been American legislation that requires operators to be able to pinpoint the position of any mobile phone from which an emergency call is made, prompting operators to add positioning technology to their networks and handsets. The commercial possibilities of the technology are beginning to cause a lot of excitement.

Even without pinpoint accuracy, existing networks can already locate a user within a particular cell (the area covered by a mobile-network base station). In urban areas, such cells are quite small, so location-based services are possible even now. A host of firms are now offering location-based services such as traffic and weather reports, driving directions, travel and entertainment information and restaurant recommendations. Many of these firms have established partnerships with operators, which control access to the location information; others have done deals with infrastructure providers, so that operators can buy support for location-based services off the shelf.

Early examples of location-based services include a tourist-information system in Italy and a traffic-information service in South Korea. According to Jean-Michel Durocher of Webraska, a location-based services firm in Paris, the most popular services are currently those that enable users to find the nearest restaurants, railway stations, car parks and petrol stations. Other uses for the technology include giving directions to pedestrians and car drivers, and providing localised weather forecasts, a service that seems to be particularly popular in Germany.

Inevitably, the most advanced implementations of location-based services can be found in Japan. In Tokyo, J-Phone's J-Navi service lets users enter a phone number, address or landmark, and then searches the area within a 500-metre radius. This makes it possible to find the subway station nearest to a particular shop, or a particular kind of restaurant within walking distance of a particular office building. Most important, users of the service can download a full-colour map. Standing on a street corner in Tokyo, Yoshitaka Noda of NS Solutions, the firm that put the J-Navi system together, calls up a detailed map of the area and points out the route to the nearest subway station. A few more clicks reveal that there are 20 sushi bars within 500m of *The Economist*'s Tokyo office.

At the time of its launch in May 2000, J-Navi was expected to handle around 100,000 hits per day, but on its third day of operation it already had 1.6m. J-Navi is now used about 2m times a day. Searching is free, but users pay for the data-transport costs, so in practice it costs about ¥5 (4 cents) a time. Downloading a map costs ¥20, because it requires the transmission of a colour graphic; about 50,000 users a day request a map.

Location-based services in Tokyo, a city with few street names, may have a unique appeal. But J-Navi's popularity seems to hold a number of lessons for would-be providers of similar services in other parts of the world. First, the search area is defined by a phone number or address, so there is no need for fancy positioning technology. Second, 30% of queries involve a location other than the user's current one; it seems that people use the service to find out in advance how to get somewhere another time. Third, although J-Navi was expected to be used mainly by business users and commuters, it has turned out to appeal to a far wider audience. Fourth, using J-Navi costs about the same as sending a text message; the service is popular because it is cheap. J-Phone gets very little revenue from the service itself, but uses it to generate traffic and to distinguish itself from its competitors.

Looking up addresses and nearby restaurants does not seem to be a particularly lucrative opportunity: a typical user is unlikely to use the service more than a couple of times a week. But there is another potential use of location-based services that could generate more traffic, and hence more revenue: mobile advertising.

A message from our sponsor

The appeal of mobile phones to advertisers is obvious: they are personal devices, they spend all day with their owners, and their whereabouts are known. Proponents of location-based marketing see all sorts of mouth-watering opportunities. Someone walking down the street might be alerted by his phone to special offers from nearby shops; or everyone attending a pop concert might have a coupon offering a discount on CDs sent to their phones. Frost & Sullivan, a consultancy, predicts that by 2005 some 37 billion advertisements and alerts will be sent to mobile devices in Europe, and that 65% of users will be prepared to receive ads in this way. Such messages are expected to be worth $7.4 billion in revenues.

It sounds like a great idea for advertisers and operators, but more like a nightmare for consumers, many of whom are already fending off a growing torrent of junk e-mails on their PCs. Research by Gartner says a surge of "spam" text-message advertising is probably inevitable in the coming months and years. To avoid offending people, mobile advertisements must be optional (ie, users must actively request them), personalised, moderate in volume (no more than five a day), and free to the recipient. One thing that might persuade people to accept mobile advertising could be the use of advertising to subsidise access costs.

One trial, carried out in Britain by the Mobile Channel, a mobile advertising company, seemed to suggest that mobile advertising has a bright future. It found that text-message advertisements had an average response rate of 10–20%, far higher than the response rate to direct mail (3%) and Internet banners (less than 1%). But novelty has a lot to do with it: early experiments with Internet advertising suggested that it, too, would be wildly successful. It would be very dangerous to read too much into these early results.

That said, it is telling that Vesku Paananen, a Finnish wireless entrepreneur who pioneered ringtone downloads in 1998, has now set up a new firm, Add2Phone, for mobile marketing. The company sent over 2m advertising messages in the first six months of 2001. In Finland, Mr Paananen notes, firms are forbidden to send unsolicited text messages, so users must opt in. Users must also, he cautions, have a means to unsubscribe, or be sure that they will receive messages only for a limited time (a week, say, or a month) when they sign up for a promotion. ZagMe, a British location-based advertising service, allows people who enter a shopping mall to specify how long they plan to stay, and sends them coupons and advertisements during that period, but not thereafter. In Japan, J-Phone's location-based advertising service, J-SkyStation, has a clever technological fix for the problem of obtrusive messaging: advertisements are sent silently into a special cache on users' phones, allowing users to look if they choose, or else ignore them.

Yet all in all, it is clear that location-based services in general, and mobile advertising in particular, are not going to be huge moneyspinners. Analysys puts the total value of the location-based services market at $18 billion by 2006; Ovum, another consultancy, suggests $20 billion. That may sound a lot, but it is modest compared with the text-messaging business, which already brings in over $30 billion a year worldwide and is still growing fast.

Communication, not commerce

So what is the killer app? Mobile operators will probably make some money on m-commerce and location-based services, if only through the associated transport revenues. Access to free content, such as online banking and train timetables, will also enhance transport revenues. And the example of i-mode suggests that users may be prepared to pay a small amount to receive news, weather, sports scores, horoscopes and so on. But the subscription revenue associated with these services is tiny; again, the real money is in the transport. What can operators do to boost traffic and maximise transport revenues?

The answer seems obvious: person-to-person communication. The success of text messaging relative to WAP shows that people like to use their phones to communicate with each other, rather than to download information from content providers. In the words of Andrew Odlyzko, a former AT&T researcher who is now at the University of Minnesota: "Content is not king – connectivity is more important." Indeed, he argues that the killer app for 3G phones might turn out to be increased voice traffic.

There is some evidence for this; for example, i-mode users make

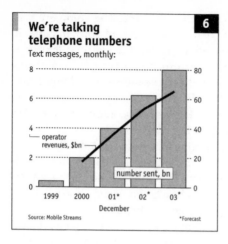

We're talking telephone numbers 6

Text messages, monthly:

operator revenues, $bn

number sent, bn

1999　2000　01*　02*　03*
December

Source: Mobile Streams

*Forecast

more voice calls than do voice-only subscribers. They may be ringing in response to incoming text messages and e-mails, or they may be taking advantage of the fact that phone numbers can be embedded as clickable hyperlinks in i-mode pages; look up a restaurant review on an i-mode phone, and you can call the restaurant with a single click. Yutaka Mizukoshi of the Boston Consulting Group in Tokyo explains that even if he has a desktop phone in front of him, if an e-mail arrives on his i-mode phone asking him to call, he will simply click and call back using his i-mode phone. The effect of combining voice and data, he says, is to concentrate communications into the mobile handset.

The key to increasing traffic, therefore, and hence revenue, is to offer new kinds of person-to-person communications services. E-mail is one obvious example, and one that accounts for a large volume of i-mode traffic; another is group-based text messaging, which allows individuals to communicate with a group of people via a single message. Kare Souru of Popsystems, a Finnish group-messaging firm, suggests that operators can make group messaging attractive by making it 15–20% cheaper to send a message to a group than to each of its members individually. The revenue per message is thus smaller, but the volume will be much higher. Popsystems' trials with a schoolgirls' basketball team found that group messaging increased message traffic per user by 60%, and overall revenue (including voice) by 20–30%. Furthermore, Popsystems found that when users are able to establish their own groups and invite others to join, each user recruits an average of 2.5 additional users per month.

Another step beyond simple text-based messaging is to add graphics and animations. There are already several competing standards to do this, notes Mr Buckingham of Mobile Streams. But a standard called EMS, short for enhanced messaging service, is pulling ahead of the pack, at least in Europe. It uses multiple text messages to send text, plus a larger payload of graphics or music. These messages are then combined on the recipient's phone; if the phone is not capable of supporting graphics or music, only the text is displayed. This multiplies message traffic. It also introduces users to the idea that more complex messages cost more to send, which is crucial if users are to be charged extra for more complex services in future.

EMS, however, is just a transitional format. The greatest scope is for multimedia messaging (MMS), which is like text messaging, but allows pictures and sound clips to be sent as well as text. Unlike EMS, MMS is a true industry standard, and has far wider support. The potential for MMS is enormous. Europeans are already using text messages in place of holiday postcards: imagine being able to send pictures too. In Japan, J-Phone's handsets with built-in cameras have been a huge success. And since MMS is based on Internet e-mail standards, it will be possible to send MMS messages between phones and PCs, which is currently hard to do. This will increase traffic further. Since MMS messages will use more bandwidth than text messages, operators will be able to charge more for them. MMS is also likely to boost voice traffic: according to research carried out by Nokia, seven out of ten picture messages generate a phone call in response.

As 2.5G and 3G networks come into wider use, EMS and MMS are likely to figure prominently in persuading users to upgrade to colour-screen handsets. Low-resolution CMOS sensors, of the kind found in cheap digital cameras, are a commodity, and can be added to handsets very cheaply. Location-based services and mobile advertising will also be able to exploit picture-messaging services, by transmitting maps or colour photographs. Pornography is one inevitable application; already, several wireless porn outfits are lining up to offer their services.

Once users have switched to fancier handsets, the hope is that text messaging and associated services will become precursors to more advanced services. Having got used to paying for information delivered to their phones by buying, say, ringing tones, goes the theory, users may be prepared to try other things. Rick Allen of Logica, a leading vendor of messaging and billing systems, calls this the "three box" strategy. The idea is to lead users from text messaging on to content, then to location-based services, and eventually on to m-commerce.

Say something

It should come as no surprise if the killer app for the mobile Internet, at least for consumers, turns out to be person-to-person communication. That, after all, has been the golden prize of all previous technologies, from telegraph to telephone to mobile phone. The Internet's killer app is e-mail, even if the web accounts for more traffic. Transmitting speech, words, pictures and graphics are all social activities, and mobile phones are primarily social devices. "Messaging is a much bigger chunk of this new market than people want to talk about," says Niklas Savander of Nokia, the world's leading handset maker. "Perhaps they think it's boring." Certainly it means that in the short term, the fancier possibilities of the mobile Internet are more likely to be adopted by businesses.

A different way of working

**All sorts of companies are finding mobile Internet technology
surprisingly useful**

FOR ALL THE hullabaloo over new business models and dotcom
start-ups, the benefits of the fixed-line Internet have accrued mostly
to existing firms that reinvented themselves around it, rather than new
firms that started from scratch. By adopting e-mail, intranets, extranets,
customer-relationship management (CRM) and enterprise-resource planning (ERP) systems, large firms have made huge investments in Internet
technology. It is now a relatively simple step, say proponents of the
mobile Internet, to extend all of these systems to mobile devices, so that
employees can access the information they need from anywhere.

Mobile devices, says Martin Dunsby of Deloitte Consulting, enable
workers on the front line to get at data in the back office. "The value is
not from cutting the cable," he says, "it's from the process change that's
enabled by the technology." He gives the example of time-and-expense
tracking, in which consultants enter the amount of time spent working
on a particular project into a handheld device. This may not be any
quicker for them than filling in a paper timesheet, but it means that the
data can be sent back to head office straight away and an invoice
issued, rather than the information being keyed into the accounting
system at head office several days later. Similarly, when visiting a client,
a salesman can use a wireless handheld device to call up the latest
inventory levels, technical support histories and so on.

It may sound mundane, but given the sorts of corporate information
systems that many firms, particularly in America, have already implemented, it is a logical next step. This means, says Mr Dunsby, that in the
next three years most of the action on the mobile Internet will be in the
corporate market. Adoption will be fastest in America: a survey by Zona
Research found that 66% of American firms are planning to deploy
mobile Internet technology by 2004, if they have not done so already.

Business use of the mobile Internet also escapes the chicken-and-egg
problem of aligning services with available hardware: a firm that wants
to provide its workforce with wireless access can specify the entire
system from top to bottom, from handsets to software to network
provider. Existing technology, in the form of 2.5G and Wi-Fi networks, is

already good enough to make this possible today: there is no need to wait for 3G. And there is a growing range of handheld computers, Internet-capable phones and intermediate devices to choose from.

The wireless workforce

At present, adoption of mobile Internet technology makes more sense in some industries than others. Not surprisingly, it holds particular appeal for firms with mobile workforces. The early adopters, says Joan Herbig of XcelleNet, a firm that provides remote management software for corporate handheld devices, are the same users that first adopted laptops in the 1990s: sales forces in financial services, health care and pharmaceuticals, field workers for utilities and so on.

Surely most mobile workers already have laptops? Yes, but wireless handheld devices have a number of advantages. Laptops have complicated operating system software, plus a whole lot of additional software loaded on top, whereas handheld computers or phones are much simpler machines, with almost nothing to go wrong. This means that support costs are lower by a factor of at least five. When Goldman Sachs employees were given BlackBerry e-mail pagers developed by Research in Motion, a Canadian wireless firm, their use of laptops fell by 45%. A fifth of BlackBerry users stopped using their laptops altogether.

The BlackBerry looks like a glorified pager. It has a small keyboard and an always-on wireless data connection, and allows users to send and receive e-mail on the move as though they were sitting in front of their desktop PCs. It is the first of a new breed of devices taking up the space between handheld computers and mobile phones. Its proponents claim that it makes employees more responsive, because they can reply to messages at any time. It can also make them more productive, because they can catch up with their e-mail on the move rather than back at the office; a BlackBerry, unlike a laptop, can be used in a taxi. According to figures from Boston Consulting, the BlackBerry costs $900 per user per year. For an employee earning $100,000 a year, the system pays for itself even if it saves only five minutes a day.

In addition to savings in support costs and improvements in productivity, handhelds have other advantages over laptops, says Richard Owen of AvantGo, a firm whose software links handhelds to corporate information systems. Handhelds are less obtrusive (salespeople, he says, hate having to boot up laptops in front of customers), and "you look like you're on top of things". Using a laptop simply to fill in on-screen forms and gather data is overkill, so salespeople are often reluctant to use laptops to collect data in the field, says Mr Owen. One of AvantGo's clients, Alcatel, a telecoms-equipment maker, reported a fivefold increase in the amount of data sent back from the field by sales representatives after introducing handhelds.

But mobile Internet technology may also hold attractions for firms without large roving workforces. Research carried out on behalf of BT Cellnet, a British mobile operator, found that employees spend an average of two hours a day away from their desks, and return to a backlog of e-mail that takes 45 minutes to deal with. "Once upon a time, people sat in an office from nine to five, and that was the boundary of their world," says Chris Matthiasson of BT Cellnet. "Now everyone in an organisation is a mobile worker." He cites trends such as the growth of flexible and off-site working, teleworking, longer commuting times, and the fact that large companies tend to be geographically spread, which requires employees to travel more than they used to. Mobile devices, he suggests, allow the best use to be made of travelling time, and of time between meetings. Wireless technology, he implies, can offer an immediate productivity gain, no matter what business you are in.

Going wireless need not be a huge investment. Most firms already have e-mail systems in place, and numerous wireless-technology firms have popped up to help firms extend these to wireless devices with the minimum of fuss. The BlackBerry system, for example, can be set up in a few hours; it simply involves installing an extra box that establishes a secure link between the wireless operator's systems and the client company's e-mail server. The wireless services of Seven, a Silicon Valley wireless start-up, do not require firms to install any new hardware at all. Instead, Seven's software resides in the network operator's systems; it queries the client firm's systems across the Internet, then reformats the resulting data to make them suitable for display on a phone.

Suck it and see

Firms that wish to go wireless, therefore, need not buy fancy new equipment, but can simply pay a network operator to provide wireless access as a subscription service. Seven's corporate e-mail system, for example, is typically resold by network operators for £3 ($4.40) per user per month; other applications cost extra. There are, of course, transport charges on top, and employees must have suitable handsets. But since wireless access is a service, not a product, it is relatively easy for firms to try it out. Tim Dunne of Nextel, a business-oriented American wireless operator, says most firms start off by using Nextel's 2.5G network to access e-mail using WAP phones, and then move on to wireless-enabling their other e-business applications. Nextel recently introduced a Motorola handset capable of running small pieces, or "applets", of software written in Java. This makes far more complex applications possible, because it allows firms to write their own software to run on the handset if they choose.

Foot in the door

So the technology is available, but the wireless industry still has work to do in convincing firms to adopt it. "Wireless people get obsessed with networks, carriers and so on, which is like airlines thinking about planes, not passengers," says Mr Owen. "The industry has been very poor at explaining why any of this is worth doing, rather than just being interesting technology." The BlackBerry has succeeded, says Mr Owen, because it is easy to explain what it does, and why it is useful. To get a foot in the door, he jokes, the mobile Internet needs a clever-sounding theory to justify it, such as "total cost of ownership" (which encouraged firms to centralise their computing systems) or "just-in-time ordering" (which encouraged them to adopt fancy inventory-management systems).

At the same time, there is the danger of overhyping the technology, as happened with WAP. "People have to understand that wireless applications are not web applications. It's completely different," says Bill Nguyen of Seven. Call up your e-mail on a PC, he explains, and you can list dozens of messages at once, sort them and manipulate them in various ways. Not so on a handset. This means, says Denise Lahey of OracleMobile, that the software that pipes information to the handset needs to be smarter: to figure out which e-mails are important, perhaps given the time of day, the user's calendar, and even the user's location. On the way to a meeting that is suddenly cancelled, for example, a user will probably want to e-mail the other people who were due to attend; so when a new mail is created, their names can appear at the top of the address-book list.

Although the mobile Internet is currently being sold as a means of improving productivity and reducing support costs, the widespread deployment of wireless technology is likely to have far more wide-ranging effects. According to Joe Manget of Boston Consulting, the use of wireless will go through three distinct stages. The first, which is currently under way, involves extending existing systems and processes to mobile devices to achieve productivity gains. For example, Nissan, a Japanese car maker, found that giving its salesmen wireless access to up-to-the-minute inventory and pricing information reduced the average number of visits required to close a sale from five to three and allowed a 40% cut in back-office staff. McKesson HBOC, America's largest drug wholesaler, has introduced wireless devices to track inventory and shipments. The company spent $52m on 1,300 handheld computers and on equipping its distribution centres with wireless-network coverage. Warehouse workers use the technology to monitor inventory, and to record and check

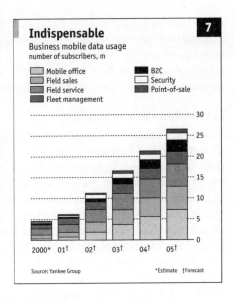

Indispensable   7

Business mobile data usage
number of subscribers, m

the contents of each shipment, thus eliminating the need to count inventory by hand and reducing errors. McKesson has already saved more than the cost of installation, having achieved an 8% productivity gain and an 80% fall in the number of incorrect shipments.

The next stage, says Mr Manget, will involve the transformation of existing business models using wireless technology. At Svenska Cellulosa Aktiebolaget, a Swedish pulp and paper company, foremen use a wireless system to send instructions to loggers in the field, specifying which trees to cut and in what order. This enables the company to co-ordinate harvesting decisions with inventory and transport requirements and match those decisions to market needs. But the transformation of business processes is not without risks; the mobile Internet, like the fixed-line Internet before it, is likely to disrupt existing power structures and decision-making processes within organisations, as frontline workers gain access to corporate information they never had before. In the final stage, entirely new business models will emerge that would not have been possible without wireless.

One development Mr Manget notes with particular interest is the rise of machine-to-machine, or "M2M", communications. In Japan, for example, NTT DoCoMo has got together with vending-machine manufacturers and soft-drink companies to equip vending machines with wireless capability. The new machines not only accept payments from wireless devices but also use wireless technology to transmit information on inventory levels and maintenance problems. Drinks companies use the data captured to improve distribution and product selection, and to reduce the number of out-of-stock items, which has increased profits per machine by as much as 70%.

Let machine talk to machine

Similarly, Toshiba is using wireless technology for remote monitoring of photocopiers, so that technicians can be dispatched as soon as there are signs of a problem. This reduces servicing costs and, since machines are out of action less often, increases usage and revenues. Caterpillar, a construction-machinery maker, is also using wireless technology to ensure that its machines are kept running smoothly; this has reduced downtime by 66%, and has resulted in a 60% reduction in maintenance costs, since worn parts can be replaced before they fail and cause damage to other components. In the long term, this approach may well be extended to all kinds of machines, including cars and household appliances; on the mobile Internet, machines might eventually outnumber people.

Such futuristic visions aside, the mobile Internet clearly has much to offer companies here and now. Corporate enthusiasm for the technology may even kick-start interest in new services in the consumer market, says Mr Nguyen; after all, employees are consumers too. Give an employee a PC, and he is connected to the Internet when he is at his desk; but give him a mobile device, and he is connected anywhere, any time. One of the mobile Internet's knock-on effects could be that the distinction between work and leisure will become increasingly blurred.

Wireless bunfight

Who would have thought that thin air could be so competitive?

IN THE LOBBY of Nokia's glistening glass-and-steel headquarters just outside Helsinki sits a table with a huge silver-and-bronze chessboard inlaid into its surface. It might have been purpose-made to symbolise the complicated manoeuvring currently going on in the industry.

Mobile operators, who act as the hub between customers, handset makers and content providers, have the hardest task. Most of them are saddled with huge debts after paying billions for their 3G licences; the agreements currently being struck to share network infrastructure are bound to lead to broader consolidation.

To make matters worse, they will find that providing complex data services is a very different business from running a voice network. Operators are having to choose partners to provide content, and decide which services to offer their customers. To attract business customers with new services such as phone-based e-mail, operators are having to get far more involved with their corporate customers' computer systems. In short, they either have to get into the business of system integration or to form partnerships with systems integrators and software firms. Unless they move into these new areas, operators risk being turned into "dumb pipes" by specialist wireless firms acting as resellers of airtime.

Operators also face potential conflicts with banks. Given their ability to charge users for small transactions, they are clearly in a position to set up their own electronic payment systems. Some customers trust their operator no less than they trust their bank. In Europe, a number of operators have applied for banking licences. Mobile phones might emerge as a popular method of payment, at least for small items. But individual operators will not be able to install equipment at the point of sale in shops, restaurants and so on; the banks and credit-card companies are much better placed to do that. Such equipment will need to communicate with phone handsets, so handset makers will also need to co-operate.

The handset makers, for their part, have an ambivalent relationship with operators. Both need to ensure that their equipment is mutually compatible. This can make for rather cosy relations between particular operators and handset makers. In Japan, for example, Panasonic and NEC are chummy with DoCoMo; Kyocera and Toshiba are close to KDDI.

In other parts of the world, where plug-in chips, called SIM cards, make it easier to switch from one operator to another on the same phone, things are not so friendly. Both the operators and the handset makers would like to "own" the customer. This tension is particularly evident in the case of Nokia, the world's largest handset maker. Because Nokia makes some of the sleekest, sexiest phones around, operators must include Nokia models among the choice of handsets they offer their customers. Nokia, for its part, wants customers to think of themselves as Nokia users. By adding unique features to its phones, it can help ensure that customers buy another Nokia model next time.

The handset makers, meanwhile, are fighting new battles of their own. At the moment, thanks to different technology standards in different parts of the world, the handset market is balkanised. But 3G phones will be a global standard (or as close to it as makes little difference). The Japanese phone makers see this as their opportunity to break into new markets, notably in Europe, where they are currently thin on the ground. Handset makers elsewhere see 3G as a chance to get into the traditionally closed Japanese market. In preparation for this global bunfight, manufacturers are already consolidating. Ericsson, a Swedish telecoms firm, has set up a joint venture to make handsets with Japan's Sony. Similarly, Toshiba, another Japanese company, has an alliance with Siemens, a German electronics firm.

Snazziness is not everything

Who will win? According to one school of thought, the Japanese have an insuperable lead, thanks to i-mode and to being the first to launch a commercial 3G service. They know how to make small, light handsets with colour screens and Internet features, and Japanese manufacturers already dominate other areas of consumer electronics, such as hi-fi equipment, video recorders, DVD players and televisions. But although they are good at making snazzy phones, Japanese manufacturers have little experience of working with operators abroad.

Nokia and Ericsson initially rose to prominence thanks to the decision by Scandinavian countries to establish a common mobile-phone standard in the 1980s. This gave Nokia an incentive to make phones that work reliably in different countries and on different networks, and are easy to use. The company is also strong in design and marketing and has a powerful brand. According to Mark Davies Jones, an analyst at Schroder Salomon Smith Barney, the biggest threat to Nokia's dominance could be from Samsung, a Korean electronics firm. Already dominant in South Korea and America, Samsung has a proven track record in building phones that work on the CDMA technology which underpins all the 3G standards, and its phones are considered as easy to use as Nokia's.

Elsewhere on the battleground, the content aggregators – or portals – are also fighting over customers. Some portals are owned by operators, such as Genie (owned by BT), zed (owned by Sonera) and T-Motion (owned by Deutsche Telekom); others, such as Yahoo!, are independent. Portals matter because they aggregate content, and access to content generates transport fees for operators. At the moment portals are quietly gathering subscribers, with a view to becoming mobile virtual network operators (MVNOs) at some point in the future. The idea is that network operators will act as wholesalers of network airtime to MVNOs, which will be better at catering for particular niches. As wholesalers, network operators will have to share transport revenues with their MVNO partners, but MVNOs should boost the overall amount of traffic. So operators will contemplate offering their networks to MVNOs only once their own revenues have levelled off and they want to reach out to new customers, says Scott Goodwin of Speedy Tomato, a Europe-wide portal.

All kinds of firms might choose to become MVNOs, but the two obvious examples are banks and media companies. If handsets were to become a significant new payment mechanism, banks might decide to team up with operators. Media firms, for their part, have lots of content to offer, and will no doubt want to get their hands on the transport revenues that go with it. That makes record companies, cinema chains and media conglomerates, notably Disney and AOL Time Warner, likely candidates for becoming MVNOs.

Who will emerge victorious from this free-for-all is still far from clear. As mobile phones are used for more and more everyday activities – and even become a "remote control for life", as some in the industry believe – it may come down to whom people are most inclined to trust.

A mobile future

There's much more to it than the Internet without wires

To GET SOME idea of how difficult it is to predict the future of the mobile Internet, talk to Marty Cooper, a veteran of the wireless industry. On April 3rd 1973, Mr Cooper stood on a street corner in Manhattan with an odd-looking white plastic brick-shaped object in his hand. It was a device called a Dyna-Tac, developed by Motorola, an American electronics firm. To curious glances from passers-by, Mr Cooper pressed a few buttons, held the brick to his ear and started talking. He was making the first-ever cellular phone call. In acknowledgment of his role in pioneering this new technology, he is now known as "the father of the cellular phone".

But that epoch-making call did not cause mobile phones to fly off the shelves. To drum up interest in these exotic new devices, says Mr Cooper, phone companies had to resort to a hard sell. "We had to tell the first users: you can use this to fill the time in your car," he says. "It was almost fraudulent, because the real value is the freedom you get – you no longer have to be in the office or the car, you can be reached anywhere. It's the removal of the chain between you and your desk. And that's the lesson it took some years to learn."

When you call a fixed telephone, you call a place; when you call a mobile phone, you call a person. Mobile phones enable people to stay in touch wherever and whenever they want. With over 800m mobile phones in use worldwide, this may seem obvious now, but in 1973 it was impossible to imagine. Even in the early 1980s, McKinsey, a consultancy, predicted that 900,000 Americans would have mobile phones by 2000; the actual figure was over 70m.

The Internet also seemed to arrive when no one was looking. Back in 1993, just before the Internet boom, who would have foreseen the rise of Napster, or eBay, or Amazon? So making forecasts about the development of the mobile Internet, the offspring of two such spectacularly unpredictable technologies, is clearly dangerous. Peering into the fog, however, this survey ventures to predict that in the short term the mobile Internet will be used by consumers mainly for interpersonal messaging, and by businesses mainly to extend corporate information systems to mobile devices. Beyond that, who knows?

From horseless to wireless

For a historical analogy, look back to the advent of the horseless carriage, or automobile, at the end of the 19th century. At first, the most notable feature of the horseless carriage seemed to be its horselessness. But few people today climb into their car and exclaim at the absence of a horse to pull it. The mobile Internet is still at that first, horseless stage: people tend to think of it simply as the Internet without the wires, rather than something entirely new.

Yet the mobile Internet is something new, because of wireless technologies' capacity to bathe the world in bandwidth and liberate people and devices from the constraints of fixed locations. Just as water, gas and electricity are ubiquitous, always-on utilities in most of the developed world, wireless technology promises to do the same for information, by making the Internet as pervasive as the air we breathe. A British mobile operator, BT Cellnet, spotted the parallel and is renaming itself O_2, the chemical symbol for oxygen.

The longer-term effect of existing technology is hard enough to imagine as it is, but there are even stranger things in the pipeline. One trend to look out for is ad-hoc networking, in which the network architecture is much more fluid than in today's systems. Handsets might, for example, double as portable base-stations, routing data to and from other nearby handsets. So far the best working example of this idea is Cybiko, a Russian-made electronic toy that allows users to exchange messages and play games with other users nearby. Teenagers lap it up. Jens Zander, of the Royal Institute of Technology in Sweden, imagines a mobile network where each user is given a six-pack of small base-stations when he buys a handset, and is asked to sprinkle them around randomly; the network has no centre, and grows virally. This sort of idea is reminiscent of the "peer-to-peer" approach used by Napster, the infamous music-swapping service. It also resonates strongly with those building guerrilla Wi-Fi networks.

Meanwhile, Mr Cooper's new firm, ArrayComm, based in San Jose, California, is pushing a high-speed wireless data system called iBurst, which relies on "smart antennas" to make more efficient use of radio spectrum than 3G networks. Time Domain, based in Huntsville, Alabama, has developed a system called "ultra-wideband" transmission that uses low-power radio pulses to transmit data at high speeds over short distances. Even now, more than a century after Guglielmo Marconi pioneered the transmission of wireless data, there is clearly vast scope for further innovation.

Little wonder, then, that so many companies, big and small, are trying to shape the development of the emerging mobile-Internet market. Given the current economic climate, life is turning out much harder for wireless firms than it was for the dotcom generation, but the builders of the mobile Internet have the advantage of hindsight. "People are trying to learn from the experience of the fixed Internet," says John Sims of 724 Solutions, a wireless software provider. "They're saying, 'let's not repeat those mistakes'."

Lessons learnt?

For example, says Mr Sims, there is far more awareness of the need for security and privacy on the mobile Internet than there was in the early days of e-commerce. And operators, he notes, are well aware of the danger of being turned into "dumb pipes", as happened to the fixed-line Internet-access providers.

Another difference, says John Little of Portal Software, a firm that provides billing systems to mobile operators, is that wireless firms expect to have to experiment in order to find business models that work. Internet start-ups, in contrast, often committed themselves to a clever new business model, and were scuppered if it failed to work.

The demise of the dotcoms has also taught mobile-Internet firms to be deeply suspicious of business models that rely on advertising revenue. The good thing about mobile phones is that they provide a means of charging money for content and services, so tangible revenues are available now, not in the distant future.

The mobile Internet is also notable for the absence of a dominant company in the industry; the nearest equivalent is Nokia, but that is no Microsoft. At every level, from network infrastructure to software to operators to handsets, the industry is still hugely fragmented. No doubt there will be much consolidation in the coming years. But so far, says Richard Owen of AvantGo, "In wireless, Microsoft's non-monopoly has created a lot of innovation, because there isn't a dominant player."

But the mobile Internet does not get top marks for everything. Two obvious weaknesses are that it still puts too much emphasis on technology for its own sake, rather than on the benefits to users; and that, until recently, it has failed to recognise the importance of person-to-person interaction, putting its money on access to centralised content instead. It should have been quicker to pick up clues from the way in which earlier communications technologies were used, and particularly from the success of text messaging.

Ringing a bell

Caution be damned: here is a final prediction. The Internet's mobile off-shoot will turn out to be an entirely new medium, as different from the Internet as the telephone was from the telegraph. It will be accessible to people beyond the reach of today's Internet, notably those in the developing world, because it will not require complex and costly PCs, and will thus bring many of the benefits of the Internet to a far wider population than is able to enjoy them at the moment.

Funny how history repeats itself. The mass-market successor to the telegraph proved to be the telephone. The successor to that modern-day telegraph, the Internet, may well be the telephone again.

The material on pages 263–303 first appeared in a survey written by Tom Standage in *The Economist* in October 2001.

VII

FOOD FOR THE MIND

Thrills and spills

The digital revolution in entertainment was expected to sweep all before it. But so far it has proved somewhere between a disappointment and a disaster

SKY DAYTON, a young man with an engaging manner and a head injury from surfing, draws a picture of a funnel on a whiteboard. It shows what the constraints of bandwidth do to the entertainment industry. The wide bit, he explains, is creative talent, and the narrow bit is the filter that determines which part of the creative people's input makes up the industry's output. The filter is the studios and the networks – "Some guys in a room with cigars, throwing darts at a schedule," says Mr Dayton, co-founder of eCompanies, an Internet incubator in Los Angeles, and previously, at 22, of EarthLink, an Internet service provider. "The stuff that gets through the filter isn't the best stuff."

Then Mr Dayton rubs out the funnel and draws two parallel lines. The filter has gone. The output is as wide as the input. This is every frustrated screenwriter's and director's dream: access to the audience without having to go through the hell that is Hollywood. This is the vision of the digital distribution of entertainment, as preached by venture capitalists and assorted seers over the past decade: a world of unlimited bandwidth in which the barriers to entry into the entertainment business will fall so dramatically that everybody could be their own studio and their own network.

The vision of unlimited bandwidth has sucked billions of dollars of investment into e-entertainment. Some of it has gone to fund the start-ups (such as eCompanies' latest venture, Icebox) that were supposed to displace the big media companies. Some of it has come from the big entertainment companies, which were terrified by the challenge they thought the start-ups posed, but also excited by the benefits the web seemed to promise.

The Internet offered much more to the media companies than it did to most consumer businesses. Clothes and grocery retailers may take orders through their website, but they still have to send round a man with a van to deliver the parcel. The entertainment business, by contrast, should be able to deliver its goods over the Internet. The factories, the distribution networks and the shops should become redundant, leading to huge cost savings for the Internet companies.

What is more, advertising on the Internet should be more valuable than in the old-media world. Now that companies can target specific groups of consumers, fewer advertising dollars should be wasted. And now that consumers can click on ads, and buy the products direct, advertising should be more easily translated into purchases. More valuable advertising should mean more cash for the companies whose revenue comes from advertising.

In your dreams

But the reality has not matched up to the vision. "The Internet is a zero-revenue business for traditional media companies," according to a recent report from Jupiter, one of the two best-regarded Internet consultancies. The two main problems Jupiter identifies are the difficulties of distributing content on the Internet, and consumers' unwillingness to pay anything for content beyond what they are already paying for Internet access.

The music, text and picture businesses on the Internet all suffer from one or other of these problems, or both. Music is quite easy to distribute on the Internet, but nobody has yet found a way of persuading people to pay for it. Text delivery is straightforward, but nobody is buying those little portable electronic-book readers into which to download the words. And pictures cannot yet be delivered through the Internet reliably enough, and to a standard that makes them worth watching.

In the old, physical world, the entertainment industry has two revenue streams: subscription and advertising. But in the new, electronic world, there are few things that consumers seem willing to pay for. They boil down to the Wall Street Journal (if it will forgive its categorisation as entertainment), some games and a good deal of pornography. As for advertising, where exponential growth was due to start any time now, the graphs are in the process of being redrawn.

Interactive television, which has got further in Europe than in America, looks like a more promising way ahead for the entertainment business than the Internet. But doubts remain even there. In Britain, for instance, which is leading the switch from analogue to digital television, the satellite and terrestrial companies pushing the changeover have been giving the set-top boxes out free, and nobody is paying for anything much yet. The entertainment industry, which had put aside its scepticism and thrown itself into this enthusiastically, is now in a state of confusion. The future is looking increasingly uncertain. "Nobody knows anything", the screenwriter William Goldman's mantra about Hollywood, is truer than ever.

What went wrong

Nicholas Negroponte has a lot to answer for. Back in the early 1980s, this media guru drummed up some money from a lot of publishing, entertainment and technology companies and set up the Media Lab at the Massachusetts Institute of Technology. A great salesman with one of the most valuable contact lists in the business, Mr Negroponte started peddling his vision to the world's media moguls.

Stop thinking about films, books, cartoons, TV shows and games, he told them. Think about information and interfaces, bits and bytes, zeros and ones. The coming switch from analogue to digital would liberate industries that had been constrained by the limit on the number of cable channels that can be bunged down a pipe, or the handful of broadcast channels that can fit into the airwaves allocated to television. In the digital age, consumers' homes could be pumped full of as much entertainment as they could possibly wish for. Everything would be on demand. People could have whatever they wanted, whenever and wherever they wanted it – on the television, the computer, the fridge or the vacuum cleaner.

To entertainment-company executives, who had watched the computer-software companies' market capitalisations outstrip theirs many times over, this was a seductive message. Now they would become part of the digital revolution. Their stock prices would rise above the mass of boring old blue chips and follow the high-flying tech stocks. They would see their faces on the front of glossy business magazines, and get to be called visionaries. Television executives seriously pondered advice from Mr Negroponte such as: "The key to the future of television is to stop thinking about television as television. TV benefits most from thinking of it in terms of bits."

Mr Negroponte has faded as a public figure, but his influence persists. His predicted switch from analogue to digital is now beginning to happen all over the world, but as yet nobody knows what it will do to the industry. In an atmosphere of such uncertainty, people hold on to their visions. The trouble with visions is that they tend to provide little guidance on the boring business of implementation, such as getting systems to work together, putting infrastructure in place, keeping costs down, producing things that people actually want, and getting the sums to add up. Yet chief executives, unlike visionaries, do have to worry about these things, because they have to make things work, and get them out to market, and sell them to people, and, eventually, turn a profit on what they sell.

The first iteration of the Negroponte vision came in the early 1990s. Hollywood began to make overtures to Silicon Valley, which produced some collaborative ventures inevitably tagged Siliwood. The entertainment companies started experimenting with wired homes. At Time Warner's Full Service Network, in Orlando, Florida, everything except the pizza was delivered to people's living rooms via the television, just as Mr Negroponte had decreed. The consumers loved it, but the sums did not add up. The Full Service Network was shut down.

For a few years after that, entertainment-company executives turned their backs on the digital vision. They became preoccupied with analogue-world mergers, putting together huge companies with a finger in every entertainment pie. The Siliwood joint ventures were shut down, and the software executives sent back to wherever they had come from. "Two-and-a-half years ago," says Jay Samit, senior vice-president of new media at EMI, who is based at the company's Los Angeles office, "Silicon Valley types couldn't get arrested in Hollywood."

Then the dotcoms appeared on the scene. Suddenly, America's biggest entertainment companies were smaller than several Internet companies that a year earlier nobody had heard of. Hollywood filled up with instant venture capitalists. Seasoned executives were leaving decent jobs to join Internet start-ups. "I'd never had so many scared billionaires inviting me over to dinner," says Mr Samit, who, unusually among Hollywood executives, has worked on both sides of the digital divide.

Wall Street began to reward the entertainment companies that invested heavily in online ventures, and to punish those that did not. Shareholders and boards badgered managers to build up their new-media divisions or buy into Internet start-ups. Even News Corporation's Rupert Murdoch, who had been notably unenthusiastic about the Internet in its early days, threw some money at it. This second iteration of the Negroponte vision is still in full flight. But, as this survey will show, some of it will prove no more durable than the first one.

Sex, news and statistics

Where entertainment on the web scores

"TO DATE," SAYS Ted Leonsis, "digital entertainment has been a failure." As the man who has been mainly responsible for content at AOL, the company that has tried hardest to meld the entertainment and Internet industries, he should know. The record, as far as most of the entertainment business goes, supports his gloom. But there are some areas where the two work well together.

The Internet's virtues – its freedom from censorship, its speed, its low distribution costs, its global reach and its interactivity – suit some parts of the industry nicely. But only some. These include:

- Pornography. A few years ago almost all the paid content sites on the web were pornographic. Even now that the big entertainment companies and the venture capitalists have tried to turn the Internet into a mainstream entertainment medium, Internet content still has a dodgy flavour. Plenty of the material on offer on many sites would never get an airing on television because it is too sexually explicit or politically incorrect.

  Joecartoon, for instance, probably the most successful cartoon artist on the web, launched his career with a cartoon of a frog being shredded in a blender. Icebox's "Mr Wong" is a slitty-eyed old Chinaman having an affair with a Caucasian girl of dubious morals. Thanks to freedom from censorship, Internet entertainment companies are able to provide "edgy" content which, they like to think, appeals to "Generation Y" kids weary of bland television fare. But many of these companies do not have much else going for them.

- News. When Excite@Home, a broadband Internet service provider, takes its product on "mall tours," news is the part of the service that most interests potential customers, says Richard Gingras, the company's senior vice-president. The Internet is rapidly becoming one of the main media for consuming news. According to a report published in June 2000 by the Pew Research Centre, a Washington-based media-research outfit, the number of people who say they log on to the Internet for news every day has risen to 15%, from 6% in 1998. A third of Americans now read news online at least once a week, compared with 20% in 1998.

News organisations have learnt to use interactivity to create content for themselves: "Do you want to see him fry tonight? Visit our website and vote." Polls used to be commissioned at great expense from specialist research organisations. Now websites can provide speedy, cheap, do-it-yourself polls that can fill up the slack moments in a 24-hour news channel's schedule. And Interactivity is good for news. News programmes come in bite-sized pieces anyway, and everyone wants a different selection of bites.

The heavier the content, it seems, the better it does on the web. Outside the pornography business, the *Wall Street Journal* is one of the few content sites charging a subscription for its web-only content and building subscriber numbers. Others rely mostly on advertising, which doesn't pull in much money on the web.

- Sports. The Internet has what it takes to make a success of this. Sports are about speed, about belonging to communities, about data – scores, distances, runs – and about serving small audiences, all of which the Internet is good at.

    Quokka Sports, for instance, one of the leading sports sites, started off with sailing. During the Whitbread Round-the-World Race in 1997–98, it broadcast live pictures which were not available on television. The site kept the armchair sailors involved with imaginative extras, such as a fantasy Whitbread game: viewers were given the same weather information as the real sailors, and had to plot their course accordingly. Their virtual progress was then shown against the sailors' real progress.

    On the basis of the Whitbread, Quokka created a sailing "vertical" – dig down deep, and you will find all the information and products you could possibly want about sailing. It has since done the same for Indy car racing, motorcycling and some action sports (through the acquisition of mountainzone.com). And in 2000, thanks to a deal with NBC, which bought the American broadcast rights to the Olympics, it covered the Games on the web. However, there were no live web broadcasts, for fear of cannibalising television revenue.

- Niche businesses. Cheap distribution and global reach allow sites to serve thinly-spread audiences. On cable, golf is about as niche

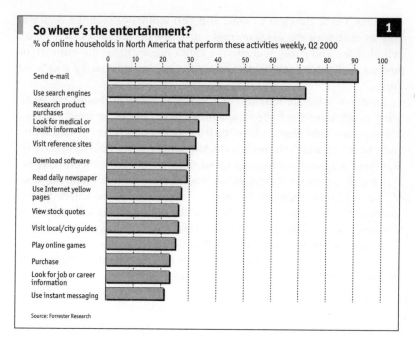

So where's the entertainment? `1`

% of online households in North America that perform these activities weekly, Q2 2000

Send e-mail
Use search engines
Research product purchases
Look for medical or health information
Visit reference sites
Download software
Read daily newspaper
Use Internet yellow pages
View stock quotes
Visit local/city guides
Play online games
Purchase
Look for job or career information
Use instant messaging

Source: Forrester Research

as they come. On the Internet, fans of sports such as cricket, spurned by sports channels around the world, are better-served than they have ever been before. At cricket.org, for instance, cricket fans get a wider range of news and information than they could on any television programme, more up-to-date than anything they could find in a magazine.

The problem for Hollywood is that most of what it produces is "linear" entertainment, otherwise known as stories, and Hollywood has become used to the idea that new technologies will offer better ways of telling, or distributing, stories. When film was first invented, nobody was sure what it was for, but it turned out to be a great way of telling stories. When talkies came along, they proved even more compelling. And television took those stories into people's living rooms.

Why stories don't work

The entertainment companies have been trying to work out how inter-activity can improve storytelling. All soaps these days, for instance, have websites where fans can touch base with their idols between

313

episodes. Go to Dawson's Creek, and you will be offered the opportunity to chat with the assistant make-up artist on the series, or vote on whether Jen and Pacey will survive either the storm, or the summer, or neither, or both. It may help promote a soap, but it is hardly a new art form.

Directors who are looking for something more profound are struggling. Some have tried allowing audiences to choose between endings. Others have used maps, where the user clicks on different locations and pulls up a clip which shows what is happening to the story in that room. But there is a good reason why these experiments do not work. Interactivity and stories are incompatible. Stories need suspense and surprise. If the audience chooses the ending, the suspense is killed. Stories demand that the audience lose itself in the telling. If it is still capable of thinking about an alternative ending, then the story has failed.

If there is the equivalent of a half-hour sitcom on the Internet, it will not come from the storytelling end of the entertainment business. More probably, it will have something to do with games.

Play to win

Games and the Internet go well together

"GAMES HAVE BEEN on a 20-year detour," says Ed Fries, Microsoft's vice-president of games publishing. Games, he points out, used to be mostly about socialising. People don't play bridge for the cards, but for the other people. Then computers turned gaming into a solitary activity. But now that the Internet has linked everybody's computers together, "Online is bringing back the social dimension into gaming," says Mr Fries.

Games are the stepchild of the entertainment industry. Until 25 years ago, they were hardly a business at all. Computer games created an industry, but the mainstream of the entertainment business still dismisses them as being for kids, geeks and other undesirables. They have no stars, so lack the glitter of celebrity. Few have intellectual pretensions. And they are part of the despised computer software industry. Hollywood has paid little attention to the games business.

But games have slowly been creeping up on the movie business. The industry got its big break when Sony's Playstation was launched in Japan in 1994 and in America and Europe in 1995. Sony pushed the Playstation in clubs and at concerts, and turned games into something adult and hip by dint of what Peter Molyneux, Britain's best-known games designer, calls "a miracle of modern marketing". The average age of a Playstation owner is 22; nearly a third are over 30.

In the past five years, from a standing start, Sony has wrested the market from Nintendo and Sega, selling 73m Playstations and 600m games. Now around half of all Japanese households, a third of American ones and a fifth of British ones have a console of one make or another.

Since the mid-1990s, the games industry has been growing faster than any other part of the entertainment business (see chart 2). In terms of revenues, it is now running neck-and-neck with movie box office. Movies still make much more money once television sales, videos and licensing deals are included, but $20 billion a year worldwide shows how far the underdog games business has come.

It still has a long way to go. It remains skewed towards the young, and its customers are almost exclusively male (except in Japan). If

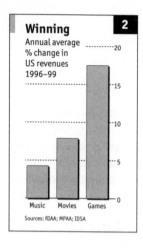

Winning 2

Annual average % change in US revenues 1996–99

Music Movies Games

Sources: RIAA; MPAA; IDSA

women could be persuaded to play, it might really take off.

Technology has driven the growth so far, and will drive it much further. Games are now at the cutting edge of several different technologies. Until recently, it was computer-aided manufacturing and design that were leading advances in graphical interfaces; now it is games.

Unlike movies, which do not change much, games are getting better every year. Remember Space Invaders, the little beasts you used to burn up in the bar? Forget them, and try one of the current crop of games. The experience is entirely different. First, 3-D imagery made the pictures better; now the application of physics is making movement more realistic; and artificial intelligence is getting players emotionally involved. The result is that, according to Peter Molyneux, games are acquiring some of the sophistication of movies. "I want you to laugh, I want you to cry, I want you to do everything the film industry wants you to do." Only, unlike films, games are naturally interactive: the player, not the director, is always in control of the story.

Virtual playmates

Now games are taking the next big leap, on to the Internet. People in the games industry are much more positive about the Internet than those in other areas of the entertainment business. They are convinced that it will be the making of them. John Riccitiello, president of Electronic Arts, says that probably two or three times as many people are playing games online as offline. "It's like a big party out there. Lots of people are wandering around, looking for something to do, and they are knocking on the door of the games business." Most of them are playing very simple stuff. The most popular is Tetris, a dull affair requiring players to stick pegs in holes. It is said to appeal to people who enjoy tidying up.

At the other end of the scale are the "massively multi-player games" (MMPGs), such as Ultima Online and EverQuest. These are role-playing games that grew out of Dungeons and Dragons, an old text-based cult game that started in the 1960s. Before computers were linked, people played by themselves. The next step was "multi-user dungeons"; then they went 2-D, and later 3-D; and now there are wild worlds where people can build alternative lives, go off on chivalrous quests, and team up with other players to fight dragons – or each other.

Players get hooked on the MMPGS. EverQuest, owned by Sony and currently the hottest, is nicknamed "Evercrack", because people find it so hard to get off it. On average, players spend 20 hours a week on it – not much less than the average viewer spends watching television. At its peak, there have been 60,000 people playing the game at once. It is, in part, the size and complexity of these worlds that keeps people in thrall: EverQuest, for example, features four continents and 25 countries.

But EverQuest has its limits. It may be one of the few entertainments on the Internet for which people are willing to pay – over 300,000 of them cough up $119 a year – but this is a pocket of obsessive, hardcore gamers, not the mainstream. Yair Landau, president of Sony Pictures Digital Entertainment, explains that the games business is polarised. "At one end we have EverQuest, at the other we have Wheel of Fortune." Wheel of Fortune is a hugely popular game show, with an interactive version which the very few Americans with interactive television (through, for instance, WebTV) can play. But what Mr Landau really needs is something that combines the power of EverQuest (and its ability to remove money from people's wallets) with the broad appeal of Wheel of Fortune.

Like the movie business, the games business is limited by the bandwidth-constraint online, but it has more flexibility to get around the problem. For some games, such as EverQuest, most of the information needed to play the game – the world that appears on screen – is delivered by post, on a CD-Rom. The only information that has to travel down the wires is about the current state of play. Occasional updates, such as a seasonal winter scene, can be downloaded.

Now the consoles are going online. Playstation 2, released in America and Europe in autumn 2000, has a plug-in hard disk and broadband connection, as well as a DVD player. And Microsoft is getting into the business too. Its X-box, with a graphical interface supposedly superior to Sony's, will also have a hard disk and Internet connection.

"The console is a Trojan horse," says Peter Molyneux. The idea is that games consoles are so powerful (the Playstation 2 has more computing power than a Pentium III) and have such sophisticated graphical interfaces that they can take over the job of the PC, the DVD and the television as well. Whether or not that happens, it looks as though Hollywood will have to start showing the once-despised games industry a little more respect.

Siren songs

Delivering music across the Internet is easy. Making money out of it isn't

THE MUSIC INDUSTRY is the canary down the digital mineshaft," says EMI's Jay Samit. Music can be easily distributed via the Internet – indeed too easily for the music business's liking.

One reason is bandwidth. In the digital world, music, pictures and text are all converted into bits. It takes many fewer bits to transmit three minutes of music than it does to transmit three minutes of video. Music can therefore travel through narrower pipes than video can. Video works only with a broadband connection.

The second reason concerns the nature of the medium. Books or magazines cannot be downloaded in the form in which they are generally consumed. Words can be read on screen or printed out, but at present neither is very convenient. By contrast, the form in which music is downloaded is the form in which it is consumed. That is why radio was one of the earliest forms of entertainment available on the Internet, and is one of the best-developed. MIT's list of online radio stations, the most comprehensive on the web, lists over 10,000. Some are offline stations that are using the Internet as a new distribution channel. The economic impact is minimal, but it gives a lot of pleasure to a lot of people – for example, those away from home who can tune in to hear their football team's results.

But there is also a bunch of companies building radio stations specifically for the medium – such as Spinner, AOL's online radio company, which gives people the option of choosing something more closely tailored to their tastes than they can find by twiddling the dial in their car. If that works as it should, it offers the listener a choice that will sound a bit like his own CD collection, but retain the serendipity and surprise element of radio.

Nobody is sure whether this model will be funded by advertising or subscription. Another question must be answered first: whether the music industry will ever be able to charge for music again.

Until quite recently, the piracy issue aroused more intellectual curiosity than terror. At the end of 1998 the record companies, in conjunction with the consumer-electronics companies, put together a group called the Secure Digital Music Initiative (SDMI) to build an anti-piracy system. Nothing much happened, but the need for a workable system increased when Napster came on the scene in late 1999. Napster allows people to share each other's files, and therefore each other's music. A case to determine whether this is legal or not is dragging through the courts. But whatever the outcome, closing down Napster is not going to solve the problem, because something worse has already emerged.

Justin Frankel, who invented Winamp, an online music player, in 1995 when he was 16 (and subsequently sold it to AOL), had a brother at college. The brother was annoyed when his campus shut down access to Napster. "Justin wrote alpha code," says Ted Leonsis at AOL, "and sent it to his brother, saying this will allow you to do what you want, and they can't shut it down." That was how Gnutella, a file-sharing system without a central server, was born. One stage worse for the music companies is Freenet, from Britain, which also allows people to share files without using a central server, and makes their movements hard to trace.

Optimists have suggested that piracy should be regarded more as a form of promotion than as theft. If people enjoyed the music on these shared files, perhaps they would go out and buy the CD. The continued rise in record sales in America suggested that this might be happening, until somebody took a closer look at record sales around campuses. The proprietor of Oliver's Records, for instance, a shop two blocks from the campus of Syracuse University, says it normally does $30,000–40,000-worth of business in April. In April 2000, it did $3,500-worth.

The music business accepts that it has to use both carrot and stick to deal with this problem, but so far neither is in evidence. The stick is still in the works, largely because the politics of SDMI have proved so difficult. All the parties have different interests. The music industry wants an ultra-secure system. The consumer-electronics people and the telecommunications companies want one that is easy to use. "The phone," explains an SDMI member, "is a stupid device. If our security could work easily on that, a 10-year-old could crack it. On the other hand, if you want the phone to have the security of a Pentium computer, people would scream because it would cost so much." He maintains that they have reached a reasonable compromise.

By late 2000, a year later than the SDMI originally predicted, the outline of a system had been drawn up. It will involve two electronic watermarks, which will be embedded in a song or piece of music. One watermark will be indestructible, and will identify the song. The other will be fragile: if the song is compressed (which it has to be in order to be sent), it will disappear. If the song has only the robust watermark, it will not play on devices sold by any of the big consumer-electronics companies, because the device will assume that it has been compressed. If it has no watermarks at all, the song could either be a home recording of the church choir or a Napster file. The device will give it the benefit of the doubt – and two watermarks, so that it cannot be copied again. The watermark scheme is half-way through being implemented: the first CDs with the first type of watermark have appeared in shops. Work on the second type is in progress.

But the companies recognise that policing will never be a complete answer. "Your main demographic [people in their teens and early 20s] is stealing," says Talal Shamoon, a senior vice-president of InterTrust, one of the companies working on developing a security system. "You don't know who they are. You don't know where they are. Your best hope is to get them to come out of the bush." The way to do that is to get some legitimate music out on the web for sale at a reasonable price.

Until summer 2000, it was impossible to buy legitimate music from the big record companies on the Internet. Plenty of independents were putting music up for sale; the majors were offering samples, but no albums. Now several have made available a few bits of downloadable music.

A handful of legitimate online retailers try to offer something more than Napster can. Napster is a poor way to build a record collection. Tracks play at different volumes; most music files are unobtainable at any one time; and Napster offers people only what they already know they like. Listen.com, by contrast, offers a hand-built, searchable directory of music by 140,000 artists, most although not all of it downloadable. Rob Reid, its founder, reckons that by making music easy to find, and beefing up the directory with editorial content about the artist and the song, he helps people expand their repertoire of favourites.

But Listen.com's success depends entirely on how much music the record companies are willing to post – and their ambivalence is evident. There was no fanfare when they started putting up music for sale, and it is not easy to find. It is hiding both from the pirates and from the retailers, who are horrified by the idea that the record business may be cutting them out.

The devil and the deep blue sea

The record companies are caught between the pirates and the retailers.

Their best way to zap the pirates would be to put all their music up on the web and price it somewhere much closer to its marginal cost (which is zero). That would greatly reduce the incentive to steal it. But it would also mean cutting out the retailers, who currently take a margin of 10–50% on record sales – and the music retail business is dominated by a few big, powerful players on whom the record companies still depend for the vast bulk of their sales, and whom they do not want to antagonise. As Bob Bowlin, chairman of Sony Music International, says, "For every action, there is an equal and opposite reaction. The bricks-and-mortar business is going to be important for the foreseeable future. It would be foolhardy to do anything which would damage that relationship."

If the music business can deal with those dangers, the Internet offers it huge benefits. Aside from cutting out the manufacturing and distribution costs of CDs, it should make marketing more efficient. At present, 95% of albums do not make money either for the artist or for the record company – not just because this is a hit-driven business, but also because music companies know so little about their consumers. Interactivity promises to change that. As Al Teller, former chief of CBS records and founder of Atomic Pop, an independent online label, says of the site he has put up for the Smashing Pumpkins: "We've got 10,000 registered users. The one thing we know about them is that they're Smashing Pumpkins fans. From a marketing point of view, that's fantastically valuable."

At the same time, the Internet should allow music companies to sell a wider range of music than ever before. Internet retailers can offer more inventory than bricks-and-mortar shops. Amazon already sells a wider range of music than any high-street record shop. Unlike their high-street counterparts, niche retailers catering for the fans of Moldovan polkas and Zimbabwean drummers can find markets all around the globe. But the business model depends on the record companies getting piracy under control, and that depends on putting a system in place that everybody can work with.

Learning to e-read

Who wants electronic books?

O N MARCH 14TH 2000, Stephen King, a horror writer, published *Riding the Bullet* on the Internet before it had appeared in print. It was an experiment with a book too short, at 66 pages, for the standard novel format. Within 24 hours, around 400,000 people had downloaded it, even though most of them also had to download the software they needed to read it. It sent servers crashing. Jack Romanos, president of Simon & Schuster, Mr King's publishers, told the *New York Times*: "I don't think anybody could have anticipated how many people were out there who are willing to accept the written word in a paperless format." It felt like the coming of age of the electronic novel.

Sending text over the Internet is easy, so the print publishing business might have been expected to feel the web's impact even before music, and long before movies. It is true that encyclopaedias in book form have been wiped out, and that textbooks are fast going electronic; but otherwise the old ways of selling printed words have barely been disrupted. That was why the Stephen King book sent the business into a tizzy. The consumers, it seemed, wanted to go somewhere the industry was reluctant to lead.

Or did they? Mr King's novel was posted free, so there was nothing to discourage surfers' natural curiosity. The word in the industry is that three-quarters of those who downloaded the book did not read it. "I am two of those people," says Youngsuk Chi, chief operating officer of Ingram Book Group, and possibly the most wired man in the publishing business. He downloaded it both at home and at work to test his equipment, and read neither copy. Horror isn't his thing.

Electronic babel

One reason why the publishing business is not yet distributing books on the Internet is that it is a lot more complicated than it might seem. For a start, most of the publishing houses' backlists are not computerised. And even for its current list, every publishing house uses its own combination of packaged and in-house software, so there are probably 2,000 different formats in use for storing books electronically.

The publishing industry needs to agree on formats for electronic storage and delivery before the mess can be sorted out. The world's big publishing houses are talking about this, but without the pressure the music industry is under, such things take a long time. And, once formats are agreed on, the industry will have to spend money on computerising, storing and downloading its lists, which only the biggest publishers will be able to afford. That is where Ingram and a handful of competitors see a market: they will buy the systems, and the publishers will out-source the work to them.

The bigger problem, though, is that people do not much like reading on their screens. The consumer-electronics industry has started to pro-duce devices designed to overcome that resistance. So far, the Rocket eBook and the SoftBook Reader – small portable screens – have not made much of an impact. Microsoft's PocketPC, which was launched in April 2000, does offer the Microsoft Reader software designed to make e-books more readable, and the PocketPC has sold well. But it is not yet clear that people are reading books on it much.

There remains a chicken-and-egg problem, which affects the elec-tronic-book people as well as others in the e-entertainment business: people will not buy the devices until lots of books are available for downloading. And once again the piracy problem is putting publishers off. Mr King's book was pirated instantly, even though downloading it was free – just to show that it could be done.

Piracy presents the publishing industry with a dilemma. On the one hand, publishers do not want to be exploited by rip-off merchants. On the other hand, a security system that prevented people from sending excerpts from books to their friends would go against any writer's instincts about the importance of shared ideas. And even from a more material standpoint, the lending of books has always been one of the most effective marketing tools.

Nevertheless, books are beginning to go online, even though for now this is mostly marginal stuff that does not easily fit into any category. It is hard to see, however, why anybody would spend hundreds of dollars on a device for reading electronic books when there are so few of them around to download – especially when there is a competing technology already on the market that is portable, open-architecture, with an acces-sible interface, and available at relatively low cost through established distribution channels: the paper book.

There's nothing on

Moving pictures are not working online

ICEBOX IS AN entertainment site created by eCompanies, Sky Dayton's venture-capital firm. It was launched in June 2000 with a tremendously hip party and a generous dose of chutzpah. "We think it has more talent than any entertainment company anywhere in the world," says Mr Dayton. Of Icebox's chief executive, Steve Stanford, who has worked both in Hollywood and in the software business, Mr Dayton says: "Steve is the most perfectly engineered hybrid of the two worlds. It's like he was created in a lab." Mr Stanford has the grace to look embarrassed.

But the content on offer at Icebox is rather less impressive than the hype. There are a few mildly amusing cartoons, three or four minutes long. Most of them are slightly more risqué, and rather less good, than the cartoons screened on television. The public is not rushing to the site: it gets too few visitors to appear on the radar of Media Metrix, which produces ratings for Internet sites.

Companies trying to sell video entertainment on the Internet face two overwhelming problems – getting anything watchable out to the viewers, and making money.

The biggest problem with distribution is bandwidth. Broadband connections allow consumers to watch video of reasonable quality, but only around 3% of American households have a broadband connection. Entertainment sites therefore either have to settle for a tiny audience, or to programme for narrowband (as Icebox does). Programming for narrowband means short cartoons. Nothing wrong with short cartoons, but several hundred sites are not going to make a living out of them.

Why has broadband deployment been so slow? Part of the blame lies with the cable industry. In countries where consumers commonly have cable-television connections (such as America), it was assumed that cable companies would swiftly be able to provide broadband Internet connections as well. And when Microsoft took an 11% stake in Comcast, a cable company, in 1997, the markets saw it as a sign that the software industry had anointed cable as the conduit of choice for the future. Cable companies' share prices shot up.

But the practicalities of getting broadband to America's surfers have proved a challenge. When the cable companies started offering people broadband connections, it took one van, one Ethernet card, two technicians and three hours. The cable industry was not geared up for the complexities of the computer industry. "We've had to take cable companies – where the greatest degree of complexity is twisted coaxial cable into the back of the set – through Ethernet cards and software-configuration issues," says Richard Gingras, senior vice-president of Excite@Home, America's biggest broadband Internet service provider. Because of the cost and complexity of all this, by the end of June 2000 fewer than half of all American homes had cable broadband available to them, according to Broadband Intelligence, an industry newsletter; and fewer than 6% of those had taken up the offer of a broadband connection. Nor have the telephone companies rushed into the business. By the end of June 2000, only a fifth of American households had access to the digital subscriber lines (DSLs) that offer broadband connections.

Both services have their limitations, which vary from country to country. In America, DSL transmission speeds range from 300 kilobits per second to 1.5 megabits. "Most video engineers think that for television-quality video you need three megabits per second," says Cynthia Brumfield, who runs Broadband Intelligence. Cable, for its part, has a theoretical limit of 10 megabits per second, but the actual speed depends on what the neighbours are doing. Telephone networks are built in a star shape, centring on the local exchange, so everybody has a separate line. Cable networks are built in a series of loops, so the wire is shared. If too many people are using it at once, transmission speeds slow down. Cable companies are rather sensitive about this.

The problems with distribution have resulted in a dearth of content, according to Ms Brumfield. Because there is little exciting broadband content available, people are reluctant to spend on broadband connections; and because there are so few broadband connections, Hollywood is unwilling to invest in broadband content. "What's needed before it becomes a mass-market medium is compelling content," she says.

Where's the money?

But even if the content could be found, revenue from entertainment on the web would still be elusive. Many companies started out with subscription as the favoured model. *Slate*, Microsoft's electronic magazine, was in the vanguard when it started charging readers in 1996. It abandoned the model in 1999, along with almost everyone else who tried it. Many sites charge for archive material; most try to boost the value of

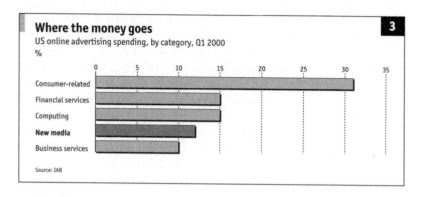

Where the money goes — **3**
US online advertising spending, by category, Q1 2000
%

Consumer-related
Financial services
Computing
New media
Business services

Source: IAB

advertising on their sites by getting visitors to fill in long registration forms; but the *Wall Street Journal* is now the only major publication, and one of the few sites outside the pornography or financial industries, that still charges a subscription for its web-only content. The problem with charging is that too many content sites are offering too much stuff for free. It may be that, as those sites burn through their money, they will be forced to start charging. But that does not mean that people will necessarily start paying.

By contrast, advertising revenue on the Internet is growing rapidly. Not much of it, however, is going to content sites (see chart 3). Total advertising revenue on the Internet is currently running at around $8 billion a year, of which around $1 billion is going to "new media" sites – hardly riches beyond the dreams of avarice. What is more, the growth of online advertising revenues has slowed sharply of late. According to a report by PricewaterhouseCoopers, online advertising revenues in America in the second quarter of 2000, at $1.9 billion, were nearly three times their level a year earlier, but only 10% higher than in the previous quarter.

Part of the reason for the increasingly sober attitude to Internet advertising is that the proportion of users who click on an advertisement is falling. In 1997, says Bryan McCarter, managing director of Zenith Interactive Solutions, the new-media arm of Zenith, a media-buying agency, click-through rates were around 1%. Since then they have fallen every year, to around 0.4% today. Advertising rates, says Mr McCarter, are also falling, though not as fast as click-through rates. "A couple of years ago," he explains, "every client thought he had to be online, but he wasn't quite sure why. Now people are thinking harder about what they want from the web."

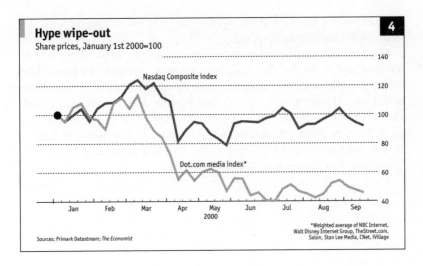

Hype wipe-out
Share prices, January 1st 2000=100

4

Nasdaq Composite index

Dot.com media index*

Jan Feb Mar Apr May Jun Jul Aug Sep
2000

*Weighted average of NBC Internet,
Walt Disney Internet Group, TheStreet.com,
Salon, Stan Lee Media, CNet, iVillage

Sources: Primark Datastream; *The Economist*

Since they cannot make a decent living from either subscriptions or advertising, many of the start-ups that were supposed to challenge the power of the big media companies are in trouble. Some have gone under. The most spectacular bomb of 2000 was Digital Entertainment Network (DEN), which had proclaimed in its manifesto that "The boob-tube zombie television is dead". But the boob-tube zombie television got the last laugh: DEN filed for bankruptcy in June, having burnt through $57m.

Pop.com folded before it even started. The site, which was to present "pops" of entertainment, was announced with great fanfare in October 1999. It was backed by Steven Spielberg and Jeffrey Katzenberg of Dreamworks, with a $50m investment from Paul Allen, Bill Gates's former partner. After a long silence, Dreamworks announced in September 2000 that all but a dozen of its 75 employees were to be laid off.

Others are finding other ways to make money. Atom Films, which has made its name by putting short films on the Internet, gets most of its revenue from selling those shorts offline, to cable companies and airlines, for instance. IFILM has shifted its strategy, from showing shorts to becoming a business-to-business portal for the film industry. Inside.com, a site devoted to the digital revolution in the entertainment industry, has admitted to having trouble generating enough subscriptions for its website – and is now going to appear as an old-fashioned print magazine.

The big media companies can at least console themselves that the start-ups no longer pose such a threat. Aside from that, there is not much on the Internet to cheer them up, for they are all struggling with the same difficulties as the start-ups.

In 1999, NBC's rocketing investments made it the envy of the rest of the business. It floated them off as a separate company in December. But by August 2000, revenues were still thin and NBCi laid off 20% of its workforce. Disney floated its Internet businesses – Go.com, its portal, and a few other properties – in late 1999. In its first nine months as a quoted company, the Walt Disney Internet Group lost nearly $1 billion. By late September 2000, the two stocks had lost around 90% and 70%, respectively, of their value since flotation.

Two new-media moguls

JOE SHIELDS of Grand Rapids, Michigan, belongs to a very select group. He makes quite a lot of money by selling content on the web. He is not keen to say how much, but it is a lot more than he was making as a T-shirt designer, which is what he did until 1998.

Mr Shields did not much like doing cartoons for the T-shirt business, so one day he "took all the rejected stuff I had lying around and threw it at the web". People quite liked it. Then somebody showed him how it could be animated. His total costs so far were pretty modest: $300 for the animation software and $50 a month for the website.

By April 1999, he was getting 2,500 hits a day. Then he put up what some *aficionados* still regard as his masterpiece: "Frog in the Blender." With this subtle combination of red and green, humour and grossness, plus a tsunami of word-of-mouth recommendations, the hits on his joecartoon.com went up to 400,000 a day. Nasty things happening to gerbils have since driven the figure up further. According to Media Metrix, an Internet ratings company, Mr Shields gets around 750,000 visitors a month. He now makes more money than any but the best-paid Internet chief executives, partly through advertising revenue, partly through licensing deals. But before cartoonists rush to the web in droves, they need to remind themselves why Mr Shields has been so successful. His pieces, unlike most of the stuff out there, are funny.

The joys of traffic

"Would you like to meet David Lynch?," asks Rob Burgess when this revered Hollywood director comes up in conversation. "We should be able to arrange it." Mr Burgess runs Macromedia, which creates software for making and viewing animation. It is a measure of these uncertain times that a Canadian software writer in San Francisco should be able to pull strings in Hollywood that once only the big agents and studio moguls were able to activate.

"We didn't set out to be an entertainment company," says Mr Burgess. "We got there because the software was so popular." So many people were coming to Macromedia's site to download Shockwave, its animation-viewing software, that Mr Burgess realised he had accidentally achieved what others were spending huge sums to create: traffic. Macromedia gets 3.9m visitors a month, according to Media

Metrix, which puts it way above any of the other entertainment start-ups, and nudging some of the big media companies' figures.

So Macromedia moved downstream. Shockwave began to show content, as well as enabling people to watch it. They have been commissioning from heavyweight old-media Hollywood types who are interested in the Internet, such as David Lynch and Tim Burton, as well as from the new children of the web. Thanks to Shockwave's high-density traffic, Mr Burgess can get a more attractive deal from content creators, and has a better chance of survival than his rivals. Even for him, however, it is a struggle: in early September 2000, Shockwave announced that it was cutting 20 of its 170 staff, to concentrate more closely on interactive content and games.

At least television works

Digital television offers the entertainment industry less exciting but more solid prospects than does the Internet

"DON'T WORRY ABOUT the difference between the TV set and the PC," Nicholas Negroponte said in 1995. In future, "there will be no distinction between the two". Yet the distinction between the two remains stubbornly clear. TVs are getting cleverer, but they are still TVs. PCs are getting cleverer, too, but they are getting no more like TVs than they were.

Part of the reason that the two are not converging is to do with human behaviour. Most people use the two electronic devices for very different purposes. Computers are for working, and televisions are for relaxing in front of.

Besides, the idea of convergence goes against the grain of technological development. If people want to watch pictures on a machine, then consumer-electronics companies will compete with each other to produce the best machine for showing pictures on, at the best price. They have no incentive to build into it the ability to do people's accounts and write their letters. Generally speaking, devices tend to become more and more specialised. The PC is now under threat, industry savants reckon, not because it can do too little but because it does too much.

The main victim of the failure of convergence is WebTV. In 1997, Microsoft bought WebTV, a Silicon Valley start-up meant to put the Internet on television. The investment was launched with great fanfare. The market, said Microsoft, was potentially vast, covering every American who did not have a PC. Microsoft's strategists thought they were making television more interesting by putting the Internet on it. "When we initially did this," says Rob Schoeben, senior director of marketing at WebTV, "we said: 'We're going to turn the television into something else'."

But Americans, it emerged, did not want their television turned into something else. Most of those who wanted to get on to the Internet, it turned out, were prepared to buy a PC, particularly as prices kept coming down. WebTV has only around 1m subscribers, or 1% of American households. It has now been junked as a mainstream brand. Microsoft's new television-related brand, UltimateTV, does not try to edge out television programming, but to make more of it. It is a box that offers television e-mail, some Internet access and a personal video recorder (PVR) – the digital sort that automatically records your favourite stuff, and allows you to pause while you are watching live television.

As a way of distributing filmed entertainment, digital television has one big advantage over the Internet: it can do it. But it is a pale version of the nirvana of unlimited distribution promised by the visionaries. It can offer many more channels than analogue television, but not the unlimited number of sites that the Internet offers. It does not, therefore, allow everybody to be a broadcaster. It does not promise to turn the media business upside down. It does, however, improve television.

Europe, the Americans concede, is an advanced civilisation in terms of its wires, as well as in wireless communications. And in digital television, Britain, with digital satellite and terrestrial platforms up and running, and digital cable and video-on-demand services over telephone lines coming in, is the furthest ahead in Europe. Competition between the platforms has speeded up the introduction of services. BSkyB, the satellite platform, has moved furthest and fastest into digital: by September 2000, it had 3.8m digital subscribers out of a total customer base of 4.5m.

Digital customers are getting more for their money than analogue ones were. The number of channels has risen, from around 40 to 200-odd. That includes a near-video-on-demand service, whereby the same movies play on several channels, with different starting times, and a number of new sports channels catering for different interests. Interactivity offers some gimmicky new features, such as five different camera angles for football games (which other broadcasters who have tested a similar product say people don't use).

Open, an e-commerce service which BSkyB set up in 1997 as a joint venture with Midland Bank, Matsushita and BT, and in which BSkyB recently increased its stake, offers e-mail and shopping. Its technology is relatively easy to use for those not familiar with the Internet, and (thanks to BSkyB's large installed base) it is getting into homes fast. In less than a year, digital television has been installed in about the same number of households as the Internet has in five years.

BSkyB says its customers love these services, and backs its claims with two figures. The churn rate – the number of customers who give up their subscriptions in any one year – is down to 3.5%, and the company's total number of subscribers has gone up markedly. Some, says Richard

Freudenstein, BSkyB's general manager, are customers won over from cable, because cable has been slower in providing a digital service; but, he says, digital has also increased the pay-TV market as a whole.

The problem, though, is turning this customer satisfaction into cash. BSkyB, which was hugely profitable in the analogue days, is now losing money, and is not even guessing at when it might break even.

Each new digital subscriber costs BSkyB £180 to acquire, but pays only slightly more than an analogue one used to. In 1998–99 (when almost all the subscribers were analogue), the average revenue per subscriber was £268; now, each digital subscriber forks out £287, not nearly enough of an increase to pay for the cost of the investment.

BSkyB evidently believes that interactivity will produce significant additional revenues in the future. Early in 2000, BSkyB's chief executive, Tony Ball, persuaded Matsushita and HSBC (a bank which had taken over Midland) to sell their BSkyB stakes. He was in a strong position: Open will prosper only if BSkyB promotes it. Open says it is very pleased with the numbers buying from its shops, although nobody is revealing any figures.

The trouble with Open is that it is a phenomenally expensive way to sell. It takes around 8% commission on sales, on top of an up-front fee. Moreover, because it is a proprietary system, not based on Internet standards, it is expensive to write programs for. So, according to an executive who has worked on developing a store for Open, the technical costs and fees to the platform add up to around £2m ($2.8m) for each retailer. For the banks and building societies that have signed up, the fees are a pinprick, and a retailer such as Woolworths, which struggles on the high street, sees being on Open as a boost to its image. But some that had planned to be on the platform, such as HMV, a music retailer, and Iceland, a food retailer, have not taken up their places.

For the moment, Open can get away with charging high fees because it is the only interactive television platform around. But in 2000 cable started, slowly, to go digital. NTL, Britain's biggest cable company (which may eventually absorb the other remaining one, Telewest), points out that it has room for many more retailers than Open, because Open's sites are broadcast on a carousel, and if there were a lot of them, the system would become unbearably slow. NTL is signing up new "content providers" hand over fist – BA, Britain's flag-carrying airline, Dorling Kindersley, a publishing firm, Argos, a discount retailer, Abbey National, a bank, Interflora and several others.

On cable and satellite, for the moment, interactive television mostly amounts to shopping. NTL would like it to be something more, but, says John Hondros, director of content at NTL, "The broadcasters aren't developing much interactive programming because they can't see the revenue model." So NTL has set up a £25m fund for independent television producers to make interactive programmes, because new content is thin on the ground. It's the broadband Internet content problem again: nobody can work out how interactivity improves content.

Instant gratification

Unlike the cable and satellite digital services, which are offering broadcast television with some add-ons, the video-on-demand services now being offered across telephone lines turn the old idea of television on its head. Video Networks, which introduced its HomeChoice brand across London in September 2000, and Yes Television, which is trying out its service in west London, are not offering more and more channels, but the chance to select what you want to watch, when you want to watch it. Many regard VOD as the holy grail of television. But there are a couple of obstacles in the way.

One is programming. Film studios are coming round to the idea of selling films in new ways, but many television producers are not. Aside from a few independent programme makers, those who produce television programmes are also those who distribute them. Since a sizeable part of the value of the business lies in distributing programmes – through the channels they own, such as ITV – the producers are not necessarily enthusiastic about handing their programmes over to the upstart VOD distributors. The BBC, one of the two main producers, is willing to sell the VOD companies quite a lot of its output, some of it up-to-the-minute; but the ITV companies will sell them only old stuff of little appeal. And unless the VOD companies can offer a wide range of programmes, people are not likely to subscribe.

The VOD companies' other problem is DSL, which brings with it some of the difficulties that have arisen with broadband Internet deployment in America. British telephone lines are more suitable conduits for video than American ones, but British VOD companies have a particular problem – BT. The near-monopoly telecom company is charging the VOD operators a £600 connection fee per subscriber, and £62 per subscriber per month. Prices were expected to come down in 2001, when competition was to be introduced into the DSL market; but in the meantime, Video Networks admitted that it was losing around £1,000 per customer per year.

A sombre ending

Most of the money spent on e-entertainment will have been wasted

IN A SURVEY that takes the failure of past predictions as its starting point, there comes an awkward moment: it has to make some predictions of its own. So here are some tentative conclusions on how the digital revolution will affect the entertainment business in the medium term.

The main determinants of change will be, first, the availability of bandwidth, and second, the extent to which the new form of distribution suits the content that is being distributed.

For music, there are clearly some big changes ahead. It uses up little bandwidth, so is easily distributed across the Internet. The question is not whether the Internet will become the main method of distributing music, but whether the music industry can make any money out of it by retaining control of copyright. The chances are that it can. The electronics, software and music businesses are now working together, albeit in a shaky alliance. They will never be able to produce an unbustable system, but they can make piracy harder. Consumers – particularly older ones – hate doing anything technically complicated. Look how people struggle with their PCs, and how few of them use their VCRs to record programmes off the air.

If the music industry manages to sort out the piracy problem, the Internet will become a hugely important source of revenue. The record companies sold their music all over again when the CD came out, and they can now sell it all over again over the Internet, again. What is more, they can sell it in more flexible packages to make it more attractive to different kinds of consumers.

The book business will start going electronic at the periphery. Textbooks, reference books and travel books – all heavy to cart around, and all much better for a search capability – suit electronic distribution. But most of the book business does not. Old-fashioned books work too well.

The games industry clearly suits the Internet. By its nature, like the web, it is interactive. Besides, the Internet brings something extra to computer games. After 20 years of enforced solitude, gaming is once more becoming the social activity it used to be. As it is changing in this and other ways, it is becoming an increasingly important part of the entertainment business. Hollywood, which has tended to look down on the games business, may find that it has to start deferring to it.

News and sports, too, are suited to the new medium. The Internet's immediacy satisfies people's appetite for the newest of news. Its ability to retrieve data allows it to satisfy the nerdiest score-watcher. Its infinite capacity allows it to target the most obscure minority interest. Its interactivity allows surfers to select only the stuff they want. For the big matches, people will always go to their televisions; but for lots of other bits of sports and news, people will look to the Internet. That will not bother the television networks. In sports, the big money is in the big matches; as for news, there never was much money in it anyway.

By contrast, the movies and television programmes that make up the bulk of the entertainment companies' output will not be distributed over the Internet for a long time yet. The Internet does nothing for them. Nobody really wants to write the ending of the story they are watching. And pictures eat up too much bandwidth. In theory, high-speed connections have plenty of that, but they are too unreliable. Nobody wants to watch a movie that starts slowing down in the car chase.

Digital television will make a bigger impact on the moving-picture entertainment business than will the Internet. Not that digital television promises to do anything as revolutionary as the web. It does not offer an unlimited number of sites, nor does it allow everybody to be their own broadcaster. But, unlike the Internet, it can deliver. Its problem will be to find a way for competing digital television platforms to make money.

Internet-enabled mobile telephony, the current big buzz in communications, does not really lend itself to providing entertainment. The screens are too small and the controls too fiddly; besides, people on the move do not usually have time to settle down for a bit of fun. They may play a few games, pick up the news headlines and see how their stock prices are doing, but no more. Phones are for phoning with.

As we were

The digital revolution is not going to overturn the power structure in the entertainment industry, or even shake it up much. As it becomes clear that the Internet will not be the main vehicle for electronic entertainment, the industry giants will, slowly, cut back their new-media investments. The losses, in the end, will not seem that enormous; after all, some of them were offset by the huge surge in advertising from the dotcoms. The period will be written out of corporate histories. And the big cheeses of the entertainment industry will, quietly and collectively, breathe a sigh of relief, because they never liked or understood the Internet anyway.

The material on pages 307–37 first appeared in a survey written by Emma Duncan in *The Economist* in October 2000.

POSTSCRIPT

THINGS HAVE BEEN going downhill faster since this survey was written. As the dotcom business slumped, so did advertising revenue, and so did media companies' spending on their digital adventures. More independent companies – including Excite@Home, Quokka Sports and Icebox – went under. More of the digital offshoots of the big media companies were folded back into the parent. Brave experiments (such as Stephen King's episodic e-book, which followed his online novella) were quietly abandoned.

Music piracy continues to grow, and the record companies continue to struggle stopping it. The courts quashed Napster's file-sharing system, but several others have sprung up in its place. The record companies' latest scheme is to launch their own file-sharing services. Music sales are falling at 5% a year.

Meanwhile, the problem of how to make money out of e-entertainment remains unsolved.

Lessons of a virtual timetable

The market for e-learning has been slow to take off. What does that say about its future?

PEARSON, A LARGE British media group that owns 50% of *The Economist*, is betting much of its future on the market for online education. Already the world's biggest "education company", Pearson plans to dominate what Marjorie Scardino, the firm's chief executive, has told analysts is a potentially vast market for electronically delivered teaching material.

There is some evidence that her strategy might be sound. In America in 2000, total spending on education was a whopping $800 billion, and a government initiative to install electronic links in schools (known as the E-Rate) has ensured that 95% of all state schools, and 63% of all the classrooms in them, have Internet access. Moreover, 40% of all college classes in America already use Internet resources.

The supporting evidence does not all come from America, though. In Britain, Warwick University is proposing to make it compulsory for its students to have a laptop computer from 2003; almost all of them will have Internet links. And there is plenty going on beyond the bricks and mortar. Mrs Scardino says that in 2001 "2m people will be seeking a degree online, and outside a campus". Even in China, Tsinghua University offers classes across the country via modem.

All that these wired pupils are waiting for, Mrs Scardino hopes, is online material to be delivered to them by Pearson, or by rivals such as Vivendi, a French media and engineering group, and Kaplan, an American career-guidance company. In February 2001, Vivendi launched education.com, a portal through which parents, teachers and children can communicate and gain electronic access to material based on the books of its Havas media subsidiary. Other firms see opportunities in supplying universities with the software and advice needed to put their lessons and administration online.

Pearson believes that it has a trump card in the competition for this market: a software company called NCS that it bought in 2000 which supplies enterprise resource planning systems to 40% of America's schools. These systems provide a platform for delivering and administering all sorts of electronic material, but in particular they are good at delivering the educational tests that promise to be a cornerstone of the Bush administration's education reform plan, with its heavy emphasis on comparative performance and accountability.

Dot college

Belief in e-learning, as it is often called, has so far weathered the downturn in the wider dotcom world. John Chambers, the influential CEO of Cisco, which supplies much of the Internet's hardware, asserts that the scale of network traffic generated by e-learning will make today's exchange of e-mail messages look like a rounding error. But his firm's business depends on an ever-rising flood of electronic data passing over the connections which it makes for electronic networks. More disinterested voices caution against confusing the obvious need to learn computer-literacy skills with the less obvious need to learn everything else via a computer.

The market for online education can be divided into three: schools, universities and business/commercial training. Since universities are where much of the early development of the Internet took place, they might have been expected to have pushed its potential furthest. But, by and large, they have concentrated on using it as a means of extending their geographic reach, through their "extension programmes". On site, their record has been more patchy.

Consider UCLA, the University of California, Los Angeles, a prime candidate for advanced electronic learning methods if ever there was one. But UCLA uses the Internet much like a big business would. And, like the average big business, it doubts whether it has saved much money as a result. It has simply been able to do a number of things better.

The first users, says Rory Hume, the university's executive vice chancellor, were the administrative staff: "The people who buy things, and who have the power to hire and fire." Next to see the opportunities were the librarians. Only late in the day did academics realise that the university's system, now being designed as a single unit, could be used as a way to distribute information about classes and, more recently, about grading. But online instruction? Lots of small experiments are running on campus, but not much more. "Even when our professors put all the material on the web," says Mr Hume, "all the students still come to lectures."

Moreover, where it is happening, online learning is a commercial turkey. The university's extension programmes are using the Internet widely – to teach from a distance, for example, a course on hotel management. But they still lose money. An immense effort by the School of Dentistry to create an online course to educate periodontists around the world cost some $750,000 and took five years to create: again, it has been a commercial failure. "People are unwilling to subscribe online for the latest information from anywhere in the world," mourns Mr Hume. "Instead, they will go to a lecture and pay much, much more. We have an enormous revenue stream from our faculty giving lectures."

The one area where the Internet is about to save the university money is in purchasing. James Davis, who came to UCLA at the end of 2000 to reorganise the way that it uses information technology, has been making it possible to combine online the institution's purchases of everything from computers to pencils, giving it more buying muscle.

Many of the other uses of information technology on the UCLA campus neither save money nor visibly enhance productivity. They simply raise the quality of the experience. One example is the websites that now exist for almost all 3,000 or so undergraduate courses. About 55–60% of them not only supply lecture notes; they also allow students to take tests online and to see their results. Another example is My.UCLA, an in-house "portal". Students can use it to search for advice, such as the entry requirements for graduate school.

Given the University of California's sprawling size, it is surprising that its nine campuses have not combined forces more, using the Internet as a bond. The main area where they have acted jointly is in setting up the California Digital Library, which drives hard bargains with the publishers of periodicals to license the use of their electronic versions. But even here, no money is saved: the university still buys paper copies. "Paper is so much more permanent than bits and bytes," explains Gloria Werner, the university's librarian.

The Internet has undoubtedly encouraged universities to reach out beyond their own campuses in order to offer more "distance learning", and at greater distances. The University of Phoenix, set up in 1989 to teach adults through a combination of old-fashioned distance learning and evening classes, is incorporating more and more e-learning into its courses, which are mostly taken by people with full-time jobs. But the extension of an institution's brand is not without risk. Increasing the number of students who claim to have studied there can damage a university's reputation if those students do not receive the level of teaching that the university's name was built on.

Prominent universities have therefore tended to band together for support in the early stages of exploring e-learning, and they have often launched their efforts under names other than their own, even though they have some of the strongest brands in education. The business schools of Columbia, in New York, the University of Chicago, the London School of Economics, Stanford in California and Carnegie Mellon in Pittsburgh, for example, have teamed up behind Cardean University, an early effort at an online institution for tertiary education. Cardean offers complete courses, mostly in business subjects, aimed at people working full-time who want to learn in the evenings, at weekends or whenever. It plans to offer full degrees eventually.

There are more than 250 firms eager to help established universities to go online. These firms build the Internet infrastructure and manage the electronic delivery of classes. Cardean, for example, is the work of UNext, an Illinois company that grew out of Knowledge Universe, an education business started by Larry Ellison, the CEO of Oracle, and Michael Milken, the developer of the junk bond market who spent 24 months in jail for fraud. Several prominent business schools – including Wharton at the University of Pennsylvania, Fuqua at Duke University, and INSEAD, near Paris – have worked with Pensare, a company based in Sunnyvale, California, to put their material online. A host of other firms, including Blackboard, Campus Pipeline, eCollege and WebCT, offer different platforms for putting course material on the Internet and for building a student community around the material.

Some firms have decided not to be the invisible force behind the e-learning efforts of established universities, but rather to become brands known in their own right as a place for students to find courses. Sometimes these are simply portals that consolidate course information from other institutions, such as Hungry Minds. Others offer courses of their own.

Business and other vocational subjects predominate. But some hope to find an audience for less utilitarian subjects among adults who feel that they missed some education when young. Mark Taylor, a sociologist at Williams College in Massachusetts, is leading an effort to offer courses in the liberal arts. Taught by professors from top universities – most of them so far in the eastern United States, such as Wellesley, Brown and Amherst – they are marketed under the name Global Education Network (GEN). The GEN project is funded by Herbert Allen, a rich alumnus of Williams, and it does not yet offer any complete courses, merely free snippets of lectures. But it was founded on the belief that there is a potential market for vigorous online intellectual stimulation.

Boxmind, with a number of Oxford University academics on its board, is another such ambitious project. By putting "star" academics at the centre of a stage away from their home institutions, websites such as GEN and Boxmind threaten (if they take off) to raise the tension between universities and their faculty over the ownership of intellectual property. With e-learning sites offering students access to the best teachers without having to call in at their institutional home, there is a danger that the universities' academic superstars may choose to go solo.

The mouse ate my homework

There is nothing new about the use of technology as a teaching tool in schools. Machines (from record-players and overhead projectors to televisions) have long been used to make lessons more vivid and engaging. The first computers in class were treated as novelties on which children could look things up in encyclopedias and play arithmetical games on multimedia CD-ROMs.

The CD-ROM has now been largely replaced by networked databases as the repository of learning material, but research continues into how to impart lessons that take advantage of a computer's capacity to present moving images and sounds as well as text, and their capacity to respond to a user's input. Pearson's Mrs Scardino thinks that the big advantage of online education is that it personalises the learning experience, allowing each student to move at his or her own pace and in his or her own way.

Online businesses aiming at the school market have tried to combine innovative "content" with building a strong "community", often through a portal. These are similar to general-interest portals such as Yahoo!, but they have specialised gateways for parents, teachers and children. Much e-learning material aims to facilitate the involvement of parents in the process. The plan is to make money by advertising and by selling goods – both electronic items, such as lesson plans to be downloaded by teachers, and more traditional supplies such as pencil sharpeners and books that can be ordered online.

Making money needs a light touch where children are concerned. Part of the appeal for parents of a portal such as MaMaMedia, which allows children to swap notes about school and to play educational games, is that it is safely ringed off from the pornography and craziness of much of the rest of the Internet. Saturating it with advertising would put off many users. And pure academic content, since much of it is defined by official syllabuses, is a commodity of sorts. For this reason, the textbook publishing industry, in effect an oligopoly of four large firms, was initially wary of the web. It was not easy to see how an online commodity could command much of a margin.

Among the most committed early users of children's educational websites were those families that teach their children at home. Since it became fully legal in all 50 states in 1993, home-schooling has taken off in America. Estimates of the number of children currently being taught at home by their parents range from 1.5m to 2m, and the number is thought to be rising at between 7% and 15% a year.

The homes in which this teaching takes place are among the most wired in the world – 94% have a computer, and almost all of those are connected to the Internet. In addition to the busy informal home-school communities in chat-rooms and on bulletin boards, a number of firms, such as Childu and School Express, offer advice, specialised bookstores and complete courses (including such specialised subjects as creationist biology for religious fundamentalists) to parents who teach their own children.

There are still plenty of sceptics about the value of online teaching in schools. Teachers are, however, gradually catching up with their screen-happy pupils. Those who enter the profession today are more likely to have used computers and the Internet in their training, and to want to bring that experience into their classrooms.

Workers' studytime

In business, network technology both makes e-learning possible and allows it to take place during the course of work. Corporations are already heavily wired, and they conduct an increasing volume of their business, both internally and externally, over the Internet. The speed at which this makes it possible to introduce new products and processes means that employees have to master more and more new information more and more quickly.

It also means that they are less likely to find someone who can explain unfamiliar material to them among their immediate co-workers, the informal kind of training that used to go on within organisations. So they look for help online. United Airlines, for example, will soon have a system that monitors what its employees are doing as they work on an aircraft, and also supplies the relevant technical data to assist them.

Firms have also embraced the notion of "knowledge" as an asset to be worked with and to be enlarged. Training is no longer treated only as a pick-me-up when fortunes are flagging, but as a necessity in order to keep up with the pack. This makes the corporate training market perhaps the most promising of all the e-learning segments – especially since corporate training, as traditionally conducted, is an expensive affair. People come together from different locations, are put up in a hotel and spend valuable time away from their desks.

More than 200 companies are fighting to offer consulting services in the promising area of e-learning for businesses. Some are conventional training companies migrating online, others are start-ups. Three that offer e-learning "platforms" have already gone public. Docent, Saba and DigitalThink, all of them based in California, already serve half the corporate e-learning market.

Pulling together all the components needed for successful e-learning is more feasible in a well-defined industrial sector. Firms that operate in the business market look for sectors that have a high degree of commonly required (and rapidly changing) technical knowledge, and formal constraints such as regulation.

Enthusiasts of e-learning claim that corporate networks in the future will move beyond the provision of courses 24 hours a day, and that they will become a growing, responsive repository of knowledge that continuously delivers to employees just what they need to know at any particular moment ... and in a form perfectly adapted to their style of learning.

No e-learning programme is close to this yet, although the United Airlines example gives a flavour of things to come. The nearest to such a "knowledge environment" is probably to be found in professional service firms like McKinsey, a consultancy where the collective expertise of the employees is almost the only asset. McKinsey consultants have developed systematic ways of pooling the results of their work into a continuously growing information resource on which all the firm's employees can draw. But they are powerfully self-motivated individuals. The same cannot be said of pupils at the likes of Beverly Hills High.

To develop the market for e-learning requires a deeper understanding of the process of learning, of how pupils respond to ideas presented by a computer rather than by a teacher or a book. The Learning Federation, a consortium of American businesses, academic institutions and government agencies, is proposing to co-ordinate research in a range of scientific disciplines that will accelerate progress in e-learning.

One proposal for funding this is to use the money raised by auctioning the radio spectrum to telecommunications companies. The plan's supporters compare today's opportunity to enlarge education through e-learning to the creation of the land-grant schools in 1862, which helped lift American agriculture and industry to international pre-eminence. Pushed by public planning and private enterprise, e-learning may yet have a comparable impact.

The material on pages 338–45 first appeared in *The Economist* in February 2001.

VIII

NUTS, BOLTS AND CLOUDS

The beast of complexity

The future of software may be a huge Internet-borne cloud of electronic offerings from which users can choose exactly what they want

STUART FELDMAN could easily be mistaken for a technology pessimist, disillusioned after more than 25 years in the business of bits and bytes. As a veteran software architect, the director of IBM's Institute for Advanced Commerce views programming as all about suffering – from ever-increasing complexity. Writing code, he explains, is like writing poetry: every word, each placement counts. Except that software is harder, because digital poems can have millions of lines which are all somehow interconnected. Try fixing programming errors, known as bugs, and you often introduce new ones. So far, he laments, nobody has found a silver bullet to kill the beast of complexity.

But give Mr Feldman a felt pen and a whiteboard, and he takes you on a journey into the future of software. Revealing his background as an astronomer, he draws something hugely complex that looks rather like a galaxy. His dots and circles represent a virtual economy of what are known as web services – anything and everything that processes information. In this "cloud", as he calls it, web services find one another automatically, negotiate and link up, creating all kinds of offerings.

Imagine, says the man from IBM, that you are running on empty and want to know the cheapest open petrol station within a mile. You speak into your cellphone, and seconds later you get the answer on the display. This sounds simple, but it requires a combination of a multitude of electronic services, including a voice-recognition and natural-language service to figure out what you want, a location service to find the open petrol stations near you and a comparison-shopping service to pick the cheapest one.

But the biggest impact of these new web services, explains Mr Feldman, will be on business. Picture yourself as the product manager of a new hand-held computer whose design team has just sent him the electronic blueprint for the device. You go to your personalised web portal and order the components, book manufacturing capacity and arrange for distribution. With the click of a mouse, you create an instant supply chain that, once the job is done, will dissolve again.

Visions, visions everywhere

All this may sound like a description of "slideware" – those glowing overhead presentations given by software salesmen that rarely deliver what they seem to promise. Yet IBM is not the only one with an ambitious vision. Hewlett-Packard, Microsoft, Oracle, Sun and a raft of start-ups are thinking along the same lines. Unless they have all got it wrong, companies, consumers and computers will one day be able to choose exactly what they want from a huge cloud of electronic offerings, via the Internet.

The reality will take a while to catch up; indeed, it may turn out to be quite different from today's vision. But there is no doubt that something big is happening in the computer industry – as big as the rise of the PC in the 1980s that turned hardware into a commodity and put software squarely at the centre of the industry. Now it looks as though software will have to cede its throne to services delivered online.

Not that software as we know it will disappear. Plenty of code will still be needed to make the new world of computing run, just as mainframe computers are still around, though in a much less dominant position. But the computer business will no longer revolve around writing big, stand-alone programs. Instead, it will concentrate on using software to create all kinds of electronic services, from simple data storage to entire business processes.

The agent of change is the Internet. For a start, it has changed the nature of software. Instead of being a static program that runs on a PC or some other piece of hardware, it turns into software that lives on a server in the network and can be accessed by an Internet browser.

But more importantly, the Internet has turned out to be a formidable promoter of open standards that actually work, for two reasons. First, the web is the ideal medium for creating standards; it allows groups to collaborate at almost no cost, and makes decision-making more transparent. Second, the ubiquitous network ensures that standards spread much faster. Moreover, the Internet has spawned institutions, such as the Internet Engineering Task Force (IETF) and the World Wide Web Consortium (W3C), which have shown that it is possible to develop robust common technical rules.

The first concrete result of all this was the open-source movement. Since the mid-1980s, thousands of volunteer programmers across the world have been collaborating, mostly via e-mail, to develop free software, often taking Internet standards as their starting point. Their flagship program is Linux, an increasingly popular operating system initially

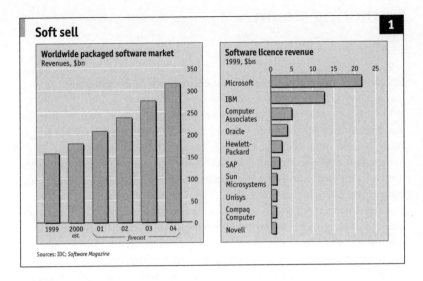

Soft sell

1

Worldwide packaged software market
Revenues, $bn

1999 | 2000 est. | 01 | 02 | 03 | 04
forecast

Software licence revenue
1999, $bn

Microsoft
IBM
Computer Associates
Oracle
Hewlett-Packard
SAP
Sun Microsystems
Unisys
Compaq Computer
Novell

Sources: IDC; *Software Magazine*

created by Linus Torvalds, a Finnish programmer.

The emergence of web services is a similar story, even though at first glance it may not look like it. The computer industry and other business sectors are collectively developing the next level of Internet standards – the common glue that will make all these web services stick together. Hence the proliferation of computer-related acronyms such as XML, RosettaNet, ebXML, XAML, SOAP, UDDI, WSDL and so on.

Why should anyone care about this geeky stuff? One good reason is that software is one of the world's largest and fastest-growing industries. In 1999 the sector sold programs worth $157 billion, according to IDC, a market-research company; and software spending, which is increasing by 15% a year, influences investments of another $800 billion in hardware and services. The changes now under way are likely to reshuffle the industry completely. The next few years may make it clear which companies will end up on top.

Moreover, the software sector could well become a model for other industries. Open-source communities, for example, are fascinating social structures. Similar communities could one day produce more than just good code. Thomas Malone, professor of information systems at the Massachusetts Institute of Technology, sees great opportunities ahead: "The Linux community is a model for a new kind of business organisation that could form the basis for a new kind of economy."

Gathering steam

Software is migrating from users' computers to the Internet

EXODUS COMMUNICATIONS' two-storey building in Santa Clara, in the heart of Silicon Valley, looks like any old warehouse. But try to enter it, and you will spot several guards behind bullet-proof glass eyeing you with suspicion. They will not even open the door until you have placed your hand in a fingerprint scanner and flashed an electronic badge. The place is protected like a high-security prison.

Once inside, it is easy to see why. It is a huge data centre, full of thousands of servers (the superfast computers that dish up web pages and other data) stacked in metal racks and locked up in chain-link cages. A powerful air-conditioning system keeps the building from overheating. Several fibre-optic cables connect to the Internet. And back-up diesel generators are ready to jump in should the power fail (a real risk in California these days).

Data centres like this are physical proof that software is becoming a service that users can access via the Internet rather than a program they run on their computers. In 2000 "application services" and their providers, known as ASPs, were all the rage in the computer industry. Venture capitalists threw money at them. Most software firms made their programs "hostable" on servers.

That initial euphoria has died down somewhat, not least because some ASPs have gone belly-up. Even the market leader, USInternet-working, recently flirted with disaster. But the software industry is not having second thoughts about the concept. Rather, it is slowly figuring out how best to use the new technology – in much the same way that, nearly a century ago, companies had to learn that as electricity took the place of steam, they no longer needed to cluster their machinery around the power source.

The software industry has to resolve several tricky questions. The most fundamental one is which sort of software – meaning essentially the computing intelligence – should reside in a user's computer and which on a server. The next one is whether companies should host applications themselves or outsource them to a service provider. And lastly, what kind of ASP will carry the day?

Processing has always been a moving target. In the early days, when computing was expensive, most applications ran on mainframes and were accessed by "dumb" terminals. Later, Moore's law – which states that the power of computer chips doubles every 18 months – turned the PC into the location of choice for software. The Internet is pulling the intelligence out into the network. But how far?

Local or remote?

The answers offered by industry heavyweights depend on which side of the business they are on. Sun Microsystems, which makes its money from powerful servers, argues that most software will move into the network and be accessed through browsers. Microsoft counters that there will always be a need for an "intelligent client", by which it means a computer running Windows.

Both of them are right and wrong – it all depends. There are still good reasons to keep software on a local computer. Most users would rather have their word-processing or spreadsheet programs on their own PCs, where they are always accessible. Online applications can be sluggish, and Internet connections are often slow and unreliable.

The rise of peer-to-peer (P2P), a much-ballyhooed new technology, has moved the argument slightly in Microsoft's favour. P2P services such as Napster, Gnutella and Aimster allow Internet users to swap music files and other data directly among themselves, but for that they need a powerful computer and a large hard disk – not just a "thin client" with a browser.

Yet the Internet is already changing PC software and making new kinds of offerings possible. For example, Eazel, a start-up, has put together an online service that monitors PCs running Linux, alerting users if there is a problem and performing automatic upgrades over the Internet. Another new service, ThinkFree, lets users work online as well as offline, downloading small office applications written in Java, a programming language, which then run in a web browser.

On the other hand, there are compelling reasons to put certain kinds of applications on a server. The strongest candidates are programs that are accessed not just by individuals or small groups of employees with a PC, but shared by many users with different devices, such as cellphones or hand-held computers. Equally, any software that needs to be updated frequently is best held on a central server.

Most new enterprise software, such as programs managing online marketplaces or customer relationships, fits these criteria. But even traditional Enterprise Resource Planning (ERP) packages stand a good chance of moving online. That is why PeopleSoft, an enterprise-software vendor, has completely rewritten the code for its human-resource and other applications for Internet use. PeopleSoft's chief executive, Craig Conway, explains that firms do not have to hire an ASP to access software on a server. But outsourcing a software service to a third party often makes sense for small and medium-sized companies that cannot afford a sophisticated IT infrastructure but want to use the most advanced products, says Mr Conway. "It's like being able to start your driving career with a brand-new Mercedes instead of a beat-up old car."

ASPs have not been a big hit with large firms so far, mostly because companies want to keep tight control of their applications and data. Still, some have been persuaded by the advantages of ASPs. Instead of having to waste time and money coming to grips with the latest technology, their customers can just hand over all the IT stuff to someone else and get on with their own business. To George Kadifa, chief executive of Corio, one of the leaders among the 500 or so ASPs that have sprung up in the United States alone, it is the worry-free implementation even more than the software itself that makes the services of companies like his so attractive.

Not everybody agrees. Bill Gurley, a general partner at Benchmark Capital, a venture-capital firm, says that companies like Corio are mere "service bureaus" that do not own much in the way of intellectual property. He also argues that most applications they host are not written specifically for the Internet, which in essence means they have to run a separate rack of servers for each customer – not a terribly efficient proposition.

That is why Mr Gurley has invested in a different kind of ASP: Employease, an Atlanta-based provider of human-resource services which has built its application entirely for the Internet. Unsurprisingly, the firm's service looks and feels much more like a big website, such as eBay, than traditional software. It runs on a single central computer system, so new customers can be easily added.

Being specially designed for the Internet is more than just a technical plus. It means that Employease is accessible not only to a company's human-resource experts (as most traditional HR applications would be) but to its employees too, as well as to third parties such as insurance companies or payroll services. "We are enabling collaboration beyond the four walls of the HR department," says Michael Seckler, Employease's co-founder and marketing chief.

When Mr Seckler and his colleagues launched their company in 1996, their offering seemed exotic, but now there are dozens of start-ups offering similar Internet-only services. NetLedger, for instance, runs a web-based accounting service. Salesforce.com can help you manage relationships with customers automatically. And ourproject.com makes it easy to manage projects online.

The awkward question

Yet these ASPs share a problem with all dotcoms: they have yet to prove that they can actually make money. In early 2000 Salesforce.com, for instance, claimed to have 2,000 customers, mostly small and medium-sized firms, which pay a monthly fee of $50 per salesperson hooked up. This is not bad for a service that went live only a year earlier, but the numbers will have to go up dramatically if Salesforce.com is to make a decent profit.

PeopleSoft's Mr Conway believes that most stand-alone ASPs lack a sustainable business model. "This is an overspawned market," he says. "A lot of fish will soon be flowing down the river." He sees applications services as another distribution channel for his company that will not need to earn its keep initially, but will over time become a large and profitable part of its business. Mr Seckler of Employease disagrees: "It's hard for incumbents to change from one model to another – you need to change the entire culture, educate the sales team and so on."

Whichever model proves the most successful, applications services are bound to have a huge impact on the software industry. But the technical challenges will pale in comparison with the new economic ones that software firms will have to adapt to. The way the sector used to operate was fundamentally inefficient. Although a service at heart (however automated), it conducted itself like a manufacturing business. It put its programs on floppy disks or CD-ROMs, marketed them heavily and left the rest to the users.

This created unhealthy incentives for software firms. Instead of developing software that works well and is easy to use, many of them concentrated on selling yet another upgrade with even more features. The capital markets reinforced this customer-unfriendly behaviour. The companies found that being first with a new product was often an unbeatable advantage, and that their valuations in financial markets depended on steep revenue growth.

When software becomes an online service, the interests of vendors and customers become better aligned because the providers have to behave more like a utility. "I only get paid if my service is running

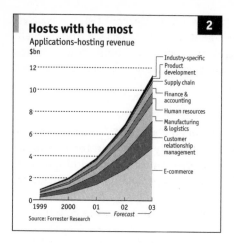

Hosts with the most `2`

Applications-hosting revenue
$bn

- Industry-specific
- Product development
- Supply chain
- Finance & accounting
- Human resources
- Manufacturing & logistics
- Customer relationship management
- E-commerce

1999 2000 01 02 03

Forecast

Source: Forrester Research

and my customers are happy," says Marc Benioff, chairman of Salesforce.com and a 13-year veteran of Oracle. Describing his years in traditional enterprise software, he recalls: "We didn't care if you were up and running, we only cared about the numbers."

Software firms must also get used to the fact that they will no longer receive the bulk of their licensing fees up front. Most ASPs charge a one-off installation fee, but much of their revenue comes from monthly subscriptions. Incumbents will find the transition hard to make, because it will bring down their growth rates for licensing revenue – and so far it has been these rates, according to a recent study by McKinsey, a consultancy, that have been the main determinant of software companies' valuations.

Among the big companies, Oracle is probably the furthest along the path from the offline to the online world. As early as 1998, the world's second-largest software firm put most of its programs online – not least because it hoped to cut support costs. If Oracle runs its programs itself, they are a lot easier to maintain, argues Tim Chou, president of Business OnLine, soon to be rebranded Oracle.com.

So far, the new unit makes up a small part of Oracle's business, contributing only a tiny share of its total revenue of $2.7 billion in the first quarter of 2001. But if Larry Ellison, the firm's chief executive, is right in predicting that by 2004 about two-thirds of all applications will be delivered by ASPs, this part should grow quickly. "Whether we can make this transition is the $50 billion question," says Mr Chou. "In any case, if we don't try, we won't exist in the future."

But becoming an ASP is not the only big challenge the Internet has posed for incumbents such as Oracle. The other, just as formidable, one is the trend towards open-source software developed online.

Out in the open

The world is taking to open source

THE WORDS MUST have shocked the global community of volunteer software developers known as the open-source movement. Volker Wiegand, president of the American subsidiary of SUSE, a leading Linux vendor, declared at the beginning of 2001 on the website Linuxgram that Linux as a business wasn't working out. His company had laid off two-thirds of its staff in early February. The free operating system was a "fallen angel" and a victim of irrational expectations, he said.

The statement, in essence, repeated what a Microsoft executive had argued shortly before. "There isn't much value in free," according to Doug Miller, group product manager for competitive strategies. He predicted that many Linux businesses would falter before the end of 2001, and that this, together with technical shortcomings, would "call into question whether Linux should be used at all" (instead of his own company's ubiquitous Windows, that is).

The Linux hype has undeniably crested. SUSE is not the only company that has had trouble building a viable business based on the free operating system. VA Linux, for example, a start-up which, in late 1999, enjoyed the most successful flotation ever – with shares gaining almost 700% on their first day of trading – said in early 2001 that it would cut a quarter of its staff and take nine months longer than planned to achieve profitability. Only the market leader, Red Hat, is doing better than expected.

But this should not be read as a sign of the imminent demise of open-source software via Linux, its standard-bearer. Most people in the software industry believe that open-source is here to stay. Steve Ballmer, Microsoft's chief executive, recently called Linux "threat number one". Steven Milunovich, a leading analyst with Merrill Lynch, an investment bank, argues that open-source is a "disruptive technology" that could topple such industry heavyweights as Microsoft and Sun.

In fact, the open-source movement is less about "world domination", which hackers often joke about, and more about an industry which, thanks to the Internet, is learning that there is value in deep co-operation as well as in hard competition. "Much more than a cause, the open-source movement is an effect of the Internet," says Tim O'Reilly, head of

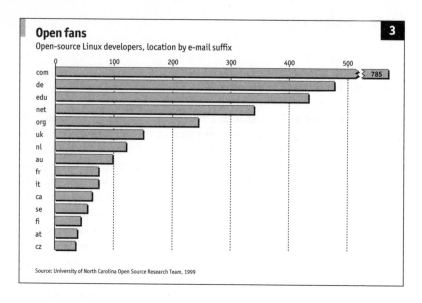

Open fans

Open-source Linux developers, location by e-mail suffix

com	785
de	
edu	
net	
org	
uk	
nl	
au	
fr	
it	
ca	
se	
fi	
at	
cz	

Source: University of North Carolina Open Source Research Team, 1999

an eponymous firm that publishes computer books, and a leading open-source thinker.

Open-source is often described as the software industry come full circle. Indeed, in the early days of computing, programs came bundled with the hardware and complete with the source code (the set of computer instructions which are then translated into binary code, the form of software that computers can understand and act on). Pioneers needed to tweak their programs, and were happy to share the improvements they made.

It was only in the 1970s, as computing spread, that firms such as Microsoft started to withhold the source code, thereby making software proprietary and turning it into a big business. Firms can sell a program without revealing the instructions that underlie it, just as Coca-Cola can market its soft drinks without giving away its secret recipe (though there have been plenty of attempts at reverse engineering, in software as well as soft drinks).

Many early hackers were horrified by the decision to withhold the source code. Proprietary software was "spiritually wasteful", they said, because it discouraged co-operation. One of them, Richard Stallman, founded the Free Software Foundation in 1983 and developed the concept of "copyleft" (as opposed to copyright), which he codified in a licence that now comes with most open-source software. It states that developers can do whatever they want with the programs, even sell their own versions, as long as they make the source code available.

Although this licence, called the General Public License (GPL), has never been enforced, it has done much to keep open-source software from splintering into competing commercial versions, open-source advocates say. The GPL, in effect, removes the incentive to turn a program into a proprietary product, they argue, because the licence is "viral": all changes to the source code automatically become part of the software commons. That is why James Allchin, who is in charge of operating systems at Microsoft, recently called the licence an "intellectual-property destroyer".

But open-source projects are best understood as a new generation of standards bodies, much like the Internet Engineering Task Force (IETF). The traditional methods for setting technical standards have turned out to be rather inefficient. Governments and industry committees are slow and often get it wrong. Monopolies are faster and do a better job, but they tend to put the brake on innovation in order to keep their dominant position.

The Internet has opened up a third way. It allows engineers in different corners of the world to collaborate at almost zero cost. Since communication is so easy, there is no need for official leaders or a big bureaucracy to keep things running. And the decision-making process is more transparent: discussions and documents, for instance, are easy to get at and to search. What counts is the quality of the argument, not the power of special interests.

The Internet pioneers were the first to create online communities for their work. They modelled their groups on organisations they had grown up in: universities and engineer corps. That is the main reason why the IETF turned out to be an interesting cross between a scientific community and a guild. It is run by elders, and its main mantra is: "We reject kings, presidents and voting. We believe in rough consensus and running code."

For love, not money

Most open-source projects are organised in much the same way. Their members are motivated mainly by fame rather than fortune; it is considered a coup to write a "patch" that passes the peer review of fellow developers and gets incorporated in the next release. And most open-source projects are governed by a "benevolent dictator", an individual with exceptional programming, organisational and communications skills – such as Linus Torvalds, the creator of Linux.

The power of these leaders, however, is not absolute. If developers are unhappy with them, they can always "fork" – take the source code and start a new development branch. To Brian Behlendorf, one of the leading developers of the Apache web server, another successful open-source project, and now chief technology officer at CollabNet, this keeps the benevolent dictators from becoming real ones.

Yet cheap communications alone are not usually enough to get open-source communities off the ground, nor to make them as successful as Apache or Linux. They need a starting point and a framework. For Apache, these were provided by a program developed by the National Centre for Supercomputing Applications (NCSA) and by the open standards of the Internet. Linux had the original, very simple version written by Mr Torvalds as a starting point and Unix, a programme created in the 1970s, as a blueprint.

To be successful, open-source software must also be designed in a modular way so that groups of programmers can work independently on different components. Most of the work on Linux, for example, is done not by a large horde of hackers all working on the same code, but by small groups of a dozen or so developers, each of which concentrates on one small part of the program.

Lastly, an open-source alternative needs a market. Linux had a ready-made one: hardware makers wanted a free alternative to Windows so as to increase their margins and gain some independence from Microsoft. The recent antitrust case against the software giant gave them a chance to support another operating system without fear of immediate retaliation. From Compaq to Dell, from Hewlett-Packard to IBM, they have all jumped on the Linux bandwagon, offering computers with the program pre-installed and investing in Linux companies.

Ironically, it is IBM, the hackers' original arch-enemy, that has gambled the most on Linux so far. Big Blue announced that it was to spend $1 billion on Linux in 2001 because it really wants the program to become a computing standard. It is doing this because it sees itself mainly as a provider of e-business solutions, and because a standard operating system would make it much easier to integrate them, explains Irving Wladawsky-Berger, IBM's man in charge of its Linux operations.

Hence IBM's efforts to make all its important programs run on Linux and transfer the operating system itself to all kinds of computers, from tiny hand-held ones to mainframes. Using special "partitioning" software, these workhorses of computing can be turned into a virtual collection of thousands of Linux servers. IBM is also trying to help the

Linux community beef up the program for heavy-duty corporate computing, where it is still a rarity.

A lot of ifs

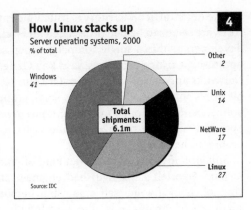

How Linux stacks up
Server operating systems, 2000
% of total

Total shipments: 6.1m

Other 2
Windows 41
Unix 14
NetWare 17
Linux 27

Source: IDC

Some people like to dismiss Linux as nothing more than a happy accident, but the program looks more like a textbook example of an emerging pattern. If a piece of software is well understood, to the point of becoming a commodity; if it can be built in a modular way; if there is a critical mass of users who are also software developers; if there are talented project managers among them; and if there is a demand for a program that is not controlled by a single vendor – then the chances are it will be suitable for an open-source project.

Admittedly, that is a lot of ifs. But Linux is by no means the only successful open-source project. Take Sendmail, which is used to route two-thirds of the world's e-mail and has spawned a start-up of the same name. It makes a living by selling a commercial product, but it also helps others to develop a free version of the program. The community of developers creates a basic product platform that Sendmail can polish and enhance to make money.

And there are many smaller, less well-known open-source projects that could gain momentum. In October, Sun Microsystems launched OpenOffice, a free alternative to Microsoft's Office suite. Forrester Research, a high-tech consultancy, forecasts that by 2004 companies will be spending 20% less on software licences than they do now because they will be using these kinds of open-source programs.

However, there are two areas of the software industry that open-source will have trouble penetrating. One is enterprise software that relates to a company's core activities. Companies will hesitate to bet their business on free software, at least until it is backed by heavy-weights such as IBM, as Linux is now. The operating system will be the test case of whether open-source software can move into heavy-duty corporate computing. Linux's most recent release is a big improvement, but it still has some way to go.

The other difficult territory is truly innovative programs. This kind of software seems to be better designed and built by a tightly knit group of developers within a single company. Open-source is extremely good at optimising existing programs, says Ray Ozzie, chief executive of Groove Networks and creator of the popular Lotus Notes. But he cannot imagine his firm's software, which allows small groups of people to collaborate online, being developed in a decentralised way. He and four colleagues operated "like one brain" for three years to solve all the engineering problems, he says.

Still, nobody can predict what kind of "mob software" may turn up. Richard Gabriel, a distinguished engineer at Sun, describes this as "a kind of semi-chaotic, self-organising behaviour in which numerous small acts of repair can lead to quickly built, complex and massive creations". Open-source is certainly a mass phenomenon, with tens of thousands of volunteer programmers across the world already taking part, and more joining in all the time, particularly in countries such as China and India. SourceForge, a website for developers, now hosts more than 18,000 open-source projects that keep 145,000 programmers busy.

Meanwhile, the corporate world is warming to open-source in unexpected ways. Firms are showing increasing interest in using this development methodology for software they create in-house or with a limited number of partners. For example, Dresdner Kleinwort Wasserstein, an investment bank, in January 2001 released the source code of an application it developed to link different types of computer systems, and has hired CollabNet to build a developer community around it.

Even Microsoft is learning to love open source. The firm has announced that it will share the source code of the latest versions of Windows with 1,000 of its best customers – but only on condition that they do not modify the program.

The official software industry, for its part, is already busy building the infrastructure for the next wave of computing: web services.

Battle of the platforms

Who will bag the market for web services?

NOBODY WOULD CALL ADP a cutting-edge software firm. The 50-year-old company, based in Roseland (New Jersey), is best known for sending pay cheques to over 29m employees worldwide. Its corporate customers transmit the relevant information to ADP electronically and the firm's computers take care of the rest. For many companies, dealing with their payroll in-house and keeping track of ever-changing labour and tax regulations would be a waste of resources.

Yet ADP is a pioneer of a phenomenon the entire industry has begun talking about: web services. These web services move one step on from web applications (which are about browsers accessing online software): they are about machines talking to each other over the Internet, or more precisely, about machines talking to machines talking to yet more machines, and so on – creating a grid or cloud of electronic offerings that feed into each other.

In this new world, a payroll service, for instance, would be a combination of several other web services. A company would outsource its human-resources system to one provider, while another would keep track of the hours employees work. The payroll firm would also be hooked up to a service that delivers information about regulatory changes and, if the money has to be transferred abroad, with a bank offering foreign-exchange services.

The network thus becomes one huge distributed computer, just as Sun Microsystems, the leading maker of network servers, has been predicting for decades. And just like a mainframe or a PC, this mother of all machines needs an operating system, a platform on which all these services can be developed. It is hardly surprising, then, that the computer industry is already gearing up for another religious war over which platform will come out on top.

On one side is Microsoft with its .NET plan. Jostling on the other are the rest of the industry heavyweights, all touting open standards, but each promoting its own, slightly different version. Hewlett-Packard and Sun recently announced initiatives called NetAction and Open Net Environment (ONE) respectively. Oracle's effort goes by the name of Dynamic Services, while IBM is touring its WebSphere platform.

And then there is a raft of fast-growing newcomers, as yet known only to insiders, that are also hoping to get a big piece of the pie. BEA Systems, a Silicon Valley firm founded in 1995, is the most ambitious, with its plan to do for web services and e-commerce "what Microsoft has done for the PC", as chief executive Bill Coleman grandly puts it.

The age of web services actually began several years ago, but only experts noticed because it happened deep down in companies' IT plumbing. Since the late 1980s, business computing has been dominated by a model called client/server, in which most processing is done on the desktop "client" (intelligent terminal) and some on the server. Peter Lewis, a noted American technology journalist, once compared this system to a restaurant where the waiter takes your order for a hamburger, goes to the kitchen and comes back with some raw meat and a bun. You get to cook the burger at your table and add your favourite condiments.

This has advantages: the service is faster and the food is cooked exactly to your liking. But there are also serious drawbacks. The restaurant has to maintain a stove at each table; every time a client application is updated, new copies have to be installed on every computer. And a lot of raw ingredients have to be carried from the kitchen to the table; large amounts of unprocessed data have to be moved from the server to the client.

What is more, the model works well only in a small restaurant and with a simple menu. Once you have lots of customers, demanding anything from sushi to steak, you will need to train your waiters in logistics. To put it in computerese, if you have many different clients – PCs, hand-held computers, cellphones – accessing many different applications running on different operating systems, you need a set of specialised programs to manage the traffic.

Cook-your-own or ready-to-eat?

Since its rise in the mid-1990s, the Internet has created such diversity and thus provided the breeding ground for a whole new class of software to tie everything together: middleware. Its central program – some already call it the operating system of the new world of software – is something called the web-application server. To stick with the metaphor, it now takes care of all the cooking. In essence, it is a big container of pre-cooked ingredients that developers can easily assemble to create new meals (ie, web applications) which are then delivered to a table (a browser).

Although web-application servers are relatively new, they are already quite a juicy market. In 2000 they generated revenues of $1.6 billion, according to Giga Information Group, a consultancy. The firm expects the figure to grow to $9 billion by 2003. The leading vendor so far has been BEA, but the established companies, which had neglected the market, are catching up fast. IBM's WebSphere, says Giga, is now in a dead heat with BEA's WebLogic; both have a market share of 24%.

What turns simple web applications into more malleable web services is something called Extensible Markup Language (XML; see "The x-files" on page 368). This new language provides the necessary standard interfaces for different web services to seek and find each other, communicate what they do, describe the data they want to exchange, and then act on it. XML has thus become the lingua franca of the web-services cloud.

But the picture would not be complete without Java. Sun, the creator of this programming language, originally touted it as a technology that would break Microsoft's lock on the PC desktop. Small programs, called applets, written in Java were meant to be distributed via the Internet and downloaded to run on any computer, regardless of its operating system.

Yet the "write once, run anywhere" claim for Java was quickly amended by cynical developers to "write once, debug everywhere". The technology was not mature, and Microsoft, to boot, had written a dialect of Java that worked well only on Windows. Yet, as with many new technologies, Java turned out to be more successful in unexpected places. Many developers started using the language to write applications for the larger systems of corporations.

More important, Java has also become what geeks call an "application-development environment". This means that Java (or, more correctly, its extension called Java 2 Platform, Enterprise Edition, or J2EE) now provides programmers with the necessary infrastructure to create new applications – just as Windows is not only an operating system, but also a platform on which to build other programs. Most web applications today, particularly corporate ones, are written in Java and run on a J2EE-compliant server program such as WebLogic or WebSphere. What companies like about this is that it allows them to develop highly customised applications quickly without having to start from scratch.

It was mainly this trend towards computing centred on networks that prodded Microsoft into launching its ambitious .NET plan in June 2000. This is an effort to establish an alternative web-services platform, albeit one with a uniquely Microsoft flavour, since the company cannot afford either to abandon its cash cow, Windows, or to let go of its proprietary past. Although the firm's managers insist that .NET is based on open XML standards, it is nonetheless tied to Windows and leaves Microsoft essentially in control.

So far, however, much of .NET exists only on colourful slide presentations to impress Wall Street and, more importantly, to lure software developers back from the Java world and keep others from jumping in. Microsoft's most precious off-balance-sheet asset has always been the huge number of programmers creating applications for Windows, thus making the platform more valuable. That is why in the past the company has always put a lot of effort into making life easy for them.

It is also why most .NET announcements so far have been aimed at making the platform attractive to developers. Microsoft has unveiled a set of "building blocks" code-named HailStorm that include key services, such as keeping track of a user's location or storing his personal information, that developers can use to build their own services.

Microsoft's recent decision to settle its four-year-old lawsuit with Sun over Java must also be seen in this light. Sun had sued Microsoft, claiming that the software giant violated a licensing agreement by creating a version of Java that would work well only on Windows. The settlement now draws a clear line between the competing camps, forcing developers to take sides. Microsoft will, in essence, abandon Java and is encouraging programmers to switch to its own Java-like language, C# (pronounced C-sharp).

It is much too early to tell which camp will win the day. The game is certainly the Java crowd's to lose. Their platform is pretty much in place, and unlike Microsoft they do not have to drag large numbers of established developers, consultants and users from the old world of PC software into the new universe of web services.

Yet Microsoft has successfully played catch-up before, even if its methods have sometimes been questionable. And it has every incentive to make .NET a success: if it fails, it risks losing its dominant position in the computer industry, just as IBM did in the 1980s. The firm has already cranked up its formidable marketing machine. Hardly a week passes without another .NET event or deal.

Microsoft might even get some help from its competitors. They too have yet to prove that they are serious about "co-operating on the standards and competing on the implementation", as Sun puts it. The history of the operating system Unix, which fragmented into incompatible versions because Microsoft's rivals could not agree on standards, is a cautionary tale. If they start bickering again, customers might opt for .NET after all. At least they can be sure that it will be fully integrated, because that is where Microsoft excels.

It is unlikely, though, that any one company will dominate web services in the way that Microsoft has ruled the world of the PC – at least as long as firms stick more or less to open standards. And more than ever, there will be opportunities for integrators of all sorts, as the next article will show.

The X-files

WHEN THE Extensible Markup Language, or XML, first appeared on the scene in 1996 as a standard pushed by the World Wide Web Consortium (W3C), most experts thought of it simply as a more sophisticated form of the Hypertext Markup Language (HTML) in which most web pages are written today. Instead of just telling the computer how to display data, as HTML does, a document written in XML also contains information about what the data represent – say, a date or a stock price.

More important, XML can be – as its name implies – extended. Unlike HTML, which has a limited set of tags to mark up a document, XML developers can invent their own tags, as long as they explain, in a standard way, what they have done and attach this information to the document.

So far, most of the work on XML has concentrated on how to use the technology to exchange data between firms – as Electronic Data Interchange (EDI) already does, albeit clumsily. There are now dozens of groups in various industries developing XML dialects for particular sectors. The best-known are RosettaNet for electronics and Acord for insurance. But there are less weighty ones too, such as ChessML and MindreadingML, using tags such as ‹thoughtsuck› and ‹deepfears›.

In 2000, the growing community of XML developers shifted its attention to another issue: how exactly to transfer XML files between parties. As it turned out, it is not enough to know what the data represent. A supplier also has to know, for instance, that a price change has been received by its retailers and whether they intend to place orders. There are now more than a dozen transport protocols vying to become the standard for this. SOAP (Simple Object Access Protocol) has so far attracted the most backers. But there are more ambitious efforts too, such as ebXML (electronic business XML), sponsored, among others, by the United Nations.

The software industry has also begun working on how to find web services in the cloud. In September 2000, Ariba, IBM and Microsoft proposed a standard called Universal Description, Discovery and Integration (UDDI). The effort, since joined by Hewlett-Packard, Intel and Oracle, aims at building a giant worldwide directory for web services. UDDI goes hand in hand with the Web Services Description Language (WSDL), used to describe the function of a web service and its provider.

The alphabet soup can only get thicker. The industry is already working on something called Transaction Authority Markup Language (XAML), which will undo a transaction involving multiple web services if one of them fails. It comes as no surprise that leading software companies are already talking about creating a standards body for XML to avoid linguistic confusion – among computers, that is. Users may have to learn yet another x-word: XMLC for XML Consortium.

All together now

The complex business of software integration

AN ARMS MERCHANT who sells weapons to all combatants? When he was still chief technology officer of Netscape, the little David fighting Microsoft's Goliath, Marc Andreessen would have dismissed the idea. But in his new job as chief executive of Loudcloud, he is quite comfortable with it.

Mr Andreessen is in the business of integration, meaning that his firm assembles all the pieces needed to build a web application – the hosting facility, the equipment, the software, the network connections – and makes sure that everything is secure and runs well. With his new hat on, he does not care which technology his customers use. He has even been into the lion's den, virtually at least. At the launch of Microsoft's .NET in June 2000, Bill Gates, the company's boss, showed a short video featuring Mr Andreessen saying nice things about Microsoft's ambitious plan.

The idea behind Loudcloud is that companies will want to outsource the building of the infrastructure for a web application because the task is hellishly complex. In due course, Mr Andreessen's firm intends to automate that difficult process completely with a piece of software called "Opsware". More than 40 customers have already signed up, including Eazel, News Corp and Nike.

Build me a cloud

Mr Andreessen is not the only one who believes that integration will be big business in the world of web services. Call them the builders of the cloud: there are dozens of firms vying to put back together all the pieces that the Internet has helped break apart – to create the web applications, web services, business processes and even entire companies.

Open standards help to make that possible, but by themselves they are not enough. They do not kill the beast of complexity that lives in every software system. And there will always be some differences in the way software firms implement standards.

In the past, much of the integration of software systems was done in-house or by hired consultants, but increasingly software packages have taken over a big part of that function. Programs by SAP of Germany, the leading vendor of enterprise resource planning (ERP) programs, have been so successful not just because they are good products, but because they come as an integrated package. And much of Microsoft's dominance is due to its skill at integrating different pieces of software.

Making everything work well together is also the sales pitch of the established software firms for their web-service platform. "Seamless e-business demands seamless software," trumpet the advertisements for Oracle. Sun is singing a similar tune. "People are sick of buying this stuff from different vendors," said Ed Zander, Sun's president and chief operating officer, at the launch of the firm's web-services initiative.

That may be true for medium-sized firms, but large companies are more open to cutting-edge ("best of breed") software to stay ahead of their rivals. More important, no one company can be the source of all innovation, explains George Gilbert, chief software strategist at Credit Suisse First Boston, an investment bank.

So there will be plenty of work for Loudcloud and dozens of other "managed service providers" (MSPS). But MSPS are not the only fish in the integration pond. Higher up in the stack of software, other new firms are also offering services to put everything together. Asera, a much-watched start-up, has probably the most innovative approach. Its software integrates the best e-business programs from different vendors and offers them as an application service. Customers rent only the bits they need. They are thus becoming insulated from the technology. Software is turning into a utility.

Jamcracker, another start-up, also offers to simplify its customers' lives. It integrates the offerings from different application service providers (ASPS) such as book-keeping or recruiting, so customers do not have to manage a relationship with each of these ASPS separately. Customers sign on only once, and get around-the-clock technical support.

Portera takes the integration theme even further by offering a bundle of services for consulting firms. The start-up aims to manage the entire business process of its clients, says Gary Steele, its chief executive. Through the firm's web portal, called ServicePort, consultants can automate much of what they do – scheduling meetings, organising training sessions, creating project teams – and keep track of it while they are on the road.

The holy grail of integration, however, will be to weave web services together. This will be hard to do but should make lots of money. "It's the air-and-water franchise," says Bob Crowley, chief executive of Bowstreet, one of the pioneers in the market. Just as with the platform for web services, incumbents and start-ups are already jockeying for position. Large software companies offer integration as part of their package of web services. Hewlett-Packard has even designed its integration framework as neutral territory to suit different kinds of software, bridging the competing worlds of .NET and Java.

So far the more specialised integrators seem to have the edge. Vitria, for example, is one of the leaders in Enterprise Application Integration (EAI), linking disparate computer systems within a company. Web-Methods has pioneered the use of XML for business-to-business integration (B2Bi), a way to interconnect, for instance, the order and inventory management systems of different firms.

But both firms' ambitions go far beyond acting as IT plumbers. Web-Methods' chief executive, Phillip Merrick, says he wants to develop his products into a "complete platform for business process automation". Software allows companies not only to connect applications easily and link up with others, but also to program the way they run themselves and deal with their partners.

Bowstreet, for its part, wants to play the part of the big web-service unifier that sits on top of everything. It has developed software similar to Vitria's, called "Business Web Factory", that lets firms mix and match all kinds of web services. The Thread.com, for example, is using Bowstreet's technology to let clothing firms create their private supply chains at the click of a mouse – buying fabric in Thailand, booking manufacturing capacity in Bangladesh and arranging for distribution in America.

Not all the integrationists' hopes will come true, and it will take years to tame the beast of complexity. The main obstacle for web services, however, may well be the way the digital software interacts with real-life people (known as wetware).

Design Darwinism

Making computers fit for people to use

WHO KILLED ALL those dotcoms? Most experts accuse greedy venture capitalists, inexperienced managers or impatient shareholders. But Jakob Nielsen, one of the leading experts on user-friendly web design, has identified another culprit: offputting websites. Many of them were just too difficult to use. "Firms drew users to their sites with expensive promotions," he explains, "and then lost them with ineffective design and bad service." If websites and computer programs were easier to use, he argues, it would not just narrow the digital divide, but also save knowledge workers about half of the hours they now spend in front of the screen.

Mr Nielsen and his colleagues of the Nielsen Norman Group are on to something. If use of the web is to become ubiquitous, "user interfaces" (geek-speak for the point where computers and people meet) have to get a whole lot better. The slow adoption of the wireless Internet in Europe and America so far should be seen as a bright red warning light. Users are sick of clicking through dozens of menus on their mobile-phone displays to find what they want (if they are lucky).

Complaints about user interfaces are as old as the concept itself. The PC's so-called WIMP (windows, icons, menus, pointer) interface was invented in the early 1970s at Xerox Parc, a famous Silicon Valley research lab. But not much has changed since Apple came up with its Macintosh computers in 1984 and Microsoft copied it with Windows a few years later – except for the addition of a lot more bells and whistles.

It is not that nobody has tried to do anything new. Start-ups regularly come up with interesting new ways to organise computer files, for example. The Brain lets you link documents, web pages and other data and displays them as a network, similar to the way people associate things in their minds. Mirror World Technologies, a company inspired by David Gelernter, a computer-science professor at Yale University, offers a program that organises files chronologically, as a stack of index cards to search and browse. It recently relaunched the product as a network-based application called Scopeware.

But none of this has really caught on, in large part because users are rather conservative, as Steven Johnson, editor of *Feed*, an online magazine, points out in his book *Interface Culture* (Harpers San Francisco, 1997): "If there is a gravitational force operating within this field, it is the force of habit."

The economics of PC software have not worked in users' favour. For a start, more than 80% of PC users are locked into Windows. Software vendors wanted their products to appeal to as many users as possible and sell them frequent upgrades, so they added ever more features, making programs more difficult to use.

The Internet is improving this warped ecosystem. To begin with, the network cuts costs, which allows software firms to pay more attention to things like ease of use. Many programs these days are downloaded from the web, so companies spend less on marketing and distribution. Technical support has become cheaper, too, because much information can be put on the web.

The Internet also increases competition, at least for PC software. Most downloaded programs are now available in a trial version that expires after 30–60 days, giving customers a chance to see how they like them. And with pure web applications, users will get even more of a choice. Since the Internet is built on open standards, they can use almost any web application, not just those written for their particular operating system.

Because the competitors are often only one click away, the interfaces of commercial websites have already improved much faster than those of standard desktop software. Yet many e-tailers, even the more popular ones, still do not get the basics right. Creative Good, a design consultancy, carried out a test in which 50 consumers each visited the sites of eight leading e-tailers. Almost half of all attempts to make a purchase failed because the users could not work out how to complete the transaction.

Net deficit

The reasons for such failures are complex. Some designers bet too much on technology and forget simplicity. Asking consumers what they want is not necessarily the answer: many of them don't know. Only expensive and time-consuming direct observation, says Mr Nielsen, will tell you where they encounter problems with a site.

That is where a number of firms see their opportunity. Vividence, for example, has put together a panel of 150,000 surfers who are willing to review client websites. Launched in early 2000, the company already has more than 160 clients. In the long run, it wants to improve the inter-

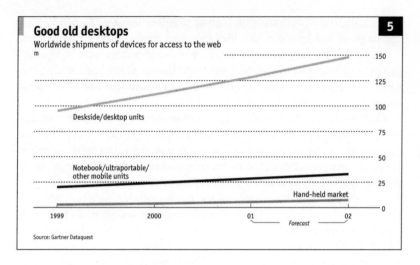

Good old desktops 5
Worldwide shipments of devices for access to the web
m

Deskside/desktop units

Notebook/ultraportable/
other mobile units

Hand-held market

1999 2000 01 02

Forecast

Source: Gartner Dataquest

faces of all kinds of electronic devices. That seems a good business to be in. According to Bruce Tognazzini, another principal of the Nielsen Norman Group, the shift from a 17-inch monitor to a cellphone's one-inch display threatens to wreak havoc with user-friendly software design. PC monitors are relatively forgiving, but with small displays designers have to anticipate what users may want to do. The Palm hand-held computer eventually succeeded because its creators talked to a lot of potential users and discovered that they wanted things like simple one-button access to basic features such as address books.

Such problems help to explain why the wireless Internet so far has not been a big hit in Europe or in America, except for sending short messages. The browsers in cellphones, based on the wireless application protocol (WAP), are a poor imitation of web browsers. Using a WAP service at present is like going through an entire deck of cards to find the right one. The next generation of WAP browsers will be a huge improvement.

Voice input will also make it easier to bypass the display limitations of wireless devices. BeVocal, a start-up operating what is known as a voice portal, already offers a combined voice and text service. Users can ask for directions and the system will talk them through to their destination or send them an e-mail. Future generations of wireless technology will allow switching between voice and text.

Yet this is still the wrong approach, argues Rich Rifredi, marketing chief at Pixo, a Silicon Valley start-up selling software to customise the user interfaces of handsets and other wireless devices. Depending on the target group for a product, they can, and should, look very different, he says. A business traveller might want one-button access to share prices; a teenager to her list of friends. "If the PC was about adding features," explains Mr Rifredi, "hand-helds and cellphones are about what to cut."

One of the products Pixo helped to build was Scout Electromedia's Modo, which enjoyed only a brief flowering before investors pulled the plug on the company in autumn 2000. It was a kind of wireless entertainment magazine, updated daily, where users would find local information on, for example, the nearest Japanese restaurant or cashpoint. The Modo provided a foretaste of the use of location as a way to select content. Hardware makers are already working on other ways to configure a device. Hewlett-Packard, for example, is experimenting with different kinds of sensors. If a device senses that it is being driven around, it could automatically call up a traffic-information service. Smell could be used to authenticate a user.

Integration of form and function will become more important, predicts Tim Brown, chief executive of Ideo, a leading design firm that developed such classics as the first mouse for the Macintosh. Once electronic devices become more pervasive, he says, the idea of a generic interface becomes obsolete. For example, elderly people are frequently overwhelmed by Windows. That is why the Health Buddy, a device developed by Ideo to monitor a person's health on a continuous basis, has a very simple interface and large buttons.

No doubt the industry will come up with many more smart, easy-to-use devices. But first it must solve another problem: how to ensure a better fit between software and the organisations that use it.

A touch of concrete

Even the best software won't work without organisational changes

WHEN CHARLIE FELD gets called in, it is almost too late. His corporate client is nearly collapsing beneath layers upon layers of information technology – different generations of hardware, uncounted pieces of software. And somehow none of it really connects. Most departments have incompatible systems. They keep their own databases and guard them jealously.

Bringing order to such chaos is the speciality of Mr Feld, chief executive of the Feld Group, and known as the Red Adair of IT. When he parachutes in, he becomes the acting chief information officer of a company for two or three years and, with a few colleagues, turns the disparate islands of computing into a coherent system. He has already sorted out the IT woes of large firms such as Delta Air Lines, Burlington Northern and Santa Fe Railway Company and Westinghouse. But fixing technology is the smaller part of what Mr Feld does. His main task is to create the right organisation for the new IT system. Unless you do that, he says, "the inertia of an organisation makes you end up with just another ugly IT animal".

This may seem pretty obvious, but when companies bring in new IT systems few of them pay much attention to the effect on their organisational structure and culture. "There is still the belief that big enough software will solve all the problems," says Rebecca Henderson, management professor at the Massachusetts Institute of Technology (MIT).

As long as software mainly replaced labour and did the boring stuff faster, such negligence might not have been too serious. But as IT penetrates every corner of an enterprise, with the Internet connecting everything, the oversight becomes potentially disastrous. It is the main reason why something goes wrong with three-quarters of software projects, according to the Standish Group, a consultancy.

That the human side of computing matters, particularly when it comes to networks, is not a new discovery. A decade ago, Wanda Orlikowski, a professor of information technologies and organisation studies at MIT, published a study on the deployment of Lotus Notes, a program known as groupware, in an international consulting firm. Her main finding: "When an organisation deploys a new technology with an intent to make substantial changes in business processes, people's technological frames and the organisation's work practices will likely require substantial change." Groupware allows employees to collaborate online, thus supposedly making an organisation more efficient. But Mrs Orlikowski found that the program was used mostly to send e-mail, transfer files or call up online news sources.

Because the program's introduction was rushed, the firm's technology group did not have the time to train people properly. More important, the organisation's reward systems and culture were at odds with the collaboration that groupware is supposed to achieve. As with many other consulting firms, the culture was competitive and individualistic. And management expected most hours to be "billable" (ie, chargeable to clients), which time spent trying out and using Notes was not.

Poor use of groupware is wasteful, but a botched implementation of enterprise resource planning (ERP) software – which handles a company's financial, manufacturing and human resources, among other things – can be downright dangerous for a company. Yet very few firms have seriously tried to figure out how these all-encompassing packages of corporate software fit into their organisation, says Thomas Davenport, director of the Institute for Strategic Change at Accenture, a consultancy formerly known as Andersen Consulting.

Do it our way

That is one of the main reasons, Mr Davenport explains, why of the 100 firms he studied for his book *Mission Critical* (Harvard Business School Press, 2000), only ten got any real value from implementing an ERP system. Most companies simply did not see the connection between information technology and organisational structure. They did not realise, for example, that once such enterprise software was in widespread use, business would have to be conducted in much the same way worldwide.

To "pour some ERP concrete into a business" (as IT managers jokingly put it) was a particular problem for American firms, which tend to give their constituent business units considerable freedom in the way they conduct their activities and how they embody this in information technology. "To switch to a centrally defined and controlled enterprise software in which everything must be common is a bracing change," Mr Davenport writes.

Often, however, the problem is simply an astonishing lack of fore-sight, in spite of all the business re-engineering that ERP systems have brought about. Mr Davenport quotes the example of a company that had introduced a sophisticated purchasing system – but forgotten to train the employees in the purchasing department. They had to learn how to use the system after the event, which seriously delayed the project.

And yet the benefits of "organisational investments" are significant, according to Erik Brynjolfsson, a management professor at the MIT Sloan School of Management. Along with two other researchers, he analysed the IT investments of 400 large firms and found that those companies that had adopted organisational changes along with the new technology tended to be more productive.

If McKinsey has its numbers right, both software vendors and their customers could be saving themselves a lot of money by paying more attention to the organisational aspects of introducing new software. In a recent study of a clothing retailer's investment in supply-chain software, the consultancy's software practice found that the program yielded sav-ings of $15m a year, but that another $55m of potential savings remained unrealised because they would have required changes in organisational structure, business processes and incentives. "Many providers don't even know how best to exploit what their software could do for com-panies," says Greg Hughes of McKinsey's software practice.

It is not only excessive faith in technology that keeps many firms from making these low-tech investments. For Mr Feld, another reason is that IT has traditionally been less well managed than other parts of a company because management considered it an expense. Mr Davenport thinks that firms were simply not willing to shell out the extra money needed to make the organisational as well as the technological changes. They also lack people who have both technical and organisational knowledge.

Now at last some vendors are starting to develop the necessary skills. I2, a leader in programs for supply-chain management based in Dallas, Texas, does not think of itself as an enterprise software company, but as a firm that "creates business value", says the company's chief executive, Sanjiv Sidhu. I2 does not just sell software, but also helps customers to transform their organisations. The firm is so confident of the benefits it offers its customers that sometimes it links part of its licensing fee to cost savings.

Web services are bound to bring further organisational challenges.

For a start, they will make the technical side of mergers and acquisitions easier. At present, many such get-togethers are frustrated because the IT systems of the intended partners cannot connect. More important, the technology will allow firms to outsource even more functions than at present and specialise in what they do best – and then, in many cases, offer that speciality as a web service. An insurance company, for example, could hone its underwriting expertise and deliver this as a web service, while buying in credit-rating, human-resource or news services from other providers.

How far will this digital division of labour go? Thomas Berquist, managing director of global investment research at Goldman Sachs, an investment bank, predicts the emergence of what he calls "industry operating systems" (I-OPS) – huge IT hubs that will take over many of the functions common to the firms in a particular sector. In future, he expects firms to outsource much of what constitutes a company today.

Mr Berquist calls these entities "operating systems" because they provide common functions for companies, much as their computer equivalents provide common functions for software applications. And in analogy to the PC world, he expects that each industry embracing an I-OPS will ultimately settle on a single big winner. But there might also be other consequences of a web-service world that require vigilance, as the conclusion of this survey will show.

As goes software ...

... so goes business – and, perhaps, even society itself

LET SOFTWARE DISAPPEAR, and life as we know it would break down, at least in developed countries. It controls most of the objects which surround us: computers, of course, but also telephones, cars, toys, TVs, much of our transport system, and so on. Yet if the vision of web services comes to pass, today's dependence on software will appear slight. Life in the cloud will mean that much of what we do, as *homo oeconomicus* at least, will be automated, from restaurant reservations to car purchases, from share trades to entire business deals.

All this is at least some years off, and may not happen at all. But the prospect raises some interesting questions. Who will write all the code needed for these services? What needs to be done to ensure that it is reliable and secure? And, last but not least, is there a way to prevent a few dominant companies or governments from controlling the cloud? Conveniently, the Internet and the institutions it has spawned may hold some answers.

Laments about a "software crisis" are almost as old as the industry itself. There are never enough skilled programmers to satisfy the demand for high-quality code. But in the years ahead this chronic imbalance could turn into a veritable "software gap", as an American presidential advisory group made up of leading computer scientists put it in a 1999 report. "This situation", the researchers wrote, "threatens to inhibit the progress of the current boom in information technology."

The group is even more concerned about the current fragility of software. Even much-tested commercial varieties are often riddled with bugs, lack security, do not perform well and are difficult to upgrade. This was a bore when most software was confined to isolated devices and networks, but it becomes a serious problem in the world of web services. Software delivered online has to be able to withstand the onslaught of millions of users, and is at risk of security attacks from myriad sources.

Whereas these technical issues have been discussed for some time, the social and political aspects have only recently come to the surface. As code increasingly penetrates daily life, it becomes de facto law that regulates behaviour, argues Lawrence Lessig, a Stanford law professor, in his book *Code and Other Laws of Cyberspace* (Basic Books, 1999). For example, code needs to be compatible with our ideals of privacy and free speech. Another pressing issue is open standards. The continuing antitrust trial against Microsoft has shown that the world needs common technical rules that are not controlled by a single company (or indeed a government). Such rules can provide a level playing field for competition. But they must not be too strict, because that would stifle innovation and diversity.

The other main regulatory issue is less obvious: it concerns directories, the digital equivalent of telephone books. Even more than open standards, they will hold the cloud together. Some directories will tell users where to find web services and what they offer; others will keep track of available hardware; and yet others will list not only the identity of users, but also where they are and whether they are online at that moment.

The reason why these directories might need to be regulated is that they are subject to strong networking effects: the more data they contain and the more users they have, the more valuable they become and the more data and users they will attract. Sometimes it will even make sense to have a single directory, as it does for the domain-name system (DNS), the current address book of the Internet. Competing domain-name systems would probably balkanise cyberspace.

Whoever controls such directories will wield potentially enormous power. If a company owned, for example, the directory for web services, it could try to make its own electronic offerings more accessible than those of its competitors. The continuing controversy about the Internet Corporation for Assigned Names and Numbers (ICANN), the body that administers the DNS, is the clearest example so far of how difficult it can be to regulate these directories.

There are other simmering disputes, too, such as whether and when AOL Time Warner should open its dominant instant-messaging (IM) system and extend it to other providers. The point about IM, a cross between a telephone call and e-mail, is that it keeps track of whether users are currently online and, in the future, will also be able to monitor where they are. This is important information for providers of smart web services. That is why the Federal Communications Commission (FCC) made its approval of the AOL/Time Warner merger contingent on the new company's promise to open up its IM system once it includes video services.

Microsoft had heavily lobbied the FCC, telling the agency that for IM to live up to its promise it must share the features of "openness and interoperability that characterise both the public telephone network and the Internet". It will be interesting to see whether Microsoft will apply the same philosophy to its own recently announced directory-like services, intended to become building blocks of the .NET world.

Luckily, the Internet is already helping to solve some of these dilemmas. Its very structure, for instance, has caused the software gap to narrow. Programmers no longer have to live in America or other developed countries, but can work from anywhere on the globe. In the future, there will increasingly be a global market for software development, just as one already exists for the manufacturing of electronics. The fast-growing software industry in India is only the beginning.

Moreover, the Internet allows for massive testing and peer review, boosting the quality of code, in particular through open-source projects. The more people look at a program, the more likely it is that mistakes will be spotted. "Given enough eyeballs, all bugs are shallow," writes Eric Raymond, another leading thinker of the open-source movement, in his influential book *The Cathedral & the Bazaar* (O'Reilly, 1999).

Finally, it is the Internet's institutions – such as the Internet Engineering Task Force (IETF) – that offer a possible solution to the regulatory issues. These consensus-building bodies are not just a good mechanism to develop robust and flexible open standards; their decision-making processes could also be applied to other issues, such as the regulation of directories. These communities are guided by respected members, known as "elders" or "benevolent dictators" (for open-source projects), who have gained their status because of the quality of their contributions.

Most of these elders are technical and social engineers who work for academic institutions or other not-for-profit organisations. Governments would do well to provide economic support for these elders instead of regulating directly, argues Paul Romer, an economics professor at Stanford University: "This would be money far better spent than on antitrust actions or agencies like the FCC."

This may be the Internet's most crucial effect on the software industry: that it has made it possible for groups akin to scientific communities, rather than market forces alone, to lay the groundwork of the digital world. That seems to be a far sounder solution than allowing a small handful of firms to become the not-so-benevolent dictators of the cloud.

The material on pages 349–83 first appeared in a survey written by Ludwig Siegele in *The Economist* in April 2001.

The Internet's new borders

Geographical lines and locations are increasingly being imposed on the Internet. Is this good or bad?

LONG, LONG AGO in the history of the Internet – way back in February 1996 – John Perry Barlow, an Internet activist, published a "Declaration of the Independence of Cyberspace". It was a well-meaning stunt that captured the spirit of the time, when great hopes were pinned on the emerging medium as a force that would encourage freedom and democracy. "Governments of the industrial world," Mr Barlow declared, "on behalf of the future, I ask you of the past to leave us alone. You are not welcome among us. You have no sovereignty where we gather. You have no moral right to rule us nor do you possess any methods of enforcement we have true reason to fear. Cyberspace does not lie within your borders."

Those were the days. At the time, it was widely believed that the Internet would help undermine authoritarian regimes, reduce governments' abilities to levy taxes, and circumvent all kinds of local regulation. The Internet was a parallel universe of pure data, an exciting new frontier where a lawless freedom prevailed. But it now seems that this was simply a glorious illusion. For it turns out that governments do, in fact, have a great deal of sovereignty over cyberspace. The Internet is often perceived as being everywhere yet nowhere, as free-floating as a cloud – but in fact it is subject to geography after all, and therefore to law.

The idea that the Internet was impossible to regulate dates back to when its architecture was far simpler than now. All sorts of new technologies have since been bolted on to the network, to speed up the delivery of content, protect networks from intruders, or target advertising depending on a user's country or city of origin. All of these technologies have mundane commercial uses. But in some cases they have also provided governments with ways to start bringing the Internet under the rule of local laws.

The same firewall and filtering technology that is used to protect corporate networks from intrusion is also, for example, used to isolate Internet users in China from the rest of the network. A report on the Internet's impact in China by the Carnegie Endowment for International Peace (CEIP), a private think-tank based in Washington, DC, found that the government has been able to limit political discourse online. Chinese citizens are encouraged to get on the Internet, but access to overseas sites is strictly controlled, and what users post online is closely monitored. The banned Falun Gong movement has had its website shut down altogether. By firewalling the whole country, China has been able to stifle the Internet's supposedly democratising influence. "The diffusion of the Internet does not necessarily spell the demise of authoritarian rule," the CEIP report glumly concluded. Similarly, Singapore and Saudi Arabia filter and censor Internet content, and South Korea has banned access to gambling websites. In Iran, it is illegal for children to use the Internet, and access-providers are required to prevent access to immoral or anti-Iranian material. In these countries, local standards apply, even on the Internet.

Local rules are also being applied in parts of the world where access to the Internet is not tightly controlled. Since the Internet consists of data sitting inside computers that are located in the real world, it turns out that legal action can be taken – and is being taken – against Internet access-providers and publishers, using old-fashioned laws, in old-fashioned courts. Libel and defamation laws have been applied to online publications within several countries, and other laws are being applied across borders too. A particularly important example concerns Yahoo!, an Internet portal which includes an auction site, and which fell foul of a French law banning the sale of Nazi memorabilia. Last November a French judge ordered Yahoo! to find a way to ban French users from purchasing such items posted on any of its sites, even sites based in America. The judge had in mind "geolocation" technology, which can work out where individual Internet users are. Although it is not perfect, it can identify a user's country of origin 70–90% of the time. Since the technology to differentiate between users from different countries exists, it is likely that local lawmakers will increasingly require websites such as Yahoo! to use it.

To American cyber-libertarians, who had hoped that the Internet would spread their free-speech gospel around the world, this is horrifying. Yahoo! is appealing against the French decision, because it sets a precedent that would require websites to filter their content to avoid breaking country-specific laws. It would also have a chilling effect on free speech, since a page posted online in one country might break the laws of another. Enforcing a judgment against the original publisher might not be possible, but EU countries have already agreed to enforce each other's laws under the Brussels Convention, and there are moves afoot to extend this scheme to other countries too, at least in the areas of civil and commercial law, under the auspices of the Hague Convention.

It is true that filtering and geolocation are not watertight, and can be circumvented by skilled users. Filters and firewalls can be defeated by dialling out to an overseas Internet access-provider; geolocation can be fooled by accessing sites via another computer in another country. E-mail can be encrypted. But while dedicated dissidents will be prepared to go to all this trouble, many Internet users are unable to change their browsers' home pages, let alone resort to these sorts of measures. So it seems unlikely that the libertarian ethos of the Internet will trickle very far down in countries with authoritarian regimes. The upshot is that local laws are already being applied on the Internet. Old-style geographical borders are proving surprisingly resilient.

Getting real

In some ways this is a shame, in others not. It is certainly a pity that the Internet has not turned out to be quite the force for freedom that it once promised to be. But in many ways, the imposition of local rules may be better than the alternatives: no regulation at all, or a single set of rules for the whole world. A complete lack of regulation gives a free hand to cheats and criminals, and expecting countries with different cultural values to agree upon even a set of lowest-common-denominator rules is unrealistic. In some areas, maybe, such as extradition and consumer protection, some countries or groups of countries may be able to agree on common rules. But more controversial matters such as free speech, pornography and gambling are best regulated locally, even if that means some countries imposing laws that cyber-libertarians object to.

Figuring out whose laws apply will not always be easy, and thrashing all of this out will take years. But it will be reassuring for consumers and businesses alike to know that online transactions are governed and protected by laws. The likely outcome is that, like shipping and aviation, the Internet will be subject to a patchwork of overlapping regulations, with local laws that respect local sensibilities, supplemented by higher-level rules governing cross-border transactions and international standards. In that respect, the rules governing the Internet will end up like those governing the physical world. That was only to be expected. Though it is inspiring to think of the Internet as a placeless datasphere, the Internet is part of the real world. Like all frontiers, it was wild for a while, but policemen always show up eventually.

The material on pages 384–6 first appeared in *The Economist* in August 2001.

POSTSCRIPT

ON NOVEMBER 7th 2001, a San Francisco judge ruled that Yahoo! was not required to comply with the French court's decision. The French organisations which had originally sued Yahoo! in a French court said they would appeal, on the ground that the American judge did not have jurisdiction over the case. The debate seems likely to continue.

Putting it in its place

The Internet is perceived as being everywhere, all at once. But geography matters in the networked world, and now more than ever

BREWSTER KAHLE unlocks the cellar door of a wooden building in San Francisco's Presidio Park. He steps inside, turns on the fluorescent lights to reveal a solid black wall of humming computers, and throws out his arm theatrically. "This", he says, "is the web." It is a seductive idea, but the web isn't really housed in a single San Francisco basement. Mr Kahle's racks of computers merely store archived copies of many of its pages which Alexa, his company, analyses to spot trends in usage. The real Internet, in contrast, is widely perceived as being everywhere, yet nowhere in particular. It is often likened to a cloud.

This perception has prompted much talk of the Internet's ability to cross borders, break down barriers and destroy distance. On the face of it, the Internet appears to make geography obsolete. But the reality is rather more complicated. If you want a high-speed digital-subscriber line (DSL) connection, for example, geographical proximity to a telephone exchange is vital, because DSL only works over relatively short distances. Similarly, go to retrieve a large software update from an online file library, and you will probably be presented with a choice of countries from which to download it; choosing a nearby country will usually result in a faster transfer. And while running an e-business from a mountain-top sounds great, it is impractical without a fast connection or a reliable source of electricity. The supposedly seamless Internet is, in other words, constrained by the realities of geography. According to Martin Dodge of University College London, who is an expert on Internet geography, "the idea that the Internet liberates you from geography is a myth".

What's more, just as there are situations where the Internet's physical geography is all too visible when it ought to be invisible, the opposite is also true. There is growing demand for the ability to determine the geographical locations of individual Internet users, in order to enforce the laws of a particular jurisdiction, target advertising, or ensure that a website pops up in the right language. These two separate challenges have spawned the development of clever tricks to obscure the physical location of data, and to determine the physical location of users –

neither of which would be needed if the Internet truly meant the end of the tyranny of geography.

Down on the farm

To see just how little the Internet resembles a cloud, it is worth taking a look at where the Internet actually is. The answer, in short, is in cities. This is partly a historical accident, says Anthony Townsend, an urban planner at the Taub Urban Research Centre at New York University. He points out that the Internet's fibre-optic cables often piggyback on old infrastructure where a right-of-way has already been established: they are laid alongside railways and roads, or inside sewers. (Engineers installing fibre-optic cables in a New York building recently unearthed a set of pneumatic tubes, along which telegrams and mail used to be sent in the 19th century.) Building the Internet on top of existing infrastructure in this way merely reinforces real-world geography. Just as cities are often railway and shipping hubs, they are also the logical places to put network hubs and servers, the powerful computers that store and distribute data.

This has led to the rise of "server farms", also known as data centres or web hotels – vast warehouses that provide floorspace, power and network connectivity for large numbers of computers, and which are located predominantly in urban areas. A typical example can be found in Santa Clara, just off California's Highway 101. It is run by Exodus Communications, a web-hosting firm which has nine server farms in Silicon Valley and another 35 around the world. From the outside, the farm is a deliberately nondescript building. A sophisticated security system, with hand scanners and video cameras, keeps out unauthorised visitors. Inside, the building resembles a jail, rather than a farm: it is packed with row upon row of computers in locked metal cages, their fans whirring and lights flashing. The air is filled with the deafening hum of air-conditioning. There are no windows and few people, and the lights are triggered by motion sensors, keeping unvisited parts of the farm in darkness. Exodus's customers house their computers inside the metal cages, which are supplied with power and network connections. Most of the world's biggest websites live in buildings like this; Exodus hosts 49 of the top 100.

As if to emphasise how physical constraints apply even to virtual spaces, server farms are still rented by the good old-fashioned square foot. According to figures published in April 2001 by Salomon Smith Barney, worldwide server-farm capacity is growing by 50% annually, and will reach 22m square feet by the end of 2001, despite the demise of the dotcoms. Cage space turns out to have other uses, too: boastful corporate logos hang from many cages, and some firms have posted job advertisements in the hope of poaching technical staff from rivals.

The signs are that the storage of information is going to become even more physically concentrated. One reason is the growth of "managed hosting" where, instead of renting space on a farm for their own servers, firms rent the computing capacity along with the power and network connectivity. In short, they simply hand over their data, and leave running the servers to the hosting company. As a result, there is no longer any need for customers to visit farms, so they need not be located in metropolitan areas, where space is limited and expensive. They can be anywhere, provided enough power and bandwidth are available.

In practice the constraint is power. A single server farm can consume as much power as a small airport, or four large hospitals. As a result, says Jon Feiber of Mohr Davidow Ventures, a venture-capital firm, the logical thing to do is to build out-of-town server farms with their own power stations. Such farms, he suggests, could be very large indeed: perhaps a dozen would be enough for the whole of the United States. Just such a facility, with a 24MW gas-fired power station, is being built just outside London by iXguardian, a British computer-services firm. It will be the largest server farm in Europe.

The combination of managed hosting and dedicated power stations means that data will be increasingly concentrated in large farms. The rise of wireless devices will drive this trend too: instead of storing data internally, such devices will store information on the network and access it when needed. But users wishing to access their data will still be spread out around the world. So centralisation will drive demand for technology that can smooth out the Internet's geographical lumpiness and speed the delivery of data; in short, technology to obscure the physical location of Internet content from its users.

First, hide the data

One way to do this is to store copies of popular lumps of content in data caches sprinkled around the world. The leader in this field, with over 11,000 caching servers in 62 countries, is Akamai, a firm based in Cambridge, Massachusetts. The geographical distribution of Akamai's infrastructure is strikingly different from that of Exodus. Broadly speaking, Akamai needs servers near the consumers of content, whereas Exodus puts its farms near the suppliers of content. Accordingly, Exodus has farms in North America, Europe, Australia and Japan, but not in Africa or South America. Akamai, on the other hand, has servers pretty much everywhere.

Akamai's customers, which include CNN and Yahoo!, are content providers who are prepared to pay to ensure that users around the world are able to access their sites smoothly and quickly. Normally, when you visit a web server, a description of the page you have requested is delivered across the network. This consists of the page's text, plus references to any graphics (or sound or film clips) associated with it. These items are then requested by your web browser and delivered across the network. Finally, the browser assembles all the components and displays the page. The problem is that while the text can be delivered quickly, the "heavy" items (such as graphics and video) are much larger and take longer to arrive. It is these items which Akamai can help to deliver more quickly.

It works like this. You request a web page in the usual way, and the page description is delivered. But the references to the page's "heavy" items are modified to fool your web browser into requesting those items from Akamai, rather than from the original web server. Taking account of your location on the network, and given the prevailing traffic conditions, Akamai then delivers the heavy items from the nearest available cache, and the page pops up much more quickly. By monitoring the demand for each item, and making more copies available in its caches when demand rises, and fewer when demand falls, Akamai's network can help to smooth out huge fluctuations in traffic. A further benefit is that the customer's web server does not have to deliver the heavy items, which reduces the load on it dramatically and makes it less likely to collapse when faced with a sudden surge of visitors.

A number of firms have followed in Akamai's footsteps by moving content to the "edges" of the Internet. But there are several other ways to speed up content delivery. One alternative approach is being taken by the Content Bridge Alliance, a group led by a California software firm called Inktomi, whose other members include AOL and Exodus. Rather than setting up a network of thousands of caches, as Akamai has done, the Content Bridge Alliance's plan is to connect existing networks and farms together more efficiently in order to speed the flow of traffic. Yet another approach is being taken by Kontiki, a firm launched in August 2001 by veterans of Netscape. It is one of several start-ups that plan to combine Akamai's approach with that of Napster, the infamous music-swapping service. Essentially, users' own computers will be used as caches, so that recently accessed content can be delivered quickly when needed to other users nearby on the network.

Now, find the users

In parallel with all this effort to obscure the physical location of data on the Internet, there is growing interest in determining the location of its users. Laws and tax regimes are based on geography, not network topology; online merchants, for example, may be allowed to sell some products in some countries but not others. The growth in interest in "geolocation" services, which attempt to pinpoint Internet users' locations based on their network addresses, also signals the realisation that traditional marketing techniques, based on geography, can be applied online too. Marie Alexander of Quova, a Silicon Valley geolocation firm, points out that goods and services exist in physical locations, and marketing is traditionally done on a geographical basis. Rather than messing around with fiddly (and privacy-invading) one-to-one marketing, she says, many firms are instead sticking with the old geographical approach, but taking it online. Thus different visitors to a website may be offered different products or special offers, depending on what is available nearby.

Quova's geolocation service, called GeoPoint, is based on a continually updated database that links Internet Protocol (IP) addresses to countries, cities and even postcodes. If you visit a website that is equipped with GeoPoint software, your IP address is relayed to Quova's servers, which look up your geographical location. This information is then used by the website to modify the page's content based on your physical location. Quova claims to be able to identify web users' country of origin with 98% accuracy, and their city of origin (at least for users in the United States) 85% of the time. Other firms, including Akamai, Digital Envoy, InfoSplit and NetGeo, offer similar services.

Once the user's location is known, existing demographic databases, which have been honed over the years to reveal what kinds of people live where, can be brought into play. But although targeted advertising is the most obvious application for geolocation, it has many other uses. It can, for example, be used to determine the right language in which to present a multilingual website. E-commerce vendors and auction houses can use geolocation to prevent the sale of goods that are illegal in certain countries; online casinos can prevent users from countries where online gambling has been outlawed from gaining access; rights-management policies for music or video broadcasts, which tend to be based on geographical territories, can also be enforced. The pharmaceuticals and financial-services industries, says Ms Alexander, which are subject to strict national regulation, can be confident that by offering goods and services for sale online they are staying within the law. Borders, she notes, are returning to the Internet.

Interest in geolocation soared after a November 2000 ruling by a French judge requiring Yahoo!, an Internet portal, to ban the auction and sale of Nazi memorabilia over the Internet to users in France. The ruling was significant because it covered sales to French users even from Yahoo!'s websites located in other countries. The implication is that to avoid breaking French law, websites around the world where such items are sold must prevent French users from gaining access – and geolocation technology allows them to do just that. Of course, the technology is far from perfect; a panel of experts, including Vinton Cerf, the networking guru who is known as the "father of the Internet", advised the judge that determining an individual user's country of origin was unlikely to be possible more than 90% of the time. But all borders are slightly porous, and the French judge decided that 90% was good enough.

Rather than adopt geolocation technology, Yahoo! responded by banning the auction of Nazi items across all of its sites, and says it has no plans to reinstate them. But it is challenging the ruling in order to avoid having other such restrictions placed on its content by other jurisdictions. The company, which is based in America, has asked a federal court in San Jose to declare the French ruling unenforceable in the United States. (Ironically, in July 2000 Yahoo! said that it would begin using Akamai's geolocation technology to target advertising and other content.)

Critics of the French ruling agree that it would set a dangerous precedent, by allowing one country to interfere with freedom of speech across the entire Internet. "If every jurisdiction in the world insisted on some form of filtering for its particular geographic territory, the web would stop functioning," Mr Cerf declared. Stanton McCandlish of the Electronic Frontier Foundation, a pressure group, says he expects other governments to adopt geolocation and other similar techniques to balkanise the Internet in coming years. But he notes that geolocation is merely the latest example in a growing trend to impose local controls on the Internet. China, for example, already filters all Internet traffic flowing into and out of the country in order to prevent its citizens from accessing particular websites.

At the same time, the French ruling is regarded in some quarters as a logical and pragmatic way forward for Internet regulation; in the real world, after all, multinational firms are used to operating under different laws in different countries. According to Lawrence Lessig, a Stanford law professor, "the notion that governments can't regulate hangs upon a particular architecture of the Net". As the Internet's architecture changes and becomes more complex, with the addition of services like filtering and geolocation, the idea that the Internet is beyond the reach of local laws and government regulation looks less and less tenable.

The revenge of geography

So much for the death of geography. And determining the location of Internet users seems likely to become even more commonplace, and even more accurate, with the rise of wireless Internet devices such as smart phones. Already, the first "location-based services" have been launched, capable of sending text messages to mobile-phone users in particular network cells. More accurate positioning will be possible in future using a number of other techniques, such as the satellite-based Global Positioning System. Advertisers are rubbing their hands at the prospect of being able to send precisely targeted offers to people near particular shops, or inside a sports arena, though privacy concerns may yet scupper their plans. Less annoyingly, users of smart phones may choose to call up location-specific information, such as maps or traffic updates, or to locate a nearby restaurant. According to a recent estimate from Analysys, a telecoms consultancy, global revenues from location-based services will reach $18 billion by 2006 – a figure that is regarded as conservative by many in the industry.

Mr Townsend notes that cities are, in a sense, vast information storage and retrieval systems, in which different districts and neighbourhoods are organised by activity or social group. A mobile Internet device, he suggests, will thus become a convenient way to probe local information and services. Location will, in effect, be used as a search parameter, to narrow down the information presented to the user. Mobile devices, he says, "reassert geography on the Internet".

At the moment, Internet users navigate a largely placeless datasphere. But in future they will want location-specific information and access to their personal data, wherever they are – and wherever it is. This will be tricky to pull off, and impossible without taking geography explicitly into account. It is undoubtedly true that the Internet means that the distance between two points on the network is no longer terribly important. But where those points are still matters very much. Distance is dying; but geography, it seems, is still alive and kicking.

The material on pages 388–94 first appeared in *The Economist* in August 2001.

Index